## Cognitive Dynamic Systems

The principles of cognition are becoming increasingly important in the areas of signal processing, communications, and control. In this ground-breaking book, Simon Haykin, a pioneer in the field and an award-winning researcher, educator, and author, sets out the fundamental ideas of cognitive dynamic systems. Weaving together the various branches of study involved, he demonstrates the power of cognitive information processing and highlights a range of future research directions.

The book begins with a discussion of the core topic, cognition, dealing in particular with the perception–action cycle. Then, the foundational topics, power spectrum estimation for sensing the environment, Bayesian filtering for environmental state estimation, and dynamic programming for action in the environment, are discussed. Building on these foundations, detailed coverage of two important applications of cognition, cognitive radar and cognitive radio, is presented.

Blending theory and practice, this insightful book is aimed at all graduate students and researchers looking for a thorough grounding in this fascinating field.

**Simon Haykin** is the Director of the Cognitive Systems Laboratory at McMaster University, Canada. He is a pioneer in adaptive signal processing theory and applications in radar and communications, areas of research that have occupied much of his professional life. For the past 10 years he has focused his entire research interests on cognitive dynamic systems: cognitive radar, cognitive radio, cognitive control, and cognition applied to the cocktail party processor for the hearing impaired. He is a Fellow of the IEEE and the Royal Society of Canada, and is the recipient of the Henry Booker Gold Medal from URSI (2002), the Honorary Degree of Doctor of Technical Sciences from ETH Zentrum, Zurich (1999), and many other medals and prizes. In addition to the seminal journal papers "Cognitive radio" and "Cognitive radar," he has also written or co-written nearly 50 books including a number of best-selling textbooks in the fields of signal processing, communications, and neural networks and learning machines.

# Cognitive Dynamic Systems

## Perception–Action Cycle, Radar, and Radio

Simon Haykin

McMaster University, Canada

CAMBRIDGE
UNIVERSITY PRESS

# CAMBRIDGE
UNIVERSITY PRESS

University Printing House, Cambridge CB2 8BS, United Kingdom

One Liberty Plaza, 20th Floor, New York, NY 10006, USA

477 Williamstown Road, Port Melbourne, VIC 3207, Australia

314-321, 3rd Floor, Plot 3, Splendor Forum, Jasola District Centre, New Delhi - 110025, India

103 Penang Road, #05-06/07, Visioncrest Commercial, Singapore 238467

Cambridge University Press is part of the University of Cambridge.

It furthers the University's mission by disseminating knowledge in the pursuit of
education, learning and research at the highest international levels of excellence.

www.cambridge.org
Information on this title: www.cambridge.org/9780521114363

First published 2012

*A catalogue record for this publication is available from the British Library*

*Library of Congress Cataloging in Publication data*
Haykin, Simon
Cognitive dynamic systems : perception–action cycle, radar, and radio / Simon Haykin.
p.   cm.
Includes bibliographical references and index.
ISBN 978-0-521-11436-3 (hardback)
1. Self-organizing systems.   2. Cognitive radio networks.   I. Title.
Q325.H39   2011
003´.7 – dc23      2011037991

ISBN  978-0-521-11436-3  Hardback

# Contents

# Preface

In my Point-of-View article, entitled "Cognitive dynamic systems," *Proceedings of the IEEE*, November 2006, I included a footnote stating that a new book on this very topic was under preparation. At long last, here is the book that I promised then, over four years later.

Just as adaptive filtering, going back to the pioneering work done by Professor Bernard Widrow and his research associates at Stanford University, represents one of the hallmarks of the twentieth century in signal processing and control, I see cognitive dynamic systems, exemplified by cognitive radar, cognitive control, and cognitive radio and other engineering systems, as one of the hallmarks of the twenty-first century.

The key question is: How do we define cognition? In this book of mine, I look to the human brain as the framework for cognition. As such, cognition embodies four basic processes:

- perception–action cycle,
- memory,
- attention, and
- intelligence,

each of which has a specific function of its own. In identifying this list of four processes, I have left out language, the fifth distinctive characteristic of human cognition, as it is outside the scope of this book. Simply put, there is no better framework than human cognition, embodying the above four processes, for the study of cognitive dynamic systems, irrespective of application.

Putting aside the introductory and epilogue chapters, the remaining six chapters of the book are organized in three main parts as follows:

- Chapter 2, entitled the "Perception–action cycle," provides an introductory treatment of the four basic processes of cognition identified above. Moreover, the latter part of the chapter presents highlights of neural networks needed for the implementation of memory.
- Chapters 3, 4, and 5 provide the fundamentals of cognitive dynamic systems, viewed from an engineering perspective. Specifically, Chapter 3 discusses power spectrum estimation as a basic tool for sensing the environment. Chapter 4 discusses the Bayesian filter as the optimal framework for estimating the state of the environment when it is hidden. In effect, Chapters 3 and 4 are devoted to how the environment is perceived

by dynamic systems, viewed in two different ways. Chapter 5 deals with dynamic programming as the mathematical framework for how the system takes action on the environment.

- Chapters 6 and 7 are devoted to two important applications of cognitive dynamic systems: cognitive radar and cognitive radio respectively; both of them are fast becoming well understood, paving the way for their practical implementation.

To conclude this Preface, it is my conviction that cognition will play the role of a software-centric information-processing mechanism that will make a difference to the theory and design of a new generation of engineering systems aimed at various applications, not just radar control, and radio.

Simon Haykin,
Ancaster, Ontario, Canada.

# Acknowledgments

In the course of writing this book, I learned a great deal about human cognition from the book *Cortex and Mind: Unifying Cognition* by Professor J. M. Fuster, University of California at Los Angeles. Just as importantly, I learned a great deal from the many lectures on cognitive dynamic systems, cognitive radio, and cognitive radar, which I had the privilege of presenting in different parts of the world.

I would like to extend my special thanks to Professor Richard Sutton and his doctoral student, Hamid Maei, University of Alberta, Canada, for introducing me to a new generation of approximate dynamic-programming algorithms, the GQ($\lambda$) and Greedy-GQ, and communicating with me by e-mail to write material presented on these two algorithms in the latter part of Chapter 5 on dynamic programming. Moreover, Hamid was gracious to read over this chapter, for which I am grateful to him.

I am also indebted to Professor Yann LeCun, New York University, and his ex-doctoral student Dr Marc'Aurelio Ranzato for highly insightful and helpful discussion on sparse coding. In a related context, clarifying concepts made by Dr Bruno Olshausen, University of California, Berkeley, are much appreciated.

I acknowledge two insightful suggestions:

(1) The notion of "fore-active radar" as the first step towards radar cognition, which was made by Professor Christopher Baker, Australian National University, Canberra.
(2) The analogy between feedback information in cognitive radar and saccade in vision, which was made by Professor José Principe, University of Florida.

I thank my ex-graduate students, Dr Ienkaran Arasaratnam, Dr Peyman Setoodeh, and Dr Yanbo Xue, and current doctoral student, Farhad Khozeimeh, for their contributions to cognitive radar and cognitive radio. I have also benefited from Dr Amin Zia, who worked with me on cognitive tracking radar as a post-doctoral fellow. I am grateful to Dr. Setoodeh for careful proof-reading of the page proofs.

Moreover, I had useful comments from Dr Terrence Sejnowski, Salk Institute, LaJolla, CA, and my faculty colleague Professor Suzanne Becker, McMaster University, Canada.

Turning to publication of the book by Cambridge University Press, I am particularly grateful to Dr Philip Meyler, Publishing Director, and his two colleagues, Sarah Marsh and Caroline Mowatt, in overseeing the book through its different stages of production. In a related context, I also wish to thank Peter Lewis for copy editing the manuscript before going into production.

The writing of this book has taken me close to four years, in the course of which I must have gone through 20 revisions of the manuscript. I am truly grateful to my Administrative Coordinator, Lola Brooks, for typing those many versions. Without her patience and dedication, completion of the manuscript would not have been possible.

Last but by no means least, I thank my wife, Nancy, for allowing me the time I needed to write this book.

Simon Haykin
Ancaster, Ontario, Canada.

# 1 Introduction

I see the emergence of a new discipline, called Cognitive Dynamic Systems, which builds on ideas in statistical signal processing, stochastic control, and information theory, and weaves those well-developed ideas into new ones drawn from neuroscience, statistical learning theory, and game theory. The discipline will provide principled tools for the design and development of a new generation of wireless dynamic systems exemplified by cognitive radio and cognitive radar with efficiency, effectiveness, and robustness as the hallmarks of performance.

This quotation[1] is taken from a point-of-view article that appeared in the *Proceedings of the Institute of Electrical and Electronics Engineers* (Haykin, 2006a). In retrospect, it is perhaps more appropriate to refer to Cognitive Dynamic Systems as an "integrative field" rather than a "discipline."

By speaking of cognitive dynamic systems as an integrative field, we mean this in the sense that its study integrates many fields that are rooted in neuroscience, cognitive science, computer science, mathematics, physics, and engineering, just to name a few. Clearly, the mixture of fields adopted in the study depends on the application of interest. However, irrespective of the application, the key question is

What is the frame of reference for justifying that a so-called cognitive dynamic system is indeed cognitive?

In this book, we adopt *human cognition* as *the frame of reference*. As for applications, the book focuses on cognitive radar and cognitive radio.

With these introductory remarks, the study begins with the next section.

## 1.1 Cognitive dynamic systems

A system, be it linear or nonlinear, is said to be *dynamic* if *time* plays a key role in its input–output behavior. In this book, we are interested in a new class of dynamic systems called *cognitive dynamic systems*, the study of which is inspired by the unique neural computational capability of the human brain[2] and the viewpoint that human cognition is a form of computation.[3]

To be specific, we say that a dynamic system, operating in an environment to be explored, is *cognitive* if it is capable of four fundamental functions (tasks) that are basic to *human cognition*:

(1) the perception–action cycle;
(2) memory;

(3) attention; and

(4) intelligence.

The perception–action cycle implies that a cognitive dynamic system has two functional parts; namely, *perceptor* and *actuator*. The cycle begins with the perceptor perceiving the environment (world) by processing the incoming stimuli, called *observables* or *measurements*. In response to *feedback information* from the perceptor about the environment, the actuator acts so as to control the perceptor via the environment, and the cycle goes on. In effect, the perceptor "guides" the actuator by virtue of what it has *learned* about the environment, and the actuator "controls" the perceptor by acting in the environment. The benefit resulting from the perception–action cycle is that of *maximizing information gained* from the environment.

Typically, the environment is *nonstationary*, which means that the underlying behavior of the environment continually changes with time. Given such an environment to deal with, we may now go on to say that a cognitive dynamic system *must also have memory*, desirably of a multiscale variety. This requirement is needed for the system to do the following:

- *learn* from the environment and store the knowledge so acquired;
- continually *update* the stored knowledge in light of environmental changes; and
- *predict* the consequences of actions taken and/or selections made by the system as a whole.

As for attention, a cognitive dynamic system must be equipped with the capability to focus its information-processing power on a target or footprint in the environment that is considered to be of special interest or strategic importance; this is done by *prioritizing* the allocation of available resources.

Turning to intelligence, among the above-mentioned four functions, it is by far the most difficult one to describe. Nevertheless, intelligence is the single most important function in a cognitive dynamic system. For the present, it suffices to say that intelligence is based on the perception–action cycle, memory, and attention for its functionality. Most importantly, it is the presence of feedback at multiple levels in the system that facilitates intelligence, which, in turn, makes it possible for the system to make *intelligent decisions* in the face of inevitable uncertainties in the environment. The feedback can itself take one of two forms:

- *global feedback*, which embodies the environment, and
- *local feedback*, which does not.

Typically, the local and global feedback loops are distributed throughout a cognitive dynamic system. The extent of feedback loops naturally varies from one cognitive dynamic system to another, depending on the application of interest.

## 1.2     The perception–action cycle

In diagrammatic form, much of what we have just described is captured so illustratively in Figure 1.1. On the right-hand side of the figure, we have the perceptor of a cognitive dynamic system that is responsible for perception of the environment. On the left-hand side of the

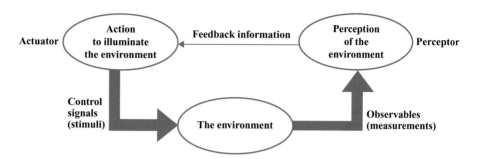

**Figure 1.1.** The perception–action cycle of a cognitive dynamic system in its most generic sense.

figure, we have the actuator of the system that is responsible for action in the environment. And, above all, we have a feedback link connecting the perceptor to the actuator.

Figure 1.1 is commonly referred to as the *perception–action cycle* of a cognitive dynamic system. This remarkable operation, in its generic form, stands out in *human cognition*, which, as mentioned at the beginning of this introductory chapter, embodies the following (Sejnowski, 2010):

- a global feedback loop, embodying perception and action so as to maximize information gain about the environment;
- multiscale memory that is organized to predict the consequences of actions;
- a memory-based attentional mechanism that prioritizes the allocation of resources; and
- a feedback-based decision-making mechanism that identifies intelligent choices in uncertain environments.

We are repeating what we described earlier in order to emphasize that these four distinctive properties of human cognition constitute the *ideal framework*, against which a dynamic system should be assessed for it to be cognitive.

## 1.3   Cognitive dynamic wireless systems: radar and radio

A new generation of engineering systems is being inspired by human cognition, the structures of which vary in details from one system to another, depending on the application of interest. In this context, two dynamic wireless systems stand out:

- *cognitive radar*, for improved performance in remote-sensing applications for system accuracy and reliability; and
- *cognitive radio*, for solving the underutilized electromagnetic spectrum problem.

### 1.3.1   Cognitive radar

There is a remarkable analogy between the *visual brain* and radar. Indeed, the perception–action cycle of Figure 1.1 applies equally well to cognitive radar by merely changing the ways in which the transmitter (actuator) and receiver (perceptor) are actually implemented.

Specifically, the function of the receiver in a radar system is to produce an *estimate of the state* of an unknown target located somewhere in the environment by processing a *sequence of observables* dependent on the target state. In effect, perception of the environment takes the form of *state estimation*. As for the transmitter in the system, its function is to adaptively select a transmitted waveform that illuminates the environment in the best manner possible. In target detection, the issue of interest is to decide as reliably as possible whether a target is present or not in the observables. In target tracking, on the other hand, the issue of interest is to estimate the target parameters (e.g. range and velocity) as accurately as possible.

With radar intended for remote-sensing applications and with its transmitter and receiver being typically collocated, much can be learned from the human brain to make a radar system cognitive.

### 1.3.2    Cognitive radio

The practical use of cognitive radio is motivated by the desire to address the *electromagnetic spectrum underutilization problem*. In today's wireless communications world, we typically find that only a small fraction of the radio spectrum assigned to legacy operators by government agencies is actually employed by primary (licensed) users. The underutilized subbands of the spectrum are commonly referred to as *spectrum holes*. The function of a cognitive radio may then be summarized as follows:

(1) The radio receiver is equipped with a *radio scene analyzer*, the purpose of which is to identify where the spectrum holes are located at a particular point in time and space.
(2) Through an external feedback link from the receiver to the transmitter, the information on spectrum holes is then passed to the radio transmitter, which is equipped with a *dynamic spectrum manager* and *transmit-power controller*. The function of the transmitter is to allocate the spectrum holes among multiple secondary (cognitive radio) users in accordance with prioritized needs.

Unlike radar, where the transmitter and receiver are ordinarily collocated, in a radio (wireless) communication system the transmitter and receiver are located in different places. Accordingly, for the receiver to send the transmitter information about the spectrum holes, we require the use of a *low-bandwidth feedback link* connecting the receiver to the transmitter.

With radio intended for wireless communications and with its transmitter and receiver being separately located, there is still a great deal we can learn from the human brain; but to make the radio cognitive, we have to use *engineering ingenuity*.

### 1.4    Illustrative cognitive radar experiment

We will now motivate the power of cognitive information-processing by considering a simple cognitive radar tracker.

The function of the receiver is to estimate the state of a target in space and thereby track its motion across time, given a set of observables (measurements) obtained on the target. With this objective in mind, a sequential state estimator suffices to take care of the perceptive needs of the receiver.

To elaborate on the sequential state estimator, for the sake of simplicity we assume that the environment is described by a *linear state-space model*, comprised of the following pair of equations:

$$\mathbf{x}(n) = \mathbf{A}\mathbf{x}(n-1) + \boldsymbol{\omega}(n), \qquad (1.1)$$

$$\mathbf{y}(n) = \mathbf{B}\mathbf{x}(n) + \mathbf{v}(\boldsymbol{\theta}_{n-1}), \qquad (1.2)$$

where $n$ denotes *discrete time*. The vector $\mathbf{x}(n)$ denotes the *state* of the environment at time $n$; the state evolution across time in (1.1) is called the *system equation*. The vector $\mathbf{y}(n)$ denotes the *measurements* recorded digitally by the receiver as input at time $n$, hence the reference to (1.2) as the *measurement equation*. Transition of the state at time $n-1$ to that at time $n$ is described by the matrix $\mathbf{A}$. Correspondingly, dependence of the measurements (observables) on the state at time $n$ is described by the *measurement matrix* $\mathbf{B}$.

Naturally, the imposition of a mathematical model on the environment, as described in (1.1) and (1.2), gives rise to *uncertainties* about the physical behavior of the environment. These uncertainties are accounted for by introducing the *system noise* $\boldsymbol{\omega}(n)$ in (1.1) and *measurement noise* $\mathbf{v}(\boldsymbol{\theta}_{n-1})$ in (1.2). The measurement noise is denoted by $\mathbf{v}$, the composition of which is dependent on the action of the transmitter. That action is controlled by a transmit-waveform parameter vector $\boldsymbol{\theta}_{n-1}$; the reason for (partially or in full) assigning time $n-1$ to this vector is to account for the propagation delay between the transmitter and receiver.

In what follows, we assume that the process noise $\omega(n)$ and measurement noise $\mathbf{v}(\boldsymbol{\theta}_{n-1})$ are both *stationary Gaussian processes* of zero mean; their respective covariance matrices are denoted by $\mathbf{Q}$ and $\mathbf{R}_{\theta}$. Invoking these assumptions and recognizing that the state-space model described in (1.1) and (1.2) is linear, it follows that the solution to the sequential state-estimation problem – that is, estimating the state $\mathbf{x}(n)$ given the sequence of observables $\{\mathbf{y}(i)\}_{i=1}^{n}$ – is to be found in the classic *Kalman filter*. The issue of sequential state estimation is discussed in Chapter 4. As such, in this introductory chapter, we will proceed on the premise that we know how to formulate the Kalman filtering algorithm. We may, therefore, go on to say that, with tracking as the issue of interest, the Kalman filter, formulated on the basis of the state-space model of (1.1) and (1.2), adequately fulfills the perceptive needs of the receiver.

## 1.4.1    The experiment

The target parameters to be estimated are the *delay* and *Doppler shift*. The delay $\tau$ is defined in terms of the range $\rho$ (i.e. distance of the target from the radar) by

$$\tau = \frac{2\rho}{c}, \qquad (1.3)$$

where $c$ is the speed of electromagnetic wave propagation (i.e. the speed of light). The Doppler shift $f_{\mathrm{D}}$ is defined in terms of the range rate $\dot{\rho}$ (i.e. velocity of the target) by

$$f_{\mathrm{D}} = \frac{2 f_{c} \dot{\rho}}{c}, \qquad (1.4)$$

where $f_c$ is the transmitted carrier frequency and the dot in $\dot{P}$ denotes differentiation with respect to time.

### 1.4.2     The environment

The unknown target is located at a distance of 3 km from the radar and it is moving at a speed of 200 m/s.

### 1.4.3     The radar

The radar is an X-band radar operating at the frequency $f_c = 10.4$ GHz; it is located at the origin. The 0 dB *signal-to-noise ratio* (SNR) at the receiver input is defined at 80 km.

### 1.4.4     State-space model

The state-space model of (1.1) and (1.2) is parameterized as follows:

$$\mathbf{A} = \begin{bmatrix} 1 & T \\ 0 & 1 \end{bmatrix},$$

$$\mathbf{B} = \begin{bmatrix} 1 & 0 \\ 0 & 1 \end{bmatrix},$$

where $T$ is the sampling period. The system noise is modeled as the target-acceleration noise, with its zero-mean covariance defined by (Bar-Shalom *et al.*, 2001)

$$\mathbf{Q} = \sigma_v^2 \begin{bmatrix} \dfrac{1}{4}T^4 & \dfrac{1}{2}T^3 \\ \dfrac{1}{2}T^3 & T^2 \end{bmatrix},$$

where the variance $\sigma_v^2 = 0.49$.

### 1.4.5     Simulation results

Figure 1.2 presents the results of Monte Carlo simulations performed to evaluate the tracking performance of three different radar configurations:[4]

(1) *Traditional active radar* with a fixed transmit waveform, in which case the measurement equation (1.2) simplifies to

$$\mathbf{y}(n) = \mathbf{Bx}(n) + v(n). \tag{1.5}$$

It is assumed that the matrices $\mathbf{A}$ and $\mathbf{B}$ and the covariance matrices $\mathbf{Q}$ and $\mathbf{R}$ are all known.

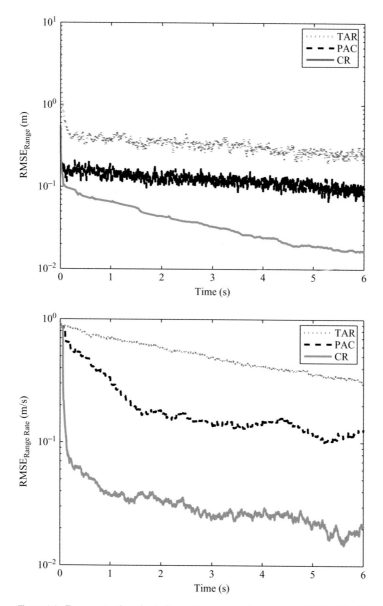

**Figure 1.2.** Demonstrating the information-processing power of global feedback and cognition in radar tracking. (a) Root mean-squared-error (RMSE) of target range, measured in meters. (b) RMSE of range rate, measured in meters per second. TAR: traditional active radar; PAC: the perception–action cycle, as in the first stage toward cognition in radar, or equivalently, the fore-active radar; CR: cognitive radar.

(2) *Perception–action cycle mechanism*, which is the first step toward radar cognition; the same mechanism is applicable to a second class referred to as the *fore-action radar*.[4] Accordingly, the measurement equation (1.2) holds.

(3) *Cognitive radar*, the transmitter of which is equipped not only with a transmit-waveform library but also another library in the receiver; the second library

makes it possible for the receiver to select appropriate values for the matrix $\mathbf{A}$ and covariance matrix $\mathbf{Q}$ in the system equation (1.1); that is, continually *model* the environment.

Figure 1.2a presents the RMSE for the target range and Figure 1.2b the RMSE for the target range-rate plotted versus time. In both parts of the figure, the top dashed curves refer to the traditional active radar, the middle dashed bold curves refer to the perception–action cycle mechanism acting alone (i.e. fore-active radar), and the bottom full curves refer to the cognitive radar.

The results presented in Figure 1.2 lead us to report two important findings:

(1) The use of global feedback in the perception–action mechanism acting alone makes a significant difference in the tracking accuracy of the radar, compared with the traditional active radar with no feedback.
(2) The addition of memory to the perception–action mechanism as in the cognitive radar brings in even more significant improvement to tracking accuracy of the radar.

## 1.5          Principle of information preservation

Having just reported findings (1) and (2) on how the fore-active radar and the cognitive radar compare with a traditional active radar in terms of tracking accuracy, we may now pose a related question:

Over and above the improvements in tracking accuracy, what else do the results of Figure 1.2 teach us?

### 1.5.1          Feedback information

Our first step in answering this fundamental question is to reiterate that the environmental state of the target consists of two parameters:

(1) *range* $\rho$, which defines how far away the target is from the radar;
(2) *range rate* $\dot{\rho}$, which defines the velocity of the target.

Since both the system and measurement equations, (1.1) and (1.2), are corrupted by additive noise processes, it follows that both the range $\rho$ and range rate $\dot{\rho}$ are *random variables*. According to Shannon's information theory (Shannon, 1948), we may, therefore, say that *information* about the target's state is contained in the sequence of measurements $\{\mathbf{y}(i)\}_{i=0}^{n}$. Moreover, in view of this statement, we may go on to speak of *feedback information*, defined in terms of the *error* between the actual state of the target and its estimate computed by the receiver as the result of operating on the sequence of measurements. It is by virtue of this feedback information passed to the transmitter by the receiver that the feedback loop around the environment is closed.

## 1.5.2   Bayesian filtering of the measurements

With state estimation as a central issue of interest in cognitive radar, we look to the *Bayesian filter as the optimal recursive data-processing algorithm*. Basically, the Bayesian filter combines all the available measurements (data) plus *prior* knowledge about the system and measuring devices and uses them all to produce the *optimal estimate* of hidden target parameters. To perform this estimation, the Bayesian filter propagates the *posterior* (i.e. probability density function of the state, conditioned on the measurements) from one estimation recursion to the other. The rationale for focusing on the posterior is that it contains *all the information* about the state that is available in the measurements. Hence, the optimal estimate is obtained by *maximizing* the posterior, which is the "best" that can be achieved (Ho and Lee, 1964).

Under the combined assumption of linearity and Gaussianity, the Bayesian filter reduces to the *Kalman filter*, hence its choice as the functional block for processing the measurements in the receiver in the motivational experiment of Section 1.4.

## 1.5.3   Information preservation through cognition

Moving on to the transmitter for action in the environment, the feedback information highlighted earlier in this section provides a basis of a *cost-to-go function* that looks to the future by one time step. Recognizing that this function is also dependent on the parameters that define the transmitted waveform, a primary function of the transmitter, therefore, is to select a set of transmit-waveform parameters that minimizes the cost-to-go function. Thus, for every cycle of the radar's perception–action cycle, the transmit-waveform parameters are selected such that perception of the radar environment in the receiver and action in the environment performed in the transmitter are optimized in a sequential manner. This process is repeated on a cycle-by-cycle basis.

To assess the overall radar performance, we need a *metric* that provides an assessment of how close the estimated state of the target is to its actual value. For the experiment, following the traditional approach in statistical signal processing, this metric was defined simply as the RMSE between the actual state and its estimated value using the Kalman filter, with both of them defined for one step into the future. What we have just described here provides the mathematical justification for improved tracking accuracy through the use of global feedback from the receiver to transmitter, reported previously under point (i) at the end of Section 1.4.

Next, through the following combination of system additions:

- *perceptual memory*, reciprocally coupled with the Kalman filter in the receiver;
- *executive memory*, reciprocally coupled with the transmit-waveform selector in the transmitter; and
- *working memory*, reciprocally coupling the executive and perceptual memories,

the transmitter and receiver are continuously matched together in their respective operations in an adaptive manner. It is through this adaptive process distributed in different parts of the radar receiver and transmitter that we are able to explain the additional significant accuracy improvement reported under point (2) at the end of Section 1.4.

Now, we are ready to answer the question that was posed at the beginning of this section. In using the Kalman filter by itself in the receiver, information about the state of the target contained in the measurements is *preserved to some extent* (Kalman, 1960). In adaptively *matching illumination* of the environment with the target through the use of feedback from the receiver to the transmitter (Gjessing, 1986), information about the state contained in the measurements is *preserved even more*. Finally, in making the radar cognitive through the provision of distributed memory, *further improvement in information preservation is achieved* by the cognitive radar system.

What we have just described here is now summed up as follows:

Cognitive information processing provides a powerful tool for fulfilling the principle of information preservation, which is aimed at preserving information about a hidden target state that is contained in the measurements.

In the statement just made on the principle of information preservation, we have emphasized the role of cognition. Information preservation may also be viewed as *information gain* in the following sense: the more we preserve information contained in the measurements about the target's state, the closer the estimated state is to its actual value. This is another way of saying that we are progressively "gaining" information about the target's state from one cycle to the next in following the perception–action cycle. By saying so, we have justified the previous use of "information gain" in describing the role of a global feedback loop in the perception–action cycle.

### 1.5.4    Concluding remarks

To conclude this discussion, it should be noted that successive improvements in information preservation, exemplified by corresponding improvements in state-estimation accuracy, are achieved at the expense of increased computational complexity. It is *apropos*, therefore, that we complement our previous statement by saying:

There is "no free lunch," in that for every gain we make in practice there is a price to be paid.

### 1.6    Organization of the book

The rest of the book is organized in seven chapters, as summarized here.

Chapter 2 is devoted to a detailed discussion of the *perception–action cycle*, which is the *baseline for the operation of every cognitive dynamic system*. This chapter also identifies three kinds of memory:

• *perceptual memory*, which is an integral part of the receiver;
• *executive memory*, which is an integral part of the transmitter; and
• *working memory*, which reciprocally couples the executive memory to perceptual memory.

These three memories, acting together, enable the radar system to *predict* the *consequences* of action taken by the radar system as a whole. As mentioned previously, unlike perception and memory, the properties of attention and intelligence are distributed across the whole dynamic system. Chapter 2 also includes discussion of neural networks for designing the memory system; *learning* from stimuli is an essential requirement of cognition.

Chapter 3 addresses *power spectrum estimation* as a means of *sensing* the environment. For reasons discussed therein, much of the material covered in that chapter is devoted to the *multitaper method* and related issues; namely, space–time processing, time–frequency analysis, and cyclostationarity. From the brief description just given, the topics covered in Chapter 3 are directly applicable to cognitive radio discussed later in Chapter 7.

Chapter 4 addresses Bayesian filtering for estimating the hidden state of the environment in a sequential manner given the observables received from the environment. Of course, the natural place for this operation is the receiver. Under the assumptions of linearity and Gaussianity, the Bayesian filter reduces to the celebrated Kalman filter. However, when the environment is nonlinear, as it is often the case in practice, we have to be content with approximate forms of the Bayesian filter. The extended Kalman filter (EKF), suitable for mild forms of nonlinearity, is one such approximation; the EKF is widely used on account of its computational simplicity. Chapter 4 includes discussion of the EKF and, most importantly, a new nonlinear filter, named the "cubature Kalman filter (CKF)." The CKF is a "method of choice" for approximating the *optimal Bayesian filter* under the Gaussian assumption in a nonlinear setting. By the nature of it, the material covered in Chapter 4 is directly applicable to cognitive radar, discussed later in Chapter 6.

Chapter 5 is devoted to another topic, dynamic programming, which, in its own way, is also basic to the study of decision-making in cognitive dynamic systems. *Dynamic programming* addresses problems on policy planning and control. As such, its implementation resides in the transmitter. Whereas state estimation deals with hidden states, formulation of dynamic programming theory assumes direct access to the state of the environment, with the challenge being that of finding an optimal policy for action in the environment to control the receiver *indirectly* on a cycle-by-cycle basis. The material covered in Chapter 5 is naturally applicable to cognitive radar in Chapter 6.

With the fundamentals of cognitive dynamic systems covered in Chapters 2–5, the stage is set to discuss practical applications. With emphasis being on new wireless dynamic systems enabled with cognition, the book focuses on two applications: cognitive radar and cognitive radio, in that order.

Cognitive radar is covered in Chapter 6. Much of the material covered in that chapter deals with the use of cognitive radar for tracking an unknown target in space (i.e. a Gaussian environment). Computer simulations are presented to demonstrate the information-processing power of cognition on the performance of a traditional radar tracker, first using global feedback only and then expanding the perception–action cycle mechanism by adding memory distributed across the entire radar. The important message to take from Chapter 6 is that, through cognitive information-processing, we now have a *transformative software technology* that is applicable to radar systems, old and new. Simply put, cognitive radar is a *game-changer*.

Next, cognitive radio is discussed in Chapter 7 to deal with the underutilized electromagnetic spectrum problem. Unlike cognitive radar, the information-processing power of cognition is exploited differently in cognitive radio. Here also, computer simulations are presented to demonstrate how radio-scene analysis in the receiver and resource allocation in the transmitter can be tackled in computationally efficient and performance-effective ways. With the electromagnetic radio spectrum being a natural resource, cognitive radio provides a *paradigm shift* in wireless communications by virtue of its information-processing power to improve utilization of this important natural resource.

Chapter 8, the last chapter entitled "Epilogue," reemphasizes the attributes that bind cognitive radar and cognitive radio to the human brain, and finishes with brief discussion of unexplored topics, which, in their own respective ways, look to how cognition can be exploited for innovative practical applications.

## Notes and practical references

1. *The predictive article on cognitive dynamic systems*
   Haykin (2006b) was emboldened to write this predictive article, emboldened by his two seminal journal papers, the first entitled "Cognitive radio: brain-empowered wireless communications," (Haykin, 2005a), and the second entitled "Cognitive radar: a way of the future" (Haykin, 2006b).
2. *Computational brain*
   For books on the computational brain, see Churchland and Sejnowski (1992) and Dayan and Abbott (2001), and Trappenberg (2010).
3. *Cognitive sciences*
   For books on cognitive science, see Pylyshyn (1984), Posner (1989), and Fuster (2003).

   In the English edition of his book entitled *On The Origins of Science: The Mechanization of the Mind*, Dupuy (2009) eloquently articulates those issues that unite cognitive science and cybernetics (dating back to Wiener, 1948) on the one hand and those other issues that have led them to move apart on the other hand. With cybernetics being the older one of the two, Dupuy argues that cybernetics should be viewed as the "parent" of cognitive science.
4. *Three classes of radar*
   In Section 6.2, three classes of radar are defined:
   • traditional active radar,
   • fore-active radar, and
   • cognitive radar.

The second class, referred to as *fore-active radar*, distinguishes itself from the traditional active radar by the use of *global feedback* from the receiver to the transmitter. In doing so, there is no distinction between a fore-active radar or the perception–action cycle of a cognitive radar. It is for this reason that we have used the terminology "perception–action cycle mechanism" for the second radar configuration in simulations described in Chapters 1 and 6.

# 2 The perception–action cycle

In Chapter 1 of his book entitled *Cortex and Mind: Unifying Cognition*, Fuster (2003) introduces a new term called the *cognit*, which is used to characterize the cognitive structure of a cortical network; it is defined as follows:

> A cognit is an item of knowledge about the world, the self, or the relations between them. Its network structure is made up of elementary representations of perception or action that have been associated with one another by learning or past experience.

Towards the end of Chapter 1 of the book, Fuster goes on to say:

> ... perception is part of the acquisition and retrieval of memory, memory stores information acquired by perception, language and memory depend on each other; language and reasoning are special forms of cognitive action; attention serves all other functions; intelligence is served by all; and so on.

These two quotes taken from Fuster's book provide us with a cohesive statement on the cortical functions of cognition in the human brain; both are important. In particular, the second quotation highlights the interrelationships between the five processes involved in cognition: perception, memory, attention, language, and intelligence. Just as language plays a distinctive role in the human brain, so it is in a cognitive dynamic system where language provides the means for effective and efficient communication among the different parts of the system. However, language is outside the scope of this book. The remaining four processes are the focus of attention henceforth.

With this brief introductory background on cognition, the stage is set in this chapter for detailed descriptions of the four cognitive processes: perception followed by action, memory, attention, and intelligence, and their respective roles in the perception–action cycle. As reiterated in the introductory chapter, the perception–action cycle is at the very heart of all cognitive dynamic systems; hence the title of this chapter.

## 2.1 Perception

For a cognitive dynamic system to perceive the environment in which it operates, the system must be equipped with an appropriate set of *sensors to learn* from the environment. For example, in cognitive radar and cognitive radio, *antennas* are placed at the front end of the receiver so as to couple it electromagnetically to the environment. By the same token, another set of antennas is placed at the output of the transmitter so as to couple it electromagnetically to the environment in its own way.

The sensors feed a functional block in the receiver called the *environmental scene analyzer*; its function is to perform *perception*, described as follows:

Perception is the sensory analysis of incoming streams of stimuli, aimed at learning the underlying physical attributes that characterize the environment in an on-line manner.

Exact composition of the environmental scene analyzer is naturally dependent on the application of interest. For example, in cognitive radar designed for target tracking, the environmental scene analyzer consists of a sequential state estimator, the function of which is to estimate the *state of the target* across time in a recursive manner. In cognitive radio for another example, the function of the environmental scene analyzer is to identify *spectrum holes* (i.e. underutilized subbands of the electromagnetic spectrum) so that they can be employed by unserviced secondary users; in effect, the spectrum holes signify the "state" of the radio environment insofar as cognitive radio is concerned.

Irrespective of the application of interest, a cognitive dynamic system's perception of the environment is continually influenced by the current data received by the system, as well as by *cognitive information* (i.e. knowledge gained from past experience) already stored in memory. In other words, every *percept* (i.e. snapshot of the perception process at every cycle in time) is made up of two components:

(1) The first component of percept refers to *recognition* and, therefore, retrieval of *relevant information about the environment* that is stored in memory for the purpose of representing past data.
(2) The other component of the percept refers to *categorization (classification)* of a new set of data that is *correlated* with the memory; in other words, the second component is an *updated* version of the first component.

Most importantly, in both cases, perceptual information processing is executed in a *self-organized and adaptive manner*.

## 2.1.1 Functional integration-across-time property of cognition

In a cognitive dynamic system, perception of the environment in the perceptor leads to *action* in the environment by the *environmental scene actuator* in the actuator. For this operation to be feasible, however, we require the use of a *feedback link* from the environmental scene analyzer to the environmental scene actuator, as depicted in Figure 2.1. Thus, whereas the perceptor perceives the environment *directly* by processing the incoming streams of stimuli, the actuator gets to know about the environment *indirectly* through the *feedback information* fed to it by the perceptor in accordance with the *perception–action cycle*.

Indeed, it is through the continuation of this cycle across time that a cognitive dynamic system acquires its *cardinal property*, enabling it to *adapt* to changes in the environment by making successive internal changes of its own through *lessons learned*

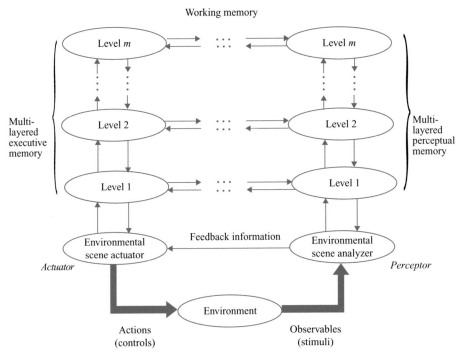

**Figure 2.1.** Directed information-flow diagram in the perception–action cycle of a cognitive dynamic system with hierarchical memory. (This figure is inspired by Fuster (2003).)

*from interactions with the environment.* To reemphasize the importance of this capability, we say that the integration of functions involved in the perception–action cycle across *time* is a distinctive characteristic of cognition, in that time plays three key roles (Fuster, 2003):

(1) Time separates the sensory input signals (stimuli) so as to guide overall system behavior.
(2) Time separates the sensory input signals responsible for perception of the environment in the perceptor from actions taken by the actuator in the environment.
(3) Time separates the occurrence of sensory feedback (i.e. feedback information) from further action.

The implication of these three points is profound, prompting us to make the following statement:

Temporal organization of the system behavior of a cognitive dynamic system requires the integration across time of three entities: percepts with other percepts, actions with other actions, and percepts with actions.

Hereafter, we refer to this property of cognitive dynamic systems as the *functional integration-across-time property*; this property plays the *cardinal role* behind the temporal organization of a cognitive dynamic system's behavior.

## 2.2 Memory

Before proceeding to discuss the important role of memory in cognitive dynamic systems, it is instructive that we differentiate between knowledge and memory:

- *Knowledge* is a memory of certain facts and relationships that exist between them, none of which changes with time; in other words, knowledge is *static* in its contents.
- *Memory*, on the other hand, is *dynamic*, in that its contents continually change over the course of time in accordance with changes in the environment.

Stated in another way, the contents of memory are subject to *time constraints*, whereas knowledge is *timeless* and, therefore, free of time constraints.

With a cognitive dynamic system typically consisting of a perceptor and actuator, it is logical to split the memory into two parts: one residing in the perceptor and the other residing in the actuator. These two parts of memory are respectively called perceptual memory and executive memory, which are discussed next in that order.

### 2.2.1 Perceptual memory

As the name implies, perceptual memory is an integral part of how, in an overall sense, the perceptor perceives the environment. To be more specific, perceptual memory provides the ability for the perceptor of a cognitive dynamic system to *interpret the incoming stimuli* so as to recognize their distinctive features and categorize the features learned accordingly in some statistical sense. We may thus make the following statement:

Perceptual memory is the experiential knowledge that is gained by the perceptor through a process of learning from the environment, such that contents of the memory continually change with time in accordance with changes in the environment; the experiential knowledge so gained through learning becomes an inextricable part of the perceptual memory.

To satisfy the objective embodied in this statement, we require two provisions:

- First, the perceptual memory is supplied with an *internal library*, the elements of which describe different realizations of a prescribed *model* of the environment; in other words, each element represents an item of knowledge about the environment or *cognit*, using Fuster's terminology.
- Second, the perceptual memory is reciprocally coupled to the environmental scene analyzer.

As illustrated in Figure 2.1, this reciprocal coupling implies the use of two links:

(1) *Bottom-up (feedforward) link.* This first link involves two operations:
  - *retrieval* of the old memory representing *features learned* about the environment *directly* from past stimuli and
  - *updating* of the old memory in light of *new* information about the environment contained in the incoming stimuli.
  In other words, new memory is formed on the old memory.

(2) *Top-down (feedback) link.* The purpose of this second link is that of *acquisition* of the new memory by the environmental scene analyzer.

We may, therefore, say that perception in the perceptor consists of *adaptive matching* of a particular environmental model retrieved from the internal library to the incoming stimuli at each cycle.

Moreover, given a *multiscale perceptual memory* in the perceptor of a cognitive dynamic system as depicted in Figure 2.1, the stimuli are processed in the perceptual memory, level by level. Hence, the *perceptual constancy* across the hierarchical structure of the memory progressively increases in *abstraction*. The net result of this abstraction is the ease of perceptual understanding of the environment by the perceptor and, hence, an improved implementation of the adaptive matching process described above.

## 2.2.2    Executive memory

Just as perceptual memory relates to perception of the environment in the perceptor, executive memory relates to the corresponding actuator's action in the environment. To be more precise, contents of the executive memory are continually changed through the transmitter's action in response to feedback information about the environment that is fed to it by the perceptor. We may thus make the following statement:

Executive memory is the experiential knowledge gained by the actuator through the lessons learned from actions taken in the environment by the actuator, with contents of the memory changing with time in accordance with how the perceptor perceives the environment; here again, the knowledge so gained through experience becomes an inextricable part of the executive memory.

To satisfy this statement, we require that the actuator be supplied with two provisions:

- First, the executive memory has an internal *library* of its own, the elements of which describe different realizations of control signal (i.e., transmit-waveforms used to illuminate the environment); in other words, each element is a *cognit* representing an item of knowledge about possible action in the environment.
- Second, the executive memory is *reciprocally* coupled to the environmental scene actuator.

Here too, the reciprocal coupling illustrated in Figure 2.1 implies the use of two links:

(1) *Bottom-up (feedforward link).* This first link involves two operations:
  - *retrieving* old memory learned about decisions made in the past and
  - *updating* the old memory in light of new information about the environment obtained in an indirect manner via the feedback link from the perceptor.
(2) *Top-down (feedback) link.* The purpose of this second link is that of *acquisition* of the new memory by the environmental scene actuator.

Moreover, with the perceptual memory made up of multiple layers (levels) as depicted in Figure 2.1, it is logical that the executive memory should have a corresponding number of layers (levels) of its own. In a manner similar to that described for the multiscale perceptual memory, we find that the output produced by the multiscale executive memory assumes the form of *abstract information* that helps the environmental scene actuator achieve its goal of decision-making in the face of environmental uncertainties with relative ease.

### 2.2.3 Final reciprocal coupling to complete the cognitive information-processing cycle

According to Fuster (2003), the two hierarchies of cortical cognits (i.e. perceptual and executive) appear to be interlinked at all of their individual layers (levels). In other words, networks of the human brain dealing with the perceptual representation and processing of incoming stimuli are reciprocally coupled with networks at a corresponding stage of executive representation and processing of feedback information.

It would, therefore, seem logical that we do the same in a cognitive dynamic system, especially where the perceptor and actuator are collocated. In other words, we seek a form of reciprocal coupling between layers of the hierarchical perceptual memory in the perceptor of the system with the corresponding layers of the hierarchical executive memory in the actuator of the system, as indicated in Figure 2.1. In so doing, the actuator and perceptor of the cognitive dynamic system are enabled to continually operate in *synchrony*.

To be more precise, this final form of reciprocal coupling of the executive and perceptual memories is required to account fully for the functional integration-across-time property of the cognitive dynamic system in its entirety. These two memories are thereby enabled to work with each other, so as to select the "best" action that can be taken by the actuator to control the perceptor via the environment in light of the *optimal* feedback information fed to it by the perceptor from one cycle to the next.

### 2.2.4 Roles of memory in cognition

With the receiver of a cognitive dynamic system functioned to perceive the environment, which, in turn, prompts the actuator to act in the environment on a cycle-by-cycle basis, the system needs a mechanism for *predicting the consequences of actions* taken by the entire system. This need is all the more essential given that the environment is nonstationary, which is always practically the case.

As already explained, there is also the need to reciprocally couple the perceptual memory with the executive memory to satisfy the functional integration-across-time property of cognition. With both of these two memories having similar multiscale structures, as indicated in Figure 2.1, this reciprocal coupling will also have to be in place on a layer-by-layer basis.

These two needs are both fulfilled by the use of a *multiscale working memory*, as shown in Figure 2.1; this new memory is discussed in the next section.

For now, it suffices for us to summarize the roles of memory in a cognitive dynamic system as follows:

(1) Through the joint activation of perceptual memory in the perceptor and executive memory in the actuator, the system acquires organized temporal behavior in accordance with the functional integration-across-time property of cognition.
(2) The memory, viewed in its entirety, predicts the consequences of actions taken by the system as a whole.

## 2.3          Working memory

In the course of describing the roles of memory in cognition, we identified the need for working memory in a cognitive dynamic system. Simply put, *working memory*[1] refers to *active memory that occupies a short span of time*. With its function being that of predicting the consequences of actions taken by the cognitive dynamic system as a whole, it follows that working memory plays a key role in the attentional mechanism focused on the internal representation of a recent event associated with a prospective action; attention is discussed in the next section.

As such, we may think of working memory as having the capacity to temporally hold a limited amount of information for two purposes: *statistical learning* in the perceptor and *probabilistic reasoning* in the actuator. The distinctive property of working memory, therefore, embodies two related points:

- the issue of dealing with a piece of information that pertains to the environment, the cognitive dynamic system itself, or the relationships between them and
- that the piece of information remains active for a relatively short period of time required to focus attention on some action to be taken by the system in the next cycle (i.e. future).

For example, the perceptor may have made a parametric selection to improve its modeling of the environment, or the actuator may have chosen a certain transmit waveform for illuminating the environment. In both cases we have an event that may have occurred at a particular cycle, and the requirement for working memory is to commit a limited amount of cognitive information-processing power to account for the consequences of that event on the next cycle.

## 2.4          Attention

In a fundamental sense, the purpose of attention is to *allocate selectively* the available computational resources to realize the execution of a goal-directed action by the transmitter. We may, therefore, think of attention as a mechanism for *prioritizing resource allocation* in terms of practical importance, which, from a practical perspective, makes a great deal of intuitive sense for the following reason. The computational (i.e. information-processing) resources of a cognitive dynamic system are naturally limited; hence, the following statement:

Attention is a mechanism that protects both the perceptual-processing power of the perceptor and the decision-making power of the actuator from the information-overload problem through prioritization of how these computational resources are allocated.

In the context of a cognitive dynamic system, the term "information overload" refers to the difficulty experienced by the system when the perceptor's task of sensing the environment and the actuator's task of decision-making are compromised by having to handle too much information.[2]

To elaborate, from the perspective of the perceptor in a cognitive dynamic system, *perceptive attention* involves focusing the computational processing power of the perceptor on a specific objective that is of special interest to the application at hand. Consider, for example, a multi-target tracking application, where there is a target of special interest that requires focused attention. With perception consisting essentially of parallel processing and adaptive matching of incoming stimuli to cognits on potential targets stored in the internal library of the perceptual memory, and with the processing continued from one cycle to the next, the bottom-up processing performed rapidly may lead to detection of the target of interest in the perceptor.

In turn, the target detection achieved through the matching process leads to top-down feedback, hence focusing and subsequent analysis. What we have just described here is basically a form of the perception–action cycle being carried out locally in the receiver with target detection as the objective of interest.

Once this objective in the perceptor has been achieved, the next issue is that of pre-serving information about the target contained in the environmental stimuli; that is, invoking the *principle of information preservation* discussed in Chapter 1. With the actuator linked to the perceptor via *feedback information*, to protect the transmitter from the information-overload problem, the *objective function* to be optimized in the actuator should be formulated in accordance with the principle of information preservation as follows:

Information about the environment fed to the actuator by the perceptor is preserved in its most effectively compressed manner.

Thus, *executive attention*, also usually viewed as *executive control*, involves a process of selecting a number of alternative cognits in the executive memory's internal library of transmit waveforms that closely match the incoming feedback information. Once this is done, the task of optimal control in the transmitter is accomplished by computing the waveform for which the objective function is optimized for action on the next cycle.

### 2.4.1 Roles of attention in cognition

Summarizing the roles of attention in cognition, we may make the following twofold statement:

(1) Based on the perceptual memory and executive memory built into a cognitive dynamic system, the attentional mechanism of the system allocates the available computational resources, including prior knowledge, and prioritizes the allocation in their order of importance.
(2) In addition to these two memories, the attentional mechanism looks to the working memory for information on the consequences of actions taken by the system, with this provision being made on a short-time basis.

## 2.5 Intelligence

Earlier on in the chapter, we identified perception, memory, attention, and intelligence as four defining processes of cognitive dynamic systems. Among these four processes, intelligence stands out as the most complex process and the hardest one to define.

We say so because the other three processes, perception, memory, and attention, in their own individual ways and varying degrees, make contributions to intelligence; hence the difficulty in defining intelligence. In the *Penguin Dictionary of Psychology*, Reber (1995: 379) makes the following point on the lack of consensus, concerning the definition of intelligence:

Few concepts in psychology have received more devoted attention and few have resisted clarification so thoroughly.

Nevertheless, we may go on to offer the following statement on intelligence in the context of cognitive dynamic systems:

Intelligence is the ability of a cognitive dynamic system to continually adjust itself through an adaptive process by making the perceptor respond to new changes in the environment so as to create new forms of action and behavior in the actuator.

As different as they are, we may say, therefore, that cognition and adaptation work hand in hand for the overall betterment of cognitive dynamic systems.

To understand the essence of intelligence, we may look to the perception–action cycle of Figure 2.1 with feedback loops distributed throughout the system. Therein, we see that *feedback control for interactions between the perceptual and executive parts of the system manifests itself both globally and locally.* To this end, we may also go on to make the following profound statement (Haykin, 2006b):

Feedback is the facilitator of computational intelligence.

Moreover, as it is with attention, intelligence is distributed across the perception–action cycle in its entirety, which, therefore, means that there is no separate structure or group of structures dedicated to intelligence as a separate function within the system.

## 2.5.1    Efficiency of processing information

For intelligence to stand for the processing of cognitive information toward the achievement of behavioral goals, the degree of intelligence is measured in terms of the *efficiency* with which that information is being processed (Fuster, 2003). The key question is:

How do we measure the efficiency of intelligence?

As observed earlier, the objective of the actuator in a cognitive dynamic system is optimal action in the environment in light of feedback information sent to the environmental scene actuator in the actuator by the environmental scene analyzer in the perceptor; the optimization is done in the actuator in some statistical sense. On this basis, therefore, we may respond to the question just raised above as follows:

Through the use of an optimal control algorithm in the actuator, a cognitive dynamic system becomes increasingly more intelligent whereby a prescribed cost-to-go function is progressively reduced and with it, information about the environment is more efficiently utilized from one cycle to the next.

In saying so, however, we should not overlook the issue of the overall computational complexity of the system.

### 2.5.2   Synchronized cognitive information processing

Looking at the perception–action cycle of Figure 2.1, we see that we have a highly complex closed-loop feedback control system, nested within numerous local feedback loops positioned alongside global feedback loops. Accordingly, the perceptor and actuator of a cognitive dynamic system would have to process information about the environment in a *self-organized, synchronized manner and on a continuous-time basis*. In cognitive radar, for example, the information of interest is about the state of a target embedded in the environment. In cognitive radio, for another example, it is about the spectrum holes (i.e. unused or partially used subbands of the radio spectrum). In both of these applications, the environment can be highly nonstationary, which makes the task of information processing a difficult proposition. It is in highly nonstationary environments that, in designing the actuator to work in synchrony with the perceptor and have them *jointly reinforce* each other on a continuous-time basis through cognition, we are enabled to make a significant practical difference in overall system performance.

### 2.5.3   The role of intelligence in cognition

In the final analysis, the cognitive role of the actuator is that of *decision-making*, in the context of which *probabilistic inference* plays a key role. The term "inference" or "reasoning" refers to a process by means of which conclusions to a problem of interest are reached. Inference may well be the outstanding characteristic of intelligence. We may, therefore, sum up the role of intelligence in cognition as follows:

The decision-making mechanism in the actuator of a cognitive dynamic system uses probabilistic inference to pick intelligent choices in the face of unavoidable uncertainties in the environment.

The uncertainties are attributed to certain physical characteristics of the environment that have been overlooked or that are difficult to account for in modeling the environment. Indeed, it may be justifiably argued that the task of decision-making in the face of environmental uncertainties is the very essence of building a robust dynamic system, which is where intelligence plays the key role. (The Bayesian perspective of inference is discussed in Chapter 4.)

### 2.6   Practical benefits of hierarchy in the perception–action cycle

The information-processing power of the perception–action cycle is enhanced by building hierarchy in the form of layers or levels into the memory distributed in the manner shown in Figure 2.1. In effect, each of the three memories depicted therein has more than one *hidden layer of neurons* (i.e. computational units) built into its structural design. To be specific, a layer of nonlinear neurons is said to be "hidden" if that layer is not reachable from the input or output (Haykin, 2009). Here, we are thinking of a *neural network* as the building block of the perceptual and executive memories.

Moreover, a neural architecture is said to be *deep* if it is composed of many hidden layers of adaptive neurons – adaptive in the sense that their synaptic weights (i.e. free parameters) are adjustable through training. According to Bengio and LeCun (2007), a deep architecture permits the representation of a wide family of functions in a more compact fashion than is possible with a shallow architecture. The rationale behind this claim is that the use of a deep architecture makes it possible to trade off space for time, while making the time–space product smaller; this rationale is indeed motivated by what goes on in the human brain. The *space–time tradeoff*, therefore, is a practical benefit of building depth into the design of memory in a cognitive dynamic system.

Moreover, the use of hierarchical depth in the memory distributed across the system results in an enlarged number of local and global feedback loops in the perception–action cycle of Figure 2.1; to be more precise, the *number of local and global feedback loops grows exponentially with hierarchical memory depth*. An immediate practical benefit of this exponential growth in feedback loops is enhanced intelligence and, hence, a more *reliable* decision-making process in the face of environmental uncertainties.

Last but by no means least, *learning features of features*, in the sense first defined by Selfridge (1958), becomes a distinct characteristic of the learning process involved in designing memory with hierarchical depth. Specifically, the features characterizing the incoming stimuli in the perceptor or those characterizing the feedback information in the actuator become *increasingly more abstract and, therefore, easier to recognize* as the hierarchical depth of memory is increased.

Therefore, we may summarize the overall benefit of hierarchical memory used in a cognitive dynamic system as follows:

In a cognitive dynamic system with hierarchical memory, increased computational complexity is traded for improved attentive and intelligent capabilities, particularly when the requirement is for reliable decision-making in the face of environmental uncertainties.

## 2.7    Neural networks for parallel distributed cognitive information processing

Thus far in this chapter, we have focused attention on the perception–action cycle, its properties, and different structural forms. In a few words, we may summarize the material covered therein as "the underlying principles of cognition in cognitive dynamic systems." However, an equally important issue that needs attention, for example, is the following question that pertains to the hierarchical perception–action cycle of Figure 2.1:

How do we construct a model of a cognitive dynamic system based on the hierarchical perception–action cycle depicted in Figure 2.1?

Clearly, there is no unique approach to address this practical issue. Nonetheless, we will follow an approach inspired by the *human brain*, which is *a complex, highly nonlinear, and distributed information-processing system*. The approach we have in mind is that of neural networks.[3]

To begin, we say: a "developing" nervous system is synonymous with a plastic brain. *Plasticity* permits the developing nervous system to *adapt* to its surrounding environment. Just as plasticity appears to be essential to the functioning of neurons as information-processing units in the human brain, so it is with neural networks made up of artificial nonlinear neurons; nonlinearity is a desirable property to have in neuronal processing. In its most general form, a *neural network* is a machine that is designed to *model* the way in which the brain performs a particular task or function of interest; the network is usually implemented by using electronic components or is simulated in software on a computer. Later on in the chapter we focus on an important class of neural networks that perform useful computations through a process of *learning*. To achieve good performance, neural networks employ a massive interconnection of "neurons" (i.e. computational units). We may, thus, offer the following definition of a neural network viewed as an adaptive machine:[4]

A neural network is a massively parallel distributed processor made up of simple but nonlinear processing units that has a natural propensity for storing experiential knowledge and making it available for use. It resembles the brain in two respects:

(1) Knowledge is acquired by the network from its environment through a learning process.
(2) Interneuron connection strengths, known as synaptic weights, are used to store the acquired knowledge.

The procedure used to perform the learning process is called a *learning algorithm*, the function of which is to modify the synaptic weights (i.e. free parameters) of the network in an adaptive fashion to attain a desired design objective. The modification of synaptic weights through training provides the traditional method for the design of neural networks.

## 2.7.1   Benefits of neural networks

It is apparent that a neural network derives its computing power through, first, its massively parallel distributed structure and, second, its ability to learn and, therefore, generalize. *Generalization* refers to the neural network's production of reasonable outputs for inputs not encountered during training (learning). These two information-processing capabilities make it possible for neural networks to find good approximate realizations of a wide range of functions.

Neural networks offer the following useful properties and capabilities:

(1) *Nonlinearity.* An artificial neuron is typically (but not always) nonlinear. A neural network, made up of an interconnection of nonlinear neurons, is itself nonlinear. Moreover, the nonlinearity is of a special kind, in the sense that it is *distributed* throughout the network. Nonlinearity is a highly important property, particularly if the underlying physics of the function of interest is inherently nonlinear.

(2) *Input–output mapping.* A popular paradigm of learning, called *learning with a teacher*, or *supervised learning*, involves modification of the synaptic weights of a neural network by applying a set of labeled *training (task) examples*. Each example consists of a unique *input signal* and a corresponding *desired (target) response*. The network is presented with an example picked at random from the training set, and the synaptic weights (free parameters) of the network are modified to minimize the

difference between the desired response and actual response of the network produced by the input signal in accordance with an appropriate statistical criterion. The training of the network is repeated for many examples in the set, until the network reaches a steady state, whereafter there are no further significant changes in the synaptic weights. The previously applied training examples may be reapplied during the training session, but it should be in a different order. Thus, the network learns from the examples by constructing an *input–output mapping* for the problem at hand. Such an approach brings to mind the study of *nonparametric statistical inference*, which is a branch of statistics dealing with model-free estimation, or, from a biological viewpoint, *tabula rasa* learning; the term "nonparametric" is used here to signify the fact that no prior assumptions are made on a statistical model for the input data.

(3) *Adaptivity.* Neural networks have a built-in capability to *adapt* their synaptic weights to changes in the surrounding environment. In particular, a neural network trained to operate in a specific environment can be easily *retrained* to deal with minor changes in the operating environmental conditions. Moreover, when it is operating in a *nonstationary* environment (i.e. one where statistics change with time), a neural network may be designed to change its synaptic weights in real time. The natural architecture of a neural network for signal processing and control applications coupled with the adaptive capability of the network makes it a useful tool in adaptive signal processing and control. As a general rule, it may be said that the more adaptive we make the system, all the time ensuring that the system remains stable, the more robust its performance will likely be when the system is required to operate in a nonstationary environment. It should be emphasized, however, that adaptivity does not always lead to robustness; indeed, it may do the very opposite. For example, an adaptive system with short-time constants may change rapidly and, therefore, tend to respond to spurious disturbances, causing a drastic degradation in system performance. To realize the full benefits of adaptivity, the principal time constants of the system should be just long enough for the system to ignore spurious disturbances, yet short enough to respond to meaningful changes in the environment; the problem described here is referred to as the *stability–plasticity dilemma* (Grossberg, 1988).

(4) *Contextual information.* Knowledge is represented by the very structure and activation state of a neural network. Every neuron in the network is potentially affected by the global activity of all other neurons in the network. Consequently, contextual information is dealt with naturally by a neural network (Rogers and McLelland, 2004).

(5) *Fault tolerance.* A neural network, implemented in hardware form, has the potential to be inherently *fault tolerant*, or capable of robust computation, in the sense that its performance degrades gracefully under adverse operating conditions. For example, if a neuron or its connecting links are damaged, recall of a stored pattern is impaired in quality. However, owing to the distributed nature of information stored in the network, the damage has to be extensive before the overall response of the network is degraded seriously. Thus, in principle, a neural network exhibits a graceful degradation in performance rather than catastrophic failure.

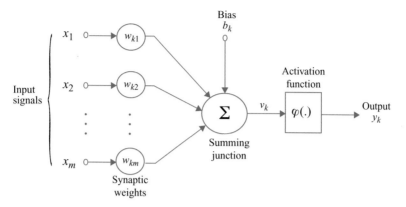

**Figure 2.2.** Nonlinear model of a neuron, labeled $k$; $v_k$ is referred to as the induced local field.

## 2.7.2   Models of a neuron

A *neuron* is an information-processing unit that is fundamental to the operation of a neural network. The block diagram of Figure 2.2 shows the *model of a neuron*, in which we identify three basic elements:

(1) A set of *synapses*, or *connecting links*, each of which is characterized by a *weight* or *strength* of its own. Specifically, a signal $x_j$ at the input of synapse $j$ connected to neuron $k$ is multiplied by the synaptic weight $w_{kj}$. It is important to make a note of the manner in which the subscripts of the synaptic weight $w_{kj}$ are written. The first subscript in $w_{kj}$ refers to the neuron in question and the second subscript refers to the input end of the synapse to which the weight refers. Unlike the weight of a synapse in the brain, the synaptic weight of an artificial neuron may lie in a range that includes negative as well as positive values.
(2) An *adder* for summing the input signals, weighted by the respective synaptic strengths of the neuron; the operations described here constitute a *linear combiner*.
(3) An *activation function* for limiting the amplitude of the output of a neuron. The activation function is also referred to as a *squashing function*, in that it squashes (limits) the permissible amplitude range of the output signal to some finite value; hence the non-linear property of the neuron. Typically, the normalized amplitude range of the output of a neuron is written as the closed unit interval [0, 1], or alternatively, [−1, 1].

The neural model of Figure 2.2 also includes an externally applied *bias*, denoted by $b_k$. The bias $b_k$ has the effect of increasing or lowering the net input of the activation function, depending on whether it is *positive (excitatory)* or *negative (inhibitory)* respectively. The overall signal $v_k$ produced at the output of the summing junction, including the bias, is referred to as the *induced local field*.

## 2.7.3   Multilayer feedforward networks

An important class of feedforward neural networks distinguishes itself by the presence of one or more *hidden layers*, whose computation nodes are correspondingly called

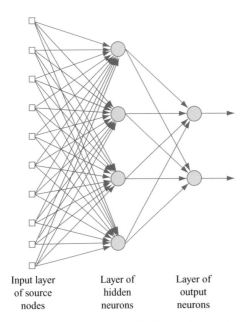

Input layer        Layer of        Layer of
of source          hidden          output
nodes              neurons         neurons

**Figure 2.3.** Fully connected feedforward network with one hidden layer and one output layer.

*hidden neurons* or *hidden units*. As mentioned previously, the term "hidden" refers to the fact that the hidden neurons are not seen directly from either the input or output of the network. The function of hidden neurons is to intervene between the external input and the network output in some useful manner. By adding one or more hidden layers, the network is enabled to extract higher order statistics from its input. In a rather loose sense, the network acquires a *global* perspective despite its local connectivity, due to the extra set of synaptic connections and the extra dimension of neural interactions (Churchland and Sejnowski, 1992).

The source nodes in the input layer of the network supply respective elements of the input vector, which constitute the input signals applied to the neurons' computation nodes in the first hidden layer. The output signals of this layer are used as inputs to the second hidden layer, and so on for the rest of the network. Typically, the neurons in each layer of the network have as their inputs the output signals of the preceding layer only; in other words, the network computes features of features. The set of output signals of the neurons in the output (final) layer of the network constitutes the overall response of the network to the input signal vector supplied by the source nodes. The architectural graph in Figure 2.3 illustrates the layout of a multilayer feedforward neural network for the case of a single hidden layer.

The neural network in Figure 2.3 is said to be *fully connected*, in the sense that every node in each layer of the network is connected to every other node in the adjacent forward layer. If, however, some of the communication links (synaptic connections) are purposely missed from the network for the purpose of reduced computational complexity and, quite possibly, improved performance, we say that the network is *partially connected*.

## 2.8 Associative learning process for memory construction

### 2.8.1 Pattern association

An *associative memory* is a brainlike distributed memory that learns by *association*, which has been known to be a prominent feature of human memory since the time of Aristotle. All models of cognition use association in one form or another as the basic operation (Anderson, 1995).

Association takes one of two forms: *autoassociation* or *heteroassociation*. In autoassociation, a neural network is required to *store* a set of patterns (vectors) by repeatedly presenting them to the network. The network is subsequently presented with a partial description or distorted (noisy) version of an original pattern stored in it, and the task is to *retrieve (recall)* that particular pattern. Heteroassociation differs from autoassociation in that an arbitrary set of input patterns is *paired* with another arbitrary set of output patterns. Autoassociation involves the use of unsupervised learning, whereas the type of learning involved in heteroassociation is supervised.

Let $x_k$ denote a *key pattern* (vector) applied to an associative memory and $y_k$ denote a *memorized pattern* (vector). The pattern association performed by the network is described by

$$x_k \rightarrow y_k, \qquad k = 1, 2, ..., q, \qquad (2.1)$$

where $q$ is the number of patterns stored in the network. The key pattern $x_k$ acts as a stimulus that not only determines the storage location of memorized pattern $y_k$, but also holds the key for its retrieval.

In an autoassociative memory, we have $y_k = x_k$; hence, the input and output (data) spaces of the network have exactly the same dimensionality. In a heteroassociative memory, on the other hand, we have $y_k \neq x_k$; hence, the dimensionality of the output space in this second case may or may not equal the dimensionality of the input space.

There are two phases involved in the operation of an associative memory:

- *storage space*, which refers to the training of the network in accordance with (2.1), and
- *recall phase*, which involves the retrieval of a memorized pattern in response to the presentation of a noisy or distorted version of a key pattern to the network.

Let the stimulus (input) $x$ represent a noisy or distorted version of a key pattern $x_j$. This stimulus produces a response (output) $y$, as indicated in Figure 2.4. For perfect recall, we should find that $y = y_j$, where $y_j$ is the memorized pattern associated with the key pattern $x_j$. When $y \neq y_j$ for $x = x_j$, the associative memory is said to have made an *error in recall*.

Figure 2.4. Input–output relation of pattern associator.

The number of patterns $q$ stored in an associative memory provides a direct measure of the *storage capacity* of the network. In designing an associative memory, the challenge is to make the storage capacity $q$ (expressed as a percentage of the total number $n$ of neurons used to construct the network) as large as possible, yet insist that a large fraction of the memorized patterns is recalled correctly.

## 2.8.2    Replicator (identity) mapping

The hidden neurons of a multilayer perceptron (MLP) play a critical role as feature detectors. A novel way in which this important property of the MLP can be exploited is in its use as a *replicator* or *identity map* (Rumelhart *et al.*, 1986). Figure 2.5 illustrates how this can be accomplished for the case of an MLP using a single

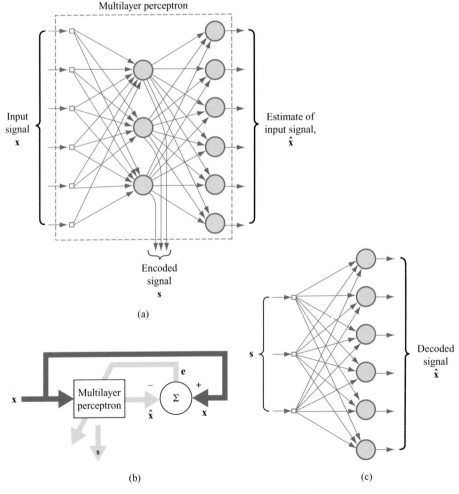

**Figure 2.5.** (a) Replicator network (identity map) with a single hidden layer used as encoder. (b) Block diagram for the supervised training of the replicator network. (c) Part of the replicator network used as decoder.

hidden layer. The network layout satisfies the following structural requirements, as illustrated in Figure 2.5a:

- the input and output layers have the same size, $m$;
- the size of the hidden layer is smaller than $m$; and
- the network is fully connected.

A given pattern $\mathbf{x}$ is simultaneously applied to the input layer as the stimulus and to the output layer as the desired response. The actual response of the output layer $\hat{\mathbf{x}}$ is intended to be an "estimate" of $\mathbf{x}$. The network is trained using a learning algorithm, with the estimation error vector, $\mathbf{x} - \hat{\mathbf{x}}$, treated as the error signal, as illustrated in Figure 2.5b. As such, the training is performed in an *unsupervised* manner (i.e. without the need for a teacher). By virtue of the special structure built into the design of the MLP, the network is *constrained* to perform identity mapping through its hidden layer. An *encoded* version of the input pattern, denoted by $\mathbf{s}$, is produced at the output of the hidden layer, as indicated in Figure 2.5a. In effect, the fully trained MLP performs the role of an "encoder." To reconstruct an estimate $\hat{\mathbf{x}}$ of the original input pattern $\mathbf{x}$ (i.e. to perform decoding), we apply the encoded signal to the hidden layer of the replicator network, as illustrated in Figure 2.5c. In effect, this latter network performs the role of a "decoder." The smaller we make the size of the hidden layer compared with the size $m$ of the input–output layers, the more effective the configuration of Figure 2.5a will be as a *data-compression system*.[5]

## 2.9    Back-propagation algorithm

There are many learning algorithms that have been devised to train an MLP for a prescribed task of interest (Haykin, 2009). Among these algorithms, the *back-propagation algorithm* is simple to implement, yet effective in performance. There are two phases to the algorithm:

(1) In the *forward phase*, the synaptic weights of the network are fixed and the input signal is propagated through the network, layer by layer, until it reaches the output. Thus, in this phase of the learning process, changes are confined to the activation potentials and outputs of the neurons in the network.

(2) In the *backward phase*, an error signal is produced by comparing the output of the network with a desired response. The resulting error signal is propagated through the network, again layer by layer, but this time the propagation is performed in the backward direction. In this second phase of the learning process, successive adjustments are made to the synaptic weights of the network. Calculation of the adjustments for the output layer is straightforward, but it is much more challenging for the hidden layers.

Usage of the term "back propagation" appears to have evolved after 1985, when the term was popularized through the publication of the seminal book entitled *Parallel Distributed Processing* (Rumelhart and McClelland, 1986).

## 2.9.1    Summary of the back-propagation algorithm

Figure 2.6a presents the architectural layout of an MLP. The corresponding signal-flow graph for back-propagation learning, incorporating both the forward and backward phases of the computations involved in the learning process, is illustrated in Figure 2.6b. The top part of the signal-flow graph accounts for the forward pass. The lower part of the signal-flow graph accounts for the backward pass, which is referred to as a *sensitivity*

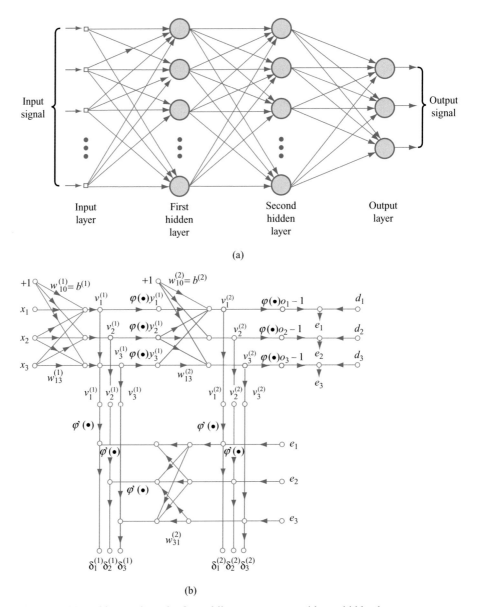

(a)

(b)

**Figure 2.6.** (a) Architectural graph of a multilayer perceptron with two hidden layers. (b) Signal-flow graphical summary of back-propagation learning. Top part of the graph refers to forward pass and bottom part of the graph refers to backward pass.

*graph* for computating the local gradients in the back-propagation algorithm (Narendra and Parthasarathy, 1990).

The *stochastic gradient method* of updating of weights is the preferred method for on-line implementation of the back-propagation algorithm. For this mode of operation, the algorithm cycles through the training sample $\{(\mathbf{x}(n), \mathbf{d}(n))\}_{n=1}^{N}$, where $\mathbf{x}(n)$ is the input signal, $\mathbf{d}(n)$ is the desired response, and $N$ is the size of the training sample. The algorithm proceeds as follows (Haykin, 2009):

(1) *Initialization.* Assuming that no prior information is available, the synaptic weights and thresholds are picked from a uniform distribution whose mean is zero and whose variance is chosen to make the standard deviation of the induced local fields of the neurons lie at the transition between the linear and standard parts of a sigmoid activation function.[6]

(2) *Presentations of training examples.* The network is presented with an epoch of training examples. For each example in the sample, ordered in some fashion, the sequence of forward and backward computations described under points (3) and (4) are performed respectively.

(3) *Forward computation.* Let a training example in the epoch be denoted by $(\mathbf{x}(n), \mathbf{d}(n))$, with the input vector $\mathbf{x}(n)$ applied to the input layer of sensory nodes and the desired response vector $\mathbf{d}(n)$ presented to the output layer of computation nodes. The induced local fields and function signals of the network are computed by proceeding forward through the network, layer by layer. The induced local field $v_j^{(l)}(n)$ for neuron $j$ in layer $l$ is

$$v_j^{(l)}(n) = \sum_i w_{ji}^{(l)}(n) y_i^{(l-1)}(n), \qquad (2.2)$$

where $y_i^{(l-1)}(n)$ is the output (function) signal of neuron $i$ in the previous layer $l-1$ at iteration $n$, and $w_{ji}^{(l)}(n)$ is the synaptic weight of neuron $j$ in layer $l$ that is fed from neuron $i$ in layer $l-1$. For $i=0$, we have $y_0^{(l-1)}(n)=+1$, and $w_{j0}^{(l)}(n)=b_j^{(l)}(n)$ is the *bias* applied to neuron $j$ in layer $l$. Assuming the use of a sigmoid function, the output signal of neuron $j$ in layer $l$ is

$$y_i^{(l)} = \varphi(v_j(n)).$$

If neuron $j$ is in the first hidden layer (i.e. $l=1$), we set

$$y_j^{(0)}(n) = x_j(n),$$

where $x_j(n)$ is the $j$th element of the input vector $\mathbf{x}(n)$. If neuron $j$ is in the output layer (i.e. $l=L$, where $L$ is referred to as the *depth* of the network), we set

$$y_j^{(L)} = o_j(n),$$

where $o_j(n)$ denotes the output signal. Correspondingly, the error signal

$$e_j(n) = d_j(n) - o_j(n) \qquad (2.3)$$

is computed for all $j$, where $d_j(n)$ is the $j$th element of the desired response vector $\mathbf{d}(n)$.

(4) *Backward computation.* The local gradients of the network, denoted by $\delta$s, are computed, using the definition

$$\delta_j^{(l)}(n) = \begin{cases} e_j^{(L)}(n)\varphi_j'(v_j^{(L)}(n)) & \text{for neuron } j \text{ in output layer } l = L, \\ \varphi_j'(v_j^{(l)}(n))\sum_k \delta_k^{(l+1)}(n)w_{kj}^{(l+1)}(n) & \text{for neuron } j \text{ in hidden layer } l < L, \end{cases} \quad (2.4)$$

where the prime in $\varphi_j'(\cdot)$ denotes differentiation with respect to the argument. The synaptic weights of the network in layer $l$ are adjusted according to the *generalized delta rule* as follows:

$$w_{ji}^{(l)}(n+1) = w_{ji}^{(l)}(n) + \alpha[w_{ji}^{(l)}(n) - w_{ji}^{(l)}(n-1)] + \eta\delta_j^{(l)}(n)y_i^{(l-1)}(n), \quad (2.5)$$

where $\eta$ is the learning-rate parameter and $\alpha$ is the momentum constant; the momentum term is included on the right-hand side of (2.5) to help the algorithm escape from local minima.

(5) *Iteration.* The forward and backward computations under points (3) and (4) are iterated by presenting new epochs of training examples to the network until the chosen stopping criterion is met.

*Notes.* The order of presentation of training examples should be randomized from epoch to epoch. The momentum and learning-rate parameter are typically adjusted (and usually decreased) as the number of training iterations increases.

## 2.10 Recurrent multilayer perceptrons

As the name implies, a *recurrent multilayer perceptron* (RMLP) is a neural network with one or more hidden layers, and with each computation layer of the network having feedback around it. Such a network is illustrated in Figure 2.7 for the case of an RMLP with two hidden layers. For the same reasons that ordinary (static) MLP are often more effective and parsimonious than neural networks with a single hidden layer, so it is with an RMLP. Moreover, the use of one or more feedback loops in the network, as illustrated in Figure 2.7, makes the RMLP not only *dynamic*, but also more computationally powerful compared with an ordinary MLP.

Let the vector $\mathbf{x}_{\mathrm{I},n}$ denote the output of the first hidden layer, $\mathbf{x}_{\mathrm{II},n}$ denote the output of the second hidden layer, and so on. Let the vector $\mathbf{x}_{o,n}$ denote the ultimate output of the output layer. Then, the general dynamic behavior of the RMLP in response to an input vector $\mathbf{u}_n$ is described by a system of coupled equations given as

$$\begin{aligned} \mathbf{x}_{\mathrm{I},n+1} &= \phi_{\mathrm{I}}(\mathbf{x}_{\mathrm{I},n}, \mathbf{u}_n), \\ \mathbf{x}_{\mathrm{II},n+1} &= \phi_{\mathrm{II}}(\mathbf{x}_{\mathrm{II},n}, \mathbf{x}_{\mathrm{I},n+1}), \\ &\vdots \\ \mathbf{x}_{o,n+1} &= \phi_o(\mathbf{x}_{o,n}, \mathbf{x}_{K,n+1}), \end{aligned} \quad (2.6)$$

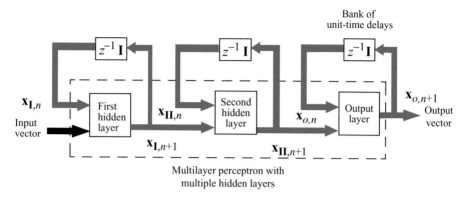

Multilayer perceptron with
multiple hidden layers

**Figure 2.7.** RMLP; feedback paths in the network are printed in color.

where $\phi_{\mathbf{I}}(.,.)$, $\phi_{\mathbf{II}}(.,.)$, …, $\phi_o(.,.)$ denote the activation functions characterizing the first hidden layer, second hidden layer, …, and output layer of the RMLP respectively, and $K$ denotes the number of hidden layers in the network; in Figure 2.7, $K = 2$. Note that in (2.6) the iteration number $n$ is used as a subscript to simplify the notation.

## 2.11 Self-organized learning

Supervised learning, exemplified by the back-propagation algorithm, requires the availability of a desired response vector, against which the actual output signal vector of the neural network is measured. This requirement, as useful as it is, limits the practical application of supervised-learning algorithms. To overcome this limitation, we look to self-organized or unsupervised learning procedures. Just as there are several ways of formulating supervised learning algorithms, so it is with unsupervised learning algorithms (Haykin, 2009). In this section, we describe an unsupervised learning algorithm called the *generalized Hebbian algorithm* (GHA), which was first described in the literature by Sanger (1989).

### 2.11.1 Hebb's postulate of learning

As the algorithm's name implies, the GHA is rooted in *Hebb's postulate of learning* (Hebb, 1949), which is the oldest and most famous of all learning rules. In signal-processing terms, this learning rule is based on a *correlational mechanism*, described as follows (Haykin, 2009):

(1) If two neurons on either side of a synapse (i.e. connecting link) are activated synchronously, then the weight of that synapse is significantly increased.
(2) If, on the other hand, the two neurons are activated asynchronously, then that synapse is selectively weakened in strength or eliminated altogether over the course of time.

Mathematically, we may thus formulate Hebb's postulate of learning simply as follows:

$$\Delta w_{kj}(n) = \eta y_k(n)x_j(n), \qquad (2.7)$$

where $\Delta w_{kj}(n)$ denotes the change in the strength (weight) of the synapse connecting node $k$ to node $j$ at time $n$, $x_j(n)$ is the input signal, $y_k(n)$ is the output signal, and $\eta$ is the learning-rate parameter. From (2.7), we find that the repeated application of the input signal $x_j(n)$ leads to an increase in the output signal $y_k(n)$ and, therefore, *exponential growth* that finally drives the synapse into saturation. At this point, no new information is stored in the synapse and selectivity is thereby lost. Some mechanism is therefore needed to stabilize the self-organized behavior of the neurons, which may be achieved through the incorporation of *competition* into the learning rule, as described next.

## 2.11.2    Generalized Hebbian algorithm

Consider the feedforward neural network of Figure 2.8, for the operation of which the following two structural assumptions are made:

(1) Each neuron in the output layer of the network is *linear*.
(2) The network has $m$ inputs and $l$ outputs, with fewer outputs than inputs; that is, $l < m$.

The only aspect of the network that is subject to training is the set of synaptic weights $\{w_{ji}\}$ connecting the source nodes $i$ in the input layer to computational nodes $j$ in the output layer, where $i = 1, 2, \ldots, m$ and $j = 1, 2, \ldots, l$. The output $y_j(n)$ of neuron $j$ at time $n$ produced in response to the set of inputs $\{x_i(n)\}_{i=1}^{m}$ is given as follows:

$$y_j(n) = \sum_{i=1}^{m} w_{ji}(n)x_i(n), \qquad j = 1, 2, \ldots, l. \tag{2.8}$$

The synaptic weight $w_{ji}(n)$ is adapted in accordance with a generalized form of Hebbian learning, as shown by

$$\Delta w_{ji}(n) = \eta \left( y_j(n)x_i(n) - y_j(n)\sum_{k=1}^{j} w_{ki}(n)y_k(n) \right), \tag{2.9}$$

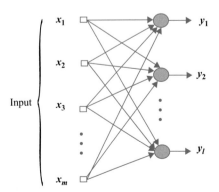

**Figure 2.8.** Feedforward network with single layer of computational nodes. The sets $\{x_i\}_{i=1}^{m}$ and $\{y_j\}_{j=1}^{l}$ constitute the input and output vectors of the network respectively.

where $i = 1, 2, \ldots, m$ and $j = 1, 2, \ldots, l$. The term $\Delta_{ji}(n)$ is the change applied to the synaptic weight $w_{ji}(n)$ at time $n$ and $\eta$ is the learning-rate parameter.

Examining (2.9), we see that the term $\eta y_j(n)x_i(n)$ on the right-hand side of the equation is attributed to Hebbian learning. As for the second term

$$\eta y_j(n) \sum_{k=1}^{j} w_{ki}(n)y_k(n),$$

we attribute it to a *competitive process* that goes on among the synapses in the network. Simply put, as a result of this process, the most vigorously growing (i.e. fittest) synapses or neurons are selected at the expenses of the weaker ones. Indeed, it is this competitive process that alleviates the exponential growth in Hebbian learning working by itself. Note that stabilization of the algorithm through competition requires the use of a minus sign on the right-hand side of (2.9).

Equations (2.8) and (2.9) sum up the GHA. The distinctive feature of this algorithm is that it operates in a self-organized manner, thereby avoiding the need for a desired response. It is this important characteristic of the algorithm that befits it for *on-line learning*.[7]

## 2.11.3 Signal-flow graph of the GHA

To develop insight into the behavior of the GHA, we rewrite (2.9) in the form

$$\Delta w_{ji}(n) = \eta y_j(n)[x'_j(n) - w_{ji}(n)y_j(n)], \qquad \begin{matrix} i = 1, 2, \ldots, m \\ j = 1, 2, \ldots, l \end{matrix}, \qquad (2.10)$$

where $x_i'(n)$ is a modified version of the $i$th element of the input vector $\mathbf{x}(n)$; it is a function of the index $j$, as shown by

$$x_i'(n) = x_i(n) - \sum_{k=1}^{j-1} w_{ki}(n)y_k(n). \qquad (2.11)$$

Next, we go one step further and rewrite (2.10) in a form that corresponds to Hebb's postulate of learning, as shown by

$$\Delta w_{ji}(n) = \eta y_j(n)x_i''(n) \qquad (2.12)$$

where

$$x_i''(n) = x_i'(n) - w_{ji}(n)y_j(n). \qquad (2.13)$$

Thus, noting that

$$w_{ji}(n+1) = w_{ji}(n) + \Delta w_{ji}(n) \qquad (2.14)$$

and

$$w_{ji}(n) = z^{-1}[w_{ji}(n+1)], \qquad (2.15)$$

where $z^{-1}$ is the unit-time delay operator, we may construct the signal-flow graph of Figure 2.9 for the GHA. From this graph, we see that the algorithm lends itself

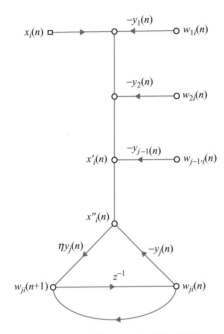

**Figure 2.9.** Signal-flow graph of the GHA.

to a *local* form of implementation, provided that it is formulated as in (2.14). Note also that $y_j(n)$, responsible for feedback in the signal-flow graph in Figure 2.9 is itself determined by (2.8); signal-flow graph representation of this latter equation is shown in Figure 2.8.

## 2.12    Summary and discussion

### 2.12.1    Cognition

This chapter presented a detailed description of the many facets of the perception–action cycle, which is of fundamental importance to the study of cognitive dynamic systems. There are four basic functions embodied in this cyclic operation:

(1) *Perception* of the environment in the perceptor, followed by *action* taken by the actuator in the environment in response to feedback information computed by the receptor.

(2) *Perceptual memory* in the perceptor and *executive memory* in the actuator, and the reciprocal coupling between them via the working memory; the function of memory is to predict the consequences of actions in the system.

(3) *Attention* to prioritize the allocation of available resources in the system.

(4) *Intelligence*, which is best described as the ability of the system to continually adjust itself through an adaptive process by responding to new changes in the environment, with new forms of action and behavior being created through a decision-making mechanism based on intelligent choices in the face of uncertainties in the environment.

## 2.12.2    Two different views of perception

With perception in the receiver looking to the environment for stimuli (observables) to process and the environment itself being stochastic, it follows that, in reality, perception is a *probabilistic process*. As such, we may formulate the problem of perception as follows:

Given a set of stimuli received from the environment, estimate the hidden state of the environment in the environmental scene analyzer as accurately as possible.

To solve this state-estimation problem, we may view perception as an *ill-posed inverse problem*. Perception is an "inverse problem" because we are trying to uncover the underlying physical laws responsible for generation of the stimuli; that is, attempting to find a *mapping from the stimuli to the state*. In practice, we typically find that inverse problems are *ill-posed* because, for one reason or another, they violate the conditions for well-posedness. According to Hadamard, for the problem of "mapping from stimuli to state" to be well posed, three conditions must be satisfied (Haykin, 2009):

- *existence,*
- *uniqueness,* and
- *continuity,* also referred to as *stability,*

which clearly speak for themselves. For example, power-spectrum estimation from a finite set of observable data, to be discussed in Chapter 3, is an ill-posed inverse problem because as we shall see in Chapter 3, the condition for uniqueness is violated. The issue of spectrum is central to the study of cognitive radio, the very purpose of which is to identify the spectrum holes (i.e. unused or partially used subbands of the radio spectrum) and exploit them for improved utilization of the radio spectrum. In effect, the spectrum holes define the "state" of the radio environment insofar as cognitive radio is concerned. In any event, the spectrum can be identified directly from the incoming radio-frequency (RF) stimuli through the application of a power-spectrum estimation procedure.

We may solve the state-estimation problem in yet another way: perception is viewed as a process of *Bayesian inference*. In this second viewpoint, information contained in the stimuli about an object of interest is represented by a conditional probability density function over the unknown state that is commonly referred to as the *posterior*. Through the adoption of Bayesian inference, the posterior is formulated as the product of two probabilistic functions: the likelihood and the prior. The likelihood is a function of the unknown state given the stimuli. As for the prior, it provides information already available about the environment "before" the estimation is performed. The perception problem is then solved through a recursive estimation procedure aimed at computing the state for which the posterior attains its maximum value. This viewpoint, adopted in Chapter 4, is well suited for cognitive radar, where the perception problem involves estimating the state of an unknown target.

The two views of perception discussed herein, namely

(1) the ill-posed inverse problem and
(2) Bayesian inference,

pave the way nicely for the next two chapters: "Power-spectrum estimation for sensing the environment" and "Bayesian filtering for state estimation of the environment" respectively.

## Notes and practical references

1. *Working memory*
   The term "working memory" is of recent origin; it was coined by Baddeley (2003). What we now call working memory was previously referred to as "short-term memory," which is the ability to remember stored information over a short period of time.
   However, there is no unanimity in the neurophysiological community on working memory. According to Fuster (2003), the term working memory is a "theoretical" construct that is essentially long term that has been activated for a short time for the processing of a prospective action.
   Nevertheless, we follow the description coined by Baddeley. From an engineering perspective, Baddeley's description befits the role described in Section 2.3 for working memory.
2. *Information overload*
   It appears that, long before the advent of the Internet, the term "information overload" was mentioned by Gross (1964) in his book on *The Managing of Organizations*. According to Wikipedia, its popularization is attributed to Alvin Toffler.
3. *Neural networks*
   For books on neural networks and learning machines, the reader is referred to:
   • *Neural Networks and Learning Machines* by Haykin (2009),
   • *Pattern Recognition and Machine Learning* by Bishop (2006),
   • *The Elements of Statistical Learning* by Hastie *et al.* (2001), and
   • *Parallel Distributed Processing, Volume 1: Foundations* by Rumelhart and McClelland (1986).
4. *Definition of neural network*
   This definition of a neural network is adapted from Aleksander and Morton (1990).
5. *Replicator neural network*
   Hecht-Nielsen (1995) describes a replicator neural network in the form of an MLP with an input layer of source nodes, three hidden layers, and an output layer:
   • The activation functions of neurons in the first and third hidden layers are defined by the hyperbolic tangent function

   $$\varphi^{(1)}(v) = \varphi^{(3)}(v) = \tanh(v),$$

   where $v$ is the induced local field of a neuron in those layers.
   • The activation function for each neuron in the second hidden layer is given by

   $$\varphi^{(2)}(v) = \frac{1}{2} + \frac{2}{2(N-1)} \sum_{j=1}^{N-1} \tanh\left[a\left(v - \frac{j}{N}\right)\right],$$

where $a$ is a gain parameter and $v$ is the induced local field of a neuron in that layer. The function $\varphi^{(2)}(v)$ describes a smooth staircase activation function with $N$ treads, thereby essentially quantizing the vector of the respective neural outputs into $K = N^n$, where $n$ is the number of neurons in the middle hidden layer.
- The neurons in the output layer are linear, with their activation functions defined by

$$\varphi^{(4)}(v) = v.$$

- Based on this neural network structure, Hecht-Nielsen describes a theorem showing that optimal data compression for arbitrary input data vector can be carried out.

6. *Activation functions*

The *sigmoid activation function* is defined by

$$\varphi(v) = \frac{1}{1 + \exp(-v)},$$

where $v$ is the input and $\varphi(v)$ is the output. Note that

$$\varphi(v) = \begin{cases} 1, & v = \infty \\ 1/2, & v = 0 \\ 0, & v = -\infty. \end{cases}$$

The input $v$ applied to the sigmoid function of a neuron is referred as the *induced local field* of the neuron.

Another activation function, representing an alternative to the sigmoid function, is the *hyperbolic tangent function*, defined by

$$\varphi(v) = \tanh(v)$$
$$= \frac{1 - \exp(-2v)}{1 + \exp(-2v)},$$

for which we now have

$$\varphi(v) = \begin{cases} 1, & \text{for } v = \infty \\ 0, & \text{for } v = 0 \\ -1, & \text{for } v = -\infty. \end{cases}$$

Whereas the sigmoid function never assumes a negative value, the hyperbolic tangent function is symmetric about the horizontal axis and may, therefore, assume both positive and negative values.

7. *Unsupervised learning*

Self-organization through the use of Hebbian learning provides one method for the formulation of unsupervised learning algorithms. Another way in which unsupervised learning can be accomplished is to look to information theory for ideas. In particular, we find that the idea of mutual information applied to a feedforward neural network provides a powerful basis for unsupervised learning. (Mutual information is

discussed in Note 8 in Chapter 6.) For example, we may start with the mutual information between the input signal and output signal of the network. By maximizing
this mutual information, viewed as the objective function of interest, the network
is enabled to extract the important features that characterize the input signal. For a
detailed treatment of information-theoretic learning models, the interested reader is
referred to Haykin (2009: Chapter 10).

Another important way in which on-line learning can be accomplished is through
*reinforcement learning*, which builds its formalism on experience gained through
continued interactions of an *agent* with its environment; discussion of this important
approach to on-line learning is taken up in Chapter 5.

# 3 Power-spectrum estimation for sensing the environment

As discussed in Chapter 2, perception of the environment is the first of four defining properties of cognition, putting aside language. To cater for perception in an engineering context, we not only need to sense the environment, but also focus attention on an underlying *discriminant* of the environment that befits the application of interest. In this chapter, we focus attention on *spectrum estimation*, which is of particular interest to sensing the environment for cognitive radio.[1] Moreover, with radar relying on the radio spectrum for its operation, spectrum sensing is also of potential interest to cognitive radar.

With spectrum estimation viewed as the discriminant for sensing the environment, there are four related issues that are of practical interest, each in its own way:

(1) *Power spectrum*, where the requirement is to estimate the average power of the incoming signal expressed as a function of frequency.
(2) *Space–time processing*, where the average power of the incoming signal is expressed as a function of frequency and spatial coordinates.
(3) *Time–frequency analysis*, in which the average power of the signal is expressed as a function of both time and frequency.
(4) *Spectral line components*, for which a special test is needed.

Throughout the presentation, the emphasis will be on the use of a finite set of observable data (measurements), which is how it is in practice.

In Section 2.12 we stated that perception in a cognitive dynamic system may be viewed as an *ill-posed inverse problem*. With power-spectrum estimation[2] being an example of perception of the environment, that statement applies equally well to each of the four issues, (1) through (4).

## 3.1 The power spectrum

We start the discussion by recognizing that the incoming signal picked up by an environmental sensor(s) is typically the *sample function of a stochastic (random) process*. As such, although it is not possible to predict the exact value of the signal at some point of time in the future, it is possible to describe the signal in terms of statistical parameters, such as the average power of the signal distributed across frequency. This parameter is called the *power spectrum* or *power spectral density* of the signal; both terms are used interchangeably in the literature, as well as in this book.

A stochastic process is said to be *wide-sense stationary* or *weakly stationary* if the following two conditions are satisfied:

(1) The *mean*, defined by the expectation of any random variable sampled from the process, is independent of time.
(2) The *autocorrelation function*, defined by the expectation of the product of any random variable sampled from the process and its delayed version, is dependent solely on the delay.

It turns out that, when the stochastic process is weakly stationary, the autocorrelation function and power spectrum of the process form a *Fourier-transform pair*. In other words, in accordance with the *Wiener–Khintchine relations*, the power spectrum is the Fourier transform of the autocorrelation function and, by the same token, the autocorrelation function is the inverse Fourier transform of the power spectrum (Haykin, 2000). Of these two functions, however, we find that, in practice, the power spectrum is the more important one of the two. Intuitively, this makes sense, considering the practical importance of average power expressed as a function of frequency.

## 3.2        Power spectrum estimation

With the power spectrum representing an important characteristic of stochastic processes, an issue of practical importance is how to *estimate* it. Unfortunately, this issue is complicated by the fact that there is a bewildering array of power-spectrum estimation procedures, with each procedure purported to have or to show some optimum property. The situation is made worse by the fact that, unless care is taken in the selection of the right method, we may end up with misleading conclusions.

### 3.2.1        Parametric methods

In parametric methods of spectrum estimation, we begin by postulating a stochastic model for the situation at hand. Depending on the specific model adopted, we may identify three different parametric approaches for spectrum estimation:

(1) *Model-identification procedures.* In this class of parametric methods, a rational function or a polynomial in the exponential $e^{-j2\pi f}$ is assumed for the transfer function of the model and a white-noise source is used to derive the model, as depicted in Figure 3.1. The power spectrum of the resulting model output provides the desired spectrum estimate. Depending on how the stochastic model is formulated, there are three *linear* models to be considered:
   (i) *autoregressive (AR) model* with an all-pole transfer function;
   (ii) *moving-average (MA) model* with an all-zero transfer function; and
   (iii) *autoregressive-moving-average (ARMA) model* with a pole–zero transfer function.

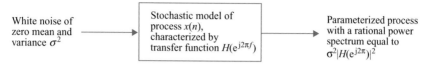

**Figure 3.1.** Rationale for model-identification procedure for power spectrum estimation.

**Figure 3.2.** Rationale of MVDR procedure for power spectrum estimation.

The resulting power spectra measured at the outputs of these models are referred to as AR, MA, and ARMA spectra respectively. The input–output relation of the model is defined by

$$S_o(f) = |H(e^{j2\pi f})|^2 \, S_i(f),$$

where $f$ denotes frequency, $H(e^{j2\pi f})$ denotes the frequency response of the model, $S_i(f)$ denotes the power spectrum of the signal at the model input, and $S_o(f)$ denotes the power spectrum of the model output. With white noise used as the driving force, the power spectrum $S_i(f)$ of the model input is equal to the white-noise variance $\sigma^2$. We then find that the power spectrum $S_o(f)$ of the model output is equal to the squared amplitude response $|H(e^{j2\pi f})|^2$ of the model, multiplied by a constant, namely, $\sigma^2$. The problem thus becomes one of estimating the model parameters that characterize the transfer function $H(e^{j2\pi f})$, such that the process produced at the model output provides an acceptable representation (in some statistical sense) of the stochastic process under study. Such an approach to power-spectrum estimation may, therefore, be viewed as a problem in *model (system) identification*.

Among the model-dependent spectra defined herein, the AR spectrum is by far the most popular. The reason for this popularity is twofold: (1) the *linear* form of the system of simultaneous equations involving the unknown AR model parameters and (2) the availability of efficient algorithms for computing the solution.

(2) *Minimum-variance distortionless response (MVDR) method.* To describe this second parametric approach for power-spectrum estimation, consider the situation depicted in Figure 3.2. The process, represented by the time series $\{x(n)\}$, is applied to a transversal filter (i.e. a discrete-time filter with an all-zero transfer function). In the *MVDR method*, the filter coefficients are chosen to minimize the variance (which is the same as the expected power for a zero-mean process) of the filter output, subject to the constraint that the frequency response of the filter is equal to unity at some desired frequency $f_0$. Under this constraint, the time series $\{x(n)\}$ is passed through the filter with no distortion at the frequency $f_0$. Moreover, signals at frequencies other than $f_0$ tend to be attenuated.

(3) *Eigendecomposition-based methods.* In this third class of parametric spectrum estimation methods, the eigendecomposition of the ensemble-average correlation matrix of the time series $\{x(n)\}$ is used to define two disjoint subspaces: *signal subspace* and *noise subspace*. This form of partitioning is then exploited to derive an appropriate algorithm for estimating the power spectrum of the process.

## 3.2.2    Nonparametric methods

In nonparametric methods of power-spectrum estimation, no assumptions are made with respect to the stochastic process under study insofar as process modeling is concerned. In other words, nonparametric methods are *model free*. Here, we may identify two different nonparametric approaches:

(1) *Periodogram-based methods.* Given an infinitely long time series $\{x(n)\}_{n=0}^{N-1}$, its power spectrum is defined by

$$S(f) = \lim_{N \to \infty} \frac{1}{N} \mathbb{E}[|X_N(f)|^2], \qquad (3.1)$$

where $X_N(f)$ is the Fourier transform of $\{x(n)\}_{n=0}^{N-1}$ and $\mathbb{E}$ is the statistical expectation operator. Mathematical justification for the definition of power spectrum of a stationary process in (3.1) is presented in Note 2 at the end of the chapter. Note also that, in (3.1), the expectation is performed first, followed by the limit operation for the data size $N$ to approach infinity; these two operations cannot, therefore, be interchanged.

According to this formula, the *periodogram*, namely $|X_N(f)|^2/N$, is adopted as the starting point for data analysis. However, the periodogram suffers from a serious limitation, in the sense that *it is not a sufficient statistic for the power spectrum*. We say so because phase information about the data $\{x(n)\}$ is ignored in computing the periodogram. Consequently, the statistical insufficiency of the periodogram is inherited by any estimator based on or equivalent to the periodogram.

(2) *Multitaper method (MTM).* A more constructive nonparametric approach is to adopt a set of special sequences known as Slepian sequences, which are fundamental to the study of time- and frequency-limited systems. The remarkable property of this family of windows is that the energy distributions of the windows (i.e. sequences) add up in a very special way that collectively defines an ideal rectangular frequency bin – ideal in the sense that the total in-bin versus out-of-bin energy concentrations is maximized. This property, in turn, allows us to trade spectral resolution for improved spectral properties (e.g. reduced variance of the spectral estimate).

When the stochastic process of interest has a purely continuous power spectrum and the underlying physical mechanism responsible for the generation of the process is unknown, then the recommended procedure is the nonparametric method of multiple windows (tapers), hereafter referred to as the *MTM*, which is considered next.

## 3.3   Multitaper method

In the older spectrum-estimation literature on nonparametric methods, it was emphasized that the estimation problem is difficult because of the *bias–variance dilemma*,
which encompasses the interplay between two conflicting statistical issues:

- *Bias* of the power-spectrum estimate of a time series due to the sidelobe-leakage phenomenon, the effect of which is *reduced by tapering* (i.e. *windowing*) the time series.
- The cost incurred by this improvement is an increase in *variance* of the estimate, which
  is due to the loss of information resulting from a reduction in the effective sample size.

How, then, can we resolve this dilemma by mitigating the loss of information due to
tapering? The answer to this fundamental question lies in the principled use of *multiple
orthonormal tapers (windows)*. Specifically, the procedure *linearly* expands the part of
the time series lying in a fixed bandwidth extending from $f - W$ to $f + W$ (centered on
some frequency $f$) in a special family of sequences known as the *Slepian sequences*;[3]
these sequences are also referred to in the literature as *discrete prolate spheroidal
sequences*. The remarkable property of Slepian sequences is that their Fourier transforms
have the *maximal energy concentration* in the bandwidth $2W$ (centered on $f$) under a
finite sample-size constraint. This property, in turn, permits trading *spectral resolution*
for improved spectral characteristics; that is, *reduced variance* of the spectral estimate
*without compromising the bias of the estimate*. In other words, the old bias–variance
tradeoff is now replaced with a *bias–resolution tradeoff*, which, once properly taken care
of, also solves the variance problem. The MTM can, therefore, produce an *accurate* estimate of the desired power spectrum, which is an important objective in its estimation.

### 3.3.1   Attributes of multitaper spectral estimation

From a practical perspective, multitaper spectral estimators have several desirable features, summarized herein (Haykin *et al.*, 2009):

(1) In multitaper spectral estimation, the *bias* is decomposed into two quantifiable
    components:
    - *local bias*, due to frequency components residing inside the user-selectable band
      from $f - W$ to $f + W$; and
    - *broadband bias*, due to frequency components found outside this band.
(2) The *resolution* of multitaper spectral estimators is naturally defined by width of the
    passband, namely $2W$.
(3) Multitaper spectral estimators offer an easy-to-quantify *tradeoff between bias and
    variance*.
(4) Direct spectrum estimation can be performed with more than just two *degrees of
    freedom* (DoFs); typically, the DoFs vary from 6 to 10, depending on the time–
    bandwidth product used in the estimation.
(5) Multitaper spectral estimation has a built-in form of *regularization*, which overcomes
    the ill-posed inverse problem associated with the power-spectrum estimation.[4]

With these highly desirable features built into the composition of a multitaper spectral estimator, it is not surprising, therefore, to find that it outperforms other well-known spectral estimators, as discussed later in the section.

## 3.3.2   Multitaper spectral estimation theory

Let $n$ denote *discrete time*, and the time series $\{x(n)\}_{n=0}^{N-1}$ represent the signal of interest. Given this time series, the MTM determines the following parameters:

- an orthonormal sequence of *Slepian tapers,* denoted by $\left\{v_n^{(k)}\right\}_{n=0}^{N-1}$; and
- a corresponding set of Fourier transforms, defined by

$$X_k(f) = \sum_{n=0}^{N-1} x(n)v_n^{(k)} \exp(-j2\pi fn), \tag{3.2}$$

where $k = 0, 1, \ldots, K-1$ and $K$ is the total number of Slepian tapers. The energy distributions of the *eigenspectra*, defined by $|X_k(f)|^2$ for varying $k$, are all concentrated inside a *resolution bandwidth* $2W$. The *time–bandwidth product*

$$C_o = NW,$$

therefore, bounds the number of tapers (windows) as shown by

$$K \le \lfloor 2NW \rfloor, \tag{3.3}$$

which, in turn, defines the DoFs available for controlling the variance of the multitaper spectral estimator. The choice of parameters $C_o$ and $K \le 2C_o$ provides a tradeoff between spectral resolution, bias, and variance. The bias of these estimates is largely controlled by the largest eigenvalue, denoted by $\lambda_0(N, W)$, which is given asymptotically by the approximate formula (Thomson, 1982)

$$1 - \lambda_0 \approx 4\pi\sqrt{C_o}\exp(-2\pi C_o).$$

This formula gives directly the fraction of the *total sidelobe energy*; that is, the total leakage into frequencies outside the interval $(-W, W)$. The total sidelobe energy decreases very rapidly with $C_o$, as may be seen in Table 3.1.

**Table 3.1.** Leakage properties of the lowest order Slepian sequence as a function of the time–bandwidth product $C_o$ (column 1). Column 2 of the table gives the asymptotic value of $1 - \lambda_0$ ($C_o$), and column 3 is the total sidelobe energy expressed in decibels (relative to total energy in the signal). Table reproduced with permission from S. Haykin, D.J. Thomson and J. Reed, 2009, Spectrum sensing for cognitive radio, *Proceedings of the IEEE*, 97, 849–877.

| $C_o = NW$ | $1 - \lambda_0$ | dB |
|:---:|:---:|:---:|
| 4 | $3.05 \times 10^{-10}$ | −95 |
| 6 | $1.31 \times 10^{-15}$ | −149 |
| 8 | $5.26 \times 10^{-21}$ | −203 |
| 10 | $2.05 \times 10^{-26}$ | −257 |

A natural spectral estimate, based on the first few eigenspectra that exhibit the least sidelobe leakage, is given by the normalized formula

$$\hat{S}(f) = \frac{\sum_{k=0}^{K-1} \lambda_k |X_k(f)|^2}{\sum_{K=0}^{K-1} \lambda_k},$$ (3.4)

where $X_k(f)$ is the Fourier transform of $x(n)$, defined in (3.2), and $\lambda_k$ is the eigenvalue associated with the $k$th eigenspectrum. It is the denominator in the MTM formula of (3.4) that makes the spectral estimator $\hat{S}(f)$ unbiased. This spectral estimator is intuitively appealing in the way it works: as the number of Slepian tapers $K$ increases, the eigenvalues decrease, causing the eigenspectra to be more contaminated by leakage. But, the eigenvalues themselves counteract by reducing the weighting applied to higher leakage eigenspectra.

### 3.3.3  Adaptive modification of multitaper spectral estimation

While the lower order eigenspectra have excellent bias properties, there is, unfortunately, some degradation as the order $K$ increases toward the limiting value defined in (3.3). To mitigate this shortcoming, a set of *adaptive weights*, denoted by $\{d_k(f)\}$, is introduced to downweight the higher order eigenspectra. Using a mean-square-error optimization procedure, the following formula for the weights is derived (Thomson, 1982):

$$d_k(f) = \frac{\sqrt{\lambda_k} S(f)}{\lambda_k S(f) + \mathbb{E}[B_k(f)]}, \quad k = 0, 1, \ldots, K-1,$$ (3.5)

where $S(f)$ is the true power spectrum, $B_k(f)$ is the broadband bias of the $k$th eigenspectrum, and $\mathbb{E}$ is the statistical expectation operator as before. Moreover, we find that

$$\mathbb{E}\,[B_k(f)] \leq (1 - \lambda_k)\sigma^2, \quad k = 0, 1, \ldots, K-1,$$ (3.6)

where $\sigma^2$ is the *process variance*, defined in terms of the time series $x(n)$ by

$$\sigma^2 = \frac{1}{N} \sum_{n=0}^{N-1} |x(n)|^2.$$ (3.7)

In order to compute the adaptive weights $d_k(f)$ using (3.5), we need to know the true spectrum $S(f)$. But if we did know it, then there would be no need to perform any spectrum estimation in the first place. Nevertheless, (3.5) is useful in setting up an *iterative procedure for computing the adaptive spectral estimator*, as shown by

$$\hat{S}(f) = \frac{\sum_{k=0}^{K-1} |d_k(f)|^2 \, \hat{S}_k(f)}{\sum_{k=0}^{K-1} |d_k(f)|^2},$$ (3.8)

where

$$\hat{S}_k(f) = |X_k(f)|^2, \quad k = 0, 1, \dots, K-1. \tag{3.9}$$

Note that if we set $|d_k(f)|^2 = \lambda_k$ for all $k$, then the estimator of (3.8) reduces to that of (3.4).

Next, setting the actual power spectrum $S(f)$ equal to the spectrum estimate $\hat{S}_k(f)$ in (3.5), then substituting the new equation into (3.8) and collecting terms, we get (after simplifications)

$$\sum_{K=0}^{K-1} \frac{\lambda_k (\hat{S}(f) - \hat{S}_k(f))}{(\lambda_k \hat{S}(f) + \hat{B}_k(f))^2} = 0, \tag{3.10}$$

where $\hat{B}_k(f)$ is an estimate of the expectation $\mathbb{E}[B_k(f)]$. Using the upper bound of (3.6), we may go on to set

$$\hat{B}_k(f) = (1 - \lambda_k)\sigma^2, \quad k = 0, 1, \dots, K-1. \tag{3.11}$$

We now have all the terms that we need to solve for the null condition of (3.10) via the *recursion*

$$\hat{S}^{(j+1)}(f) = \left[ \sum_{k=0}^{K-1} \frac{\lambda_k \hat{S}_k^{(j)}(f)}{\left( \lambda_k \hat{S}^{(j)}(f) + \hat{B}_k(f) \right)^2} \right] \left[ \sum_{k=0}^{K-1} \frac{\lambda_k}{\left( \lambda_k \hat{S}^{(j)}(f) + \hat{B}_k(f) \right)^2} \right]^{-1}, \tag{3.12}$$

where $j$ denotes an iteration step; that is, $j = 0, 1, 2, \dots$. To initialize the recursion of (3.12), for a reasonably good starting point, we may set $\hat{S}^{(j)}(0)$ equal to the average of the two lowest order eigenspectra. Convergence of the recursion described in (3.12) is usually rapid, with successive spectral estimates differing by less than 5% in 5 to 20 iterations (Drosopoulos and Haykin, 1992). The result obtained from (3.12) is substituted into (3.5) to obtain the desired weights, $d_k(f)$.

A useful by-product of the adaptive spectral estimation procedure is a *stability measure of the estimates*, given by

$$v(f) = 2 \sum_{k=0}^{K-1} |d_k(f)|^2, \tag{3.13}$$

which is the approximate DoF for the estimator $\hat{S}_k(f)$, expressed as a function of frequency $f$. If $\bar{v}$, denoting the average of $v(f)$ over frequency $f$, is significantly less than $2K$, then the result is an indication that either the bandwidth $W$ is too small or additional prewhitening of the time series $x(n)$ should be used.

The importance of *prewhitening* cannot be stressed enough, particularly for RF data. In essence, prewhitening reduces the dynamic range of the spectrum by filtering the data, prior to processing. The resulting residual spectrum is nearly flat or "white." In particular, leakage from strong components is reduced, so that the fine structure of weaker components is more likely to be resolved. In actual fact, most of the theory behind spectral estimation is smooth, almost white-like spectra to begin with, hence the need for "prewhitening."

### 3.3.4   Summarizing remarks on the MTM

(1) Estimation of the power spectrum based on (3.4) is said to be *incoherent*, because the $k$th eigenspectrum $|X_k(f)|^2$ ignores phase information for all values of the index $k$.

(2) For the parameters needed to compute the multitaper spectral estimator (3.4), recommended values (within each section of data) are as follows:
   - parameter $C_o = 4$, possibly extending up to 10;
   - number of Slepian tapers $K = 10$, possibly extending up to 16.

   These values are needed, especially when the dynamic range of the data being analyzed is large.

(3) If, and when, the number of tapers is increased toward the limiting value $2NW$, then the adaptive multitaper spectral estimator should be used.

(4) Whenever possible, prewhitening of the data prior to processing should be applied.

### 3.3.5   Comparison of the MTM with other spectral estimators

Now that we understand the idea of multitaper spectral estimation, we are ready to compare its performance against other spectral-estimation algorithms. The results described herein summarize previous experimental work that was originally reported by Drosopoulos and Haykin (1992) and reproduced in Haykin (2007). The test dataset used in that previous work was *Marple's classic synthetic dataset*, the analytic spectrum of which is known exactly (Marple, 1987). Specifically, the dataset is composed of the following components:

- two complex sinusoids of fractional frequencies 0.2 and 0.21 that are included to test the resolution capability of a spectral estimator;
- two weaker complex sinusoids of 20 dB less power at fractional frequencies −0.15 and 0.1;
- colored noise process, generated by passing two independently zero-mean real white-noise processes through identical moving-average filters; each filter has an identical raised-cosine frequency response centered at ±0.35 and occupying a bandwidth of 0.3.

Following Marple (1987), the experimental study by Drosopoulos and Haykin (1992) started with two spectral estimators:

- the periodogram with a 4096-point fast Fourier transform (FFT) and
- a tapered version of the same periodogram, using a Hamming window.

With spectral line components featuring prominently in the dataset, the experimental study also included two eigendecomposition-based spectral estimators:

- *Multiple SIgnal Classification (MUSIC) algorithm* (Schmidt, 1981) and
- *Modified Forward–Backward Linear Prediction (MFBLP) algorithm* (Kumarasan and Tufts, 1983).

The two classical spectral estimators failed in resolving the line components and also failed in correctly estimating the continuous parts of Marple's synthetic spectrum. On the other hand, the two eigendecomposition-based algorithms appeared to resolve the line components reasonably well, but failed completely to estimate the continuous parts of the spectrum correctly.

Next, the MTM formula (3.4) was tested with Marple's synthetic data, using a time–bandwidth product $C_o = 4$ and $K = 8$ Slepian tapers. The resulting composite spectrum appeared to estimate the continuous parts of the synthetic spectrum reasonably well, correctly identify the locations of the line components at –0.15 and 9.1, but lumped the more closely spaced line components at 0.2 and 0.21 into a composite combination around 0.21. With additional complexity featuring the inclusion of the *harmonic F-test for spectral line components*, to be discussed later in Section 3.7, the composite spectrum computed by the MTM did reproduce the synthetic spectrum fully and accurately; the *F*-test, included with the MTM, is intended to check for an estimate of line components existing in the power spectrum.

In light of the experimental results summarized herein, it can be said that the basic formula of the MTM in (3.4) did outperform the periodogram and its Hamming-tapered version, which is not surprising. Moreover, it outperformed the MUSIC and MFBLP algorithms insofar as the continuous parts of the spectrum are concerned, but did not perform as well in dealing with the line components. However, when the composite MTM spectrum was expanded to include the *F*-test for line components, Marple's synthetic spectrum was reconstructed fully and accurately.

It is also noteworthy that in Drosopoulos and Haykin (1992) and Haykin (2007), a comparative evaluation of the MTM and the traditional *maximum-likelihood estimation procedure,* known for its *optimality,* is presented for angle-of-arrival estimation in the presence of multipath using real-life radar data. The results presented therein for low grazing angles show that these two methods are close in performance.

In another comparative study, Bronez (1992) compared the MTM with *weighted overlapped segment averaging (WOSA),* which was originally proposed by Welch (1967). To make this comparison, theoretical measures were derived by Bronez in terms of leakage, variance, and resolution. The comparison was performed by evaluating each one of these three measures in turn, while keeping the other two measures equal. The results of the comparison demonstrated that the MTM always performed better than WOSA. For example, given the same variance and resolution, it was found that the MTM had an advantage of 10 to 20 dB over WOSA.

## 3.4    Space–time processing

As already discussed, the MTM is theorized to provide a reliable and accurate method of estimating the power spectrum of environmental stimuli as a function of frequency. As such, in the MTM we have a reliable method for identifying spectrum holes and estimating their average-power contents for cognitive radio applications; as mentioned previously in Chapters 1 and 2, spectrum holes refer to underutilized subbands of the radio spectrum. In analyzing the radio scene in the local neighborhood of a cognitive radio receiver, for example, we also need to have a *sense of direction*, so that the cognitive radio is enabled to *listen* to incoming interfering signals from unknown directions. What we are signifying here is the need for *space–time processing*. For such an application, we may *employ a set of sensors to properly "sniff" the RF environment along different directions.*

To elaborate on the underlying theory of space–time processing, consider an array of $M$ antennas sensing the environment. For the $k$th Slepian taper, let $X_k^{(m)}(f)$ denote the complex-valued Fourier transform of the input signal $x(n)$ computed by the $m$th sensor in accordance with (3.2), and $m = 0, 1, \ldots, M - 1$. With $k = 0, 1, \ldots, K - 1$, where $K$ is the number of Slepian sequences, we may then construct the $M$-by-$K$ *spatio-temporal complex-valued matrix* (Haykin, 2005a)

$$\mathbf{A}(f) = \begin{bmatrix} a_0^{(0)}X_0^{(0)}(f) & a_1^{(0)}X_1^{(0)}(f) & \cdots & a_{K-1}^{(0)}X_{K-1}^{(0)}(f) \\ a_0^{(1)}X_0^{(1)}(f) & a_1^{(1)}X_1^{(1)}(f) & \cdots & a_{K-1}^{(1)}X_{K-1}^{(1)}(f) \\ \vdots & \vdots & & \vdots \\ a_0^{(M-1)}X_0^{(M-1)}(f) & a_1^{(M-1)}X_1^{(M-1)}(f) & \cdots & a_{K-1}^{(M-1)}X_{K-1}^{(M-1)}(f) \end{bmatrix}, \quad (3.14)$$

where each row of the matrix is produced by environmental stimuli sensed at a different gridpoint, and each column is computed using a different Slepian taper; the $a_k^{(m)}$ represent coefficients accounting for different localized areas around the gridpoints.

To proceed further, we make two necessary assumptions:

(1) The number of Slepian tapers $K$ is *larger* than the number of sensors $M$; this requirement is needed to avoid "spatial undersampling" of the environment.
(2) Except for being synchronously sampled, the $M$ sensors operate *independently* of each other; this second requirement is needed to ensure that the *rank* of the matrix $\mathbf{A}(f)$ (i.e. the number of linearly independent rows) is exactly equal to $M$.

In physical terms, each entry in the matrix $\mathbf{A}(f)$ is produced by two contributions: one due to additive ambient noise at the front end of the sensor and the other due to the incoming interfering signals. Insofar as spectrum sensing is concerned, however, the primary contribution of interest is that due to the environmental stimuli. In this context, an effective tool for denoising is *singular-value decomposition* (SVD).

SVD is a generalization of *principal-components analysis*, or *eigendecomposition*. Whereas eigendecomposition involves a single orthonormal matrix, SVD involves a pair of orthonormal matrices, which we denote as follows:

- $M$-by-$M$ matrix $\mathbf{U}(f)$ defined by the $M$ sensors and
- $K$-by-$K$ matrix $\mathbf{V}(f)$ defined by the $K$ Slepian sequences.

Thus, applying SVD to the spatio-temporal matrix $\mathbf{A}(f)$ defined in (3.14), we may express the resulting decomposition as follows (Golub and Van Loan, 1996):

$$\mathbf{U}^\dagger(f)\mathbf{A}(f)\mathbf{V}(f) = [\mathbf{\Sigma}(f) \vdots \mathbf{0}], \quad (3.15)$$

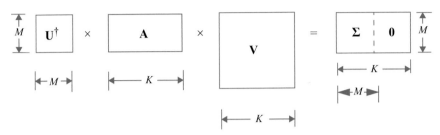

**Figure 3.3.** Diagrammatic depiction of SVD applied to a rectangular matrix **A**; left-hand side of the figure represents the matrix product $\mathbf{U}^\dagger\mathbf{A}\mathbf{V}$ and the right-hand side represents the resulting SVD.

where $\Sigma(f)$ is an $M$-by-$M$ diagonal matrix, the $k$th element of which is denoted by $\sigma_k(f)$, and $\mathbf{0}$ is a null matrix; the superscript $\dagger$ denotes *Hermitian transposition*. Figure 3.3 shows a graphical depiction of this decomposition; to simplify the depiction, dependence on the frequency $f$ has been ignored in the figure.

Henceforth, the system described by the spatio-temporal matrix $\mathbf{A}(f)$ of (3.14), involving $K$ Slepian tapers, $M$ sensors, and decomposition of the matrix in (3.15), is referred to as the *MTM-SVD processor*. Note that with the spatio-temporal matrix $\mathbf{A}(f)$ being frequency dependent, and likewise for the unitary matrices $\mathbf{U}(f)$ and $\mathbf{V}(f)$, the MTM-SVD processor is actually performing *tensor analysis*.

### 3.4.1   Physical interpretation of the action performed by the MTM-SVD processor

To understand the underlying signal operations embodied in the MTM-SVD processor, we begin by reminding ourselves of the *orthonormal properties* of matrices $\mathbf{U}$ and $\mathbf{V}$ that hold for all $f$, as shown by

$$\mathbf{U}(f)\mathbf{U}^\dagger(f) = \mathbf{I}_M$$

and

$$\mathbf{U}(f)\mathbf{V}^\dagger(f) = \mathbf{I}_K,$$

where $\mathbf{I}_M$ and $\mathbf{I}_K$ are $M$-by-$M$ and $K$-by-$K$ identity matrices respectively. Using this pair of relations in (3.15), we obtain the following decomposition of the matrix $\mathbf{A}(f)$, after some straightforward manipulations:

$$\mathbf{A}(f) = \sum_{m=0}^{M-1} \sigma_m(f)\mathbf{u}_m(f)\mathbf{v}^\dagger_m(f). \tag{3.16}$$

$\sigma_m(f)$ is the $m$th *singular value* of the matrix $\mathbf{A}(f)$, $\mathbf{u}_m(f)$ is the *left-singular vector*, and $\mathbf{v}_m(f)$ is the *right-singular vector*. In analogy with *principal-components analysis*, the decomposition of (3.16) may be viewed as one of *principal modulations* produced by the incoming time–frequency stimuli. According to this decomposition, the singular value $\sigma_m(f)$ provides a scaling of the $m$th principal modulation computed by the MTM-SVD processor. Note also that the $M$ singular values, constituting the diagonal matrix $\Sigma(f)$ in (3.15), are all real numbers. The higher order singular values, namely, $\sigma_m(f)$, ..., $\sigma_{K-1}(f)$, are all zero; they constitute the null matrix $\mathbf{0}$ in (3.15).

Using (3.16) to form the matrix product $\mathbf{A}(f)\mathbf{A}^{\dagger}(f)$ and invoking the orthonormal property of the unitary matrix $\mathbf{V}(f)$, we have the eigendecomposition

$$\mathbf{A}(f)\mathbf{A}^{\dagger}(f) = \sum_{m=0}^{M-1} \sigma_m^2(f)\mathbf{u}_m(f)\mathbf{u}^{\dagger}_m(f), \tag{3.17}$$

where $\sigma_m^2(f)$ is the $m$th eigenvalue of the eigendecomposition. Similarly, forming the other matrix product $\mathbf{A}^{\dagger}(f)\mathbf{A}(f)$ and invoking the orthonormal property of the unitary matrix $\mathbf{U}(f)$, we have the alternative eigendecomposition

$$\mathbf{A}^{\dagger}(f)\mathbf{A}(f) = \sum_{k=0}^{K-1} \sigma_k^2(f)\mathbf{v}_k(f)\mathbf{v}^{\dagger}_k(f). \tag{3.18}$$

With $\sigma_m(f) = \sigma_{m+1}(f) = \cdots = \sigma_{M-1}(f) = 0$, both eigendecompositions have exactly the *same nonzero eigenvalues*, but they have entirely *different eigenvectors*. Note that the outer products $\mathbf{u}_m(f)\mathbf{u}^{\dagger}_m(f)$ and $\mathbf{v}_m(f)\mathbf{v}^{\dagger}_m(f)$ are both of rank one.

Recalling that the index $m$ signifies a sensor and the index $k$ signifies a Slepian taper, we may now make three statements on the multiple operations being performed by the MTM-SVD processor:

(1) The $m$th eigenvalue $\sigma_m^2(f)$ is defined by

$$\sigma_m^2(f) = \sum_{k=0}^{K-1} |a_k^{(m)} X_k^{(m)}(f)|^2, \qquad m = 0, 1, \ldots, M-1. \tag{3.19}$$

Setting $|a_k^{(m)}|^2 = \lambda_k^{(m)}$ and dividing $\sigma_m^2$ by this same factor, we get

$$\hat{S}^{(m)}(f) = \frac{\displaystyle\sum_{k=0}^{K-1} \lambda_k^{(m)} |X_k^{(m)}(f)|^2}{\displaystyle\sum_{K=0}^{K-1} \lambda_k^{(m)}}, \qquad m = 0, 1, \ldots, M-1, \tag{3.20}$$

which is a rewrite of (3.4), specialized for sensor $m$. We may, therefore, make the following statement:

The eigenvalue $\sigma_m^2(f)$, except for the constant scaling factor $\left( \sum_{k=0}^{K-1} \lambda_k^{(m)} \right)^{-1}$, provides the desired multitaper spectral estimate for the incoming interference picked up by the $m$th sensor.

(2) Since the index $m$ refers to the $m$th sensor, we make the second statement:

The left singular vector $\mathbf{u}_m(f)$ defines the direction of the interfering signal picked up by the $m$th sensor at frequency $f$.

(3) The index $k$ refers to the $k$th Slepian taper; moreover, since $\sigma_k^2(f) = \sigma_m^2(f)$ for $k = 0, 1, \ldots, M-1$, we may make the third and last statement:

The right singular vector $\mathbf{v}_m(f)$ defines the multitaper coefficients for the $m$th interferer's waveform.

Most importantly, with no statistical assumptions on the additive ambient noise in each sensor or the incoming interferers, we may go on to state that the nonparametric MTM-SVD processor is indeed *robust*.

The enhanced signal-processing capability of the MTM-SVD processor just described is achieved at the expense of increased computational complexity. To elaborate, with $N$ data points and signal bandwidth $2W$, there are $N$ different frequencies with spectral resolution $2W/N$ to be considered. Accordingly, the MTM-SVD processor has to perform a total of $N$ SVDs on matrix $\mathbf{A}(f)$. Note, however, that the size of the wavenumber spectrum (i.e. the spatial distribution of the interferers) is determined by the number of sensors $M$, which is considerably smaller than the number of data points $N$. Most importantly, the wavenumber is computable in *parallel*. With the computation being performed at each frequency $f$, each of the $M$ sensors sees the full spectral footprint of the interferer pointing along its own direction; the footprint is made up of $N$ frequency points with a spectral resolution of $2W/N$.

Summing up, the MTM-SVD processor has the capability to sense the surrounding environment both in frequency and direction, the resolutions of which are respectively determined by the number of data points and the number of sensors deployed. Although by itself the MTM-SVD processor cannot provide for the attentional requirement of a cognitive radio, it provides the dynamic-spectrum manager valuable spatial information about the radio environment to resolve this requirement.

## 3.5    Time–frequency analysis

The MTM-SVD processor rests its signal-processing capability on two coordinates of processing:

- *frequency*, which is necessary for displaying the spectral content of environmental data along the frequency axis, and
- *space*, which provides the means for estimating wavenumber spectra of the environment.

However, for the spectral analysis of environmental data to be complete, there is a third coordinate that is just as important in its own way; that coordinate is *time*. Indeed, it is the temporal characterization of a stochastic process that permits us to describe the process as *stationary* or *nonstationary*.

The statistical analysis of nonstationary processes has had a rather mixed history. Although the general second-order theory of spectral analysis was described by Loève (1946), it has not been applied nearly as extensively as the theory of stationary processes published only slightly earlier by Wiener and Kolmogorov. There were, at least, four distinct reasons for this neglect, as summarized by Haykin and Thomson (1998):

(1) Loève's theory was *probabilistic*, not statistical, and there do not appear to have been successful attempts to find a statistical version of the theory until some time later.
(2) At the time of those early publications, over six decades ago, the mathematical training of most engineers and physicists in signals and stochastic processes was minimal and, recalling that even Wiener's delightful book was referred to as "The

Yellow Peril" on account of the color of its covers, it is easy to imagine the reception that a general nonstationary theory of stochastic processes would have received.

(3) Even if the theory had been commonly understood at the time and good statistical estimation procedures had been available, the computational burden would probably have been overwhelming. This was the era when Blackman–Tukey estimates of the power spectrum of a stationary process were developed. This was not because they were great estimates, but, primarily, because they were simple to understand in mathematical terms and before the (re)invention of the FFT algorithm, with the latter being computationally more efficient than other forms.

(4) Finally, it cannot be denied that the Loève theory of nonstationary processes was harder to grasp than the *Wiener–Kolmogorov theory* of stationary processes.

Confronted with the notoriously unreliable nature of a wireless channel for radio communication, for example, we have to find some way to account for the nonstationary behavior of a signal at the channel output across *time* (implicitly or explicitly) in describing the signal picked up by the receiver. Given the desirability of working in the frequency domain for well-established reasons, we may include the effect of time by adopting a *time–frequency description* of the signal. During the last three to four decades, many papers have been published on various estimates of time–frequency distributions; see, for example, the book by Cohen (1995) and the references therein. In most of that early work on time–frequency analysis, the signal is assumed to be *deterministic*. In addition, many of the proposed estimators are constrained to match continuous time $t$ and continuous frequency $f$ *marginal* density conditions. For a continuous-time signal, denoted by $x(t)$, the *time marginal* is required to satisfy the condition

$$\int_{-\infty}^{\infty} D(t, f)\, df = |x(t)|^2,  \tag{3.21}$$

where $D(t, f)$ is the *time–frequency distribution* of the signal. Similarly, if $X(f)$ is the Fourier transform of $x(t)$, then the *frequency marginal* must satisfy the second condition

$$\int_{-\infty}^{\infty} D(t, f)\, dt = |X(f)|^2.  \tag{3.22}$$

Given the large differences observed between waveforms collected on sensors spaced short distances apart, the time marginal requirement is a rather strange assumption. Worse, the frequency marginal is, except for the factor $1/N$, just the periodogram of the signal, which is known to be badly biased and inconsistent.[5] Thus, we do not consider matching marginal distributions, as commonly defined in the literature, to be important.

## 3.5.1  Theoretical background of nonstationarity

*Nonstationarity* is an inherent characteristic of most, if not all, of the stochastic processes encountered in practice. Yet, despite its highly pervasive nature and practical importance, not enough attention is paid in the literature to the characterization of nonstationary processes in a mathematically satisfactory manner.

To this end, consider a complex continuous stochastic process, a sample function of which is denoted by $x(t)$. We assume that the process is *harmonizable* (Loève, 1946, 1963), so that it permits the *Cramér representation*, described by

$$x(t) = \int_{-1/2}^{1/2} e^{j2\pi vt} dX(v), \tag{3.23}$$

where $dX(v)$ is the *generalized Fourier transform* of the process, also referred to as the *increment process*; the dummy variable $v$ has the same dimension as frequency. The bandwidth of $x(t)$ has been normalized to *unity* for convenience of presentation; consequently, as indicated in (3.23), the integration extends with respect to $v$ over the interval $[-1/2, +1/2]$. As before, it is assumed that the process $x(t)$ has zero mean; that is, $\mathbb{E}[x(t)] = 0$ for all time $t$. Correspondingly, we have $\mathbb{E}[X(v)] = 0$ for all $v$. Moreover, the power spectrum $S(f)$ is defined in terms of the generalized Fourier transform $dX(f)$ by the equation

$$\mathbb{E}[dX(f)\, dX^*(v)] = S(f)\delta(f - v)\, df dv,$$

where the asterisk denotes complex conjugation and $\delta(f)$ is the Dirac delta function in the frequency domain. Note that (3.23) is also the starting point in formulating the MTM. (See Note 2 at the end of the chapter, described in the discrete-time domain.)

To set the stage for introducing the statistical parameters of interest, we introduce the *covariance function* defined in the time domain as follows (Thomson, 2000):

$$\begin{aligned}\Gamma_L(t_1, t_2) &= \mathbb{E}[x(t_1)x^*(t_2)] \\ &= \int_{-\infty}^{\infty}\int_{-\infty}^{\infty} e^{j2\pi(t_1 f_1 - t_2 f_2)}\gamma_L(f_1, f_2)\, df_1\, df_2.\end{aligned} \tag{3.24}$$

Hereafter, the generalized two-frequency spectrum $\gamma_L(f_1, f_2)$ in the integrand of the second line in (3.24) is referred to as the *Loève spectrum*.[6] With $X(f)$ denoting the Fourier transform of $x(t)$, the Loève spectrum is itself formally defined by

$$\gamma_L(f_1, f_2)\, df_1\, df_2 = \mathbb{E}[dX(f_1)\, dX^*(f_2)]. \tag{3.25}$$

Equation (3.25) highlights the underlying feature of a nonstationary process by describing the *correlation* between the spectral elements $X(f_1)$ and $X(f_2)$ of the process at two different frequencies $f_1$ and $f_2$ respectively.

If the process is stationary, then, by definition, the covariance function $\Gamma_L(t_1, t_2)$ depends only on the time difference $t_1 - t_2$, and the Loève spectrum $\gamma_L(f_1, f_2)$ collapses to $\delta(f_1 - f_2)S(f_1)$, where $\delta(f)$ is the Dirac delta function in the frequency domain and $S(f)$ is the power spectrum of the process $x(t)$. Similarly, for a white nonstationary process, the covariance function $\Gamma_L(t_1, t_2)$ becomes $\delta(t_1 - t_2)P(t_1)$, where $\delta(t)$ is the Dirac delta function in the time domain and $P(t)$ is the expected (average) power of the process at time $t$. Thus, as both the spectrum and covariance functions include delta-function discontinuities in simple cases, neither should be expected to be "smooth"; *continuity properties* of the process, therefore, depend on direction in the $(f_1, f_2)$-plane or $(t_1, t_2)$-plane. The continuity problems are more easily dealt with by rotating both the time and frequency coordinates of the covariance function in (3.24) and Loève spectrum in (3.25) respectively by 45°. In the time domain, we may now define the new coordinates to be a "center" $t_0$ and a delay $\tau$, as shown by

$$t_1 + t_2 = 2t_0, \tag{3.26}$$

$$t_1 - t_2 = \tau. \tag{3.27}$$

Correspondingly, we may write

$$t_1 = t_0 + \frac{\tau}{2}, \tag{3.28}$$

$$t_2 = t_0 - \frac{\tau}{2}. \tag{3.29}$$

Thus, denoting the covariance function in the *rotated coordinates* by $\Gamma(t_0, \tau)$, we may go on to write

$$\Gamma(t_0, \tau) = \Gamma_L\left(t_0 + \frac{\tau}{2}, t_0 - \frac{\tau}{2}\right). \tag{3.30}$$

In a similar manner, we may define new frequency coordinates, $f$ and $g$, by writing

$$f_1 + f_2 = 2f, \tag{3.31}$$

$$f_1 - f_2 = g, \tag{3.32}$$

and go on correspondingly to write

$$f_1 = f + \frac{g}{2}, \tag{3.33}$$

$$f_2 = f - \frac{g}{2}. \tag{3.34}$$

The *rotated two-frequency spectrum* is thus defined in terms of the Loève spectrum as shown by

$$\gamma(f, g) = \gamma_L\left(f + \frac{g}{2}, f - \frac{g}{2}\right). \tag{3.35}$$

Substituting the definitions of (3.28) and (3.29) into (3.24) shows that the difference term $(t_1 f_1 - t_2 f_2)$ in the exponent of the Fourier transform becomes the sum term $(t_0 g + \tau f)$, setting the stage for us to write

$$\Gamma(t_0, \tau) = \int_{-\infty}^{\infty} \left\{ \int_{-\infty}^{\infty} e^{j2\pi(t_0 g + \tau f)} \right\} \gamma(f, g) \, df \, dg. \tag{3.36}$$

Because $f$ is associated with the time difference $\tau$, it corresponds to the ordinary frequency of stationary processes; therefore, we may refer to it as the "stationary" frequency. Similarly, because $g$ is associated with the average time $t_0$, it describes the behavior of the spectrum over long time spans; hence, we refer to $g$ as the "nonstationary" frequency.

   Consider next the continuity of the *generalized spectral density* $\gamma$, reformulated as a function of $f$ and $g$. On the line $g = 0$, the generalized spectral density $\gamma$ is just the ordinary spectrum with the usual continuity (or lack thereof) conditions normally applied to stationary spectra. As a function of $g$, however, we expect to find a $\delta$-function discontinuity at $g = 0$ if, for no other reason that almost all data contain some stationary additive noise. Consequently, smoothers in the $(f, g)$ plane or, equivalently, the $(f_1, f_2)$ plane, should not be isotropic, but require much higher resolution along the nonstationary frequency coordinate $g$ than along the stationary frequency coordinate $f$.

A slightly less arbitrary way of handling the $g$ coordinate is to apply the inverse Fourier transform to $\gamma(f, g)$ with respect to the nonstationary frequency $g$, obtaining the new formula (Thomson, 2000)

$$D(t_0, f) = \int_{-\infty}^{\infty} e^{j2\pi t_0 g} \gamma(f, g) \, dg \tag{3.37}$$

as the *dynamic spectrum* of the process; the $D(t_0, f)$ in (3.37) is *not* to be confused with the time–frequency distributions in (3.21) and (3.22). The motivation behind (3.37) is to transform very rapid variations expected around $g = 0$ on the right-hand side of the equation into a slowly varying function of $t_0$ on the left-hand side of the equation, while, at the same time we leave the usual dependence on $f$ intact. From Fourier-transform theory, we know that the Dirac delta function in the frequency domain is transformed into a constant in the time domain. It follows, therefore that, in a stationary process, $D(t_0, f)$ does not depend on $t_0$ and assumes the simple form of a power spectrum. Thus, recalling the definition of $\Gamma_L(t_1, t_2)$ in (3.24), we may redefine the dynamic spectrum as follows:

$$D(t_0, f) = \int_{-\infty}^{\infty} e^{-j2\pi\tau f} \mathbb{E}\left\{ x\left(t_0 + \frac{\tau}{2}\right) x^*\left(t_0 - \frac{\tau}{2}\right) \right\} d\tau \tag{3.38}$$

According to (3.38), the dynamic spectrum is an *expected value and nonnegative definite*, which, being a form of power spectrum, is how it should be for all $t_0$.

## 3.5.2    Spectral coherences of nonstationary processes based on the Loève transform

From an engineering perspective, we usually like to have estimates of second-order statistics of the underlying physics responsible for the generation of a nonstationary process. Moreover, it would be desirable to compute the estimates using the MTM. With this twofold objective in mind, let $X_k(f_1)$ and $X_k(f_2)$ denote the multitaper Fourier transforms of the sample function $x(t)$; these two estimates are based on the $k$th Slepian taper and are defined at two different frequencies, $f_1$ and $f_2$, in accordance with (3.2). To evaluate the *spectral correlation* of the process at $f_1$ and $f_2$, the traditional formulation is to consider the product $X_k(f_1)X_k^*(f_2)$, where, as before, the asterisk in $X_k^*(f_2)$ denotes complex conjugation.

Unfortunately, we often find that such a formulation is *insufficient* in capturing the underlying second-order statistics of the process, particularly so in the case of several communication signals that are of interest in cognitive-radio applications.[7] To complete the second-order statistical characterization of the process, we need to consider unusual products of the form $X_k(f_1)X_k(f_2)$, which do *not* involve the use of complex conjugation. However, in the literature on stochastic processes, statistical parameters involving products like $X_k(f_1)X_k(f_2)$ are frequently not named and, therefore, hardly used; and when they are used, not only are different terminologies adopted for them, but also some of the terminologies are misleading.

To put matters right, in this chapter we follow the terminology first described by Mooers (1973), and use the subscripts "inner" and "outer" to distinguish between spectral correlations based on products involving such terms as $X_k(f_1)X^*(f_2)$ and $X_k(f_1)X(f_2)$ respectively.[8] Hereafter, estimates of spectral correlations so defined are referred to as estimates of the *first* and *second kinds* respectively, and likewise for related matters.

With the matter of terminology settled, taking the complex demodulates of a nonstationary process at two different frequencies, $f_1$ and $f_2$, and invoking the inherent orthogonality property of Slepian sequences, we may now formally define the estimate of the Loève spectrum of the first kind as

$$\hat{\gamma}_{L,\text{inner}}(f_1, f_2) = \frac{1}{K} \sum_{k=0}^{K-1} X_k(f_1) X_k^*(f_2),$$
(3.39)

where, as before, $K$ is the total number of Slepian tapers. The estimate of the Loève spectrum of the second kind is correspondingly defined as

$$\hat{\gamma}_{L,\text{outer}}(f_1, f_2) = \frac{1}{K} \sum_{k=0}^{K-1} X_k(f_1) X_k(f_2),$$
(3.40)

where, on the right-hand side, there is no complex conjugation.

Thus, given a stochastic process with the complex demodulates $X_k(f_1)$ and $X_k(f_2)$, the *Loève spectral coherences of the first and second kinds* are respectively defined as

$$C_{\text{inner}}(f_1, f_2) = \frac{\hat{\gamma}_{L,\text{inner}}(f_1, f_2)}{(\hat{S}(f_1)\hat{S}(f_2))^{1/2}}$$
(3.41)

and

$$C_{\text{outer}}(f_1, f_2) = \frac{\hat{\gamma}_{L,\text{outer}}(f_1, f_2)}{(\hat{S}(f_1)\hat{S}(f_2))^{1/2}},$$
(3.42)

where, in both equations, the multitaper spectral estimate $\hat{S}(f)$ is naturally real valued. Note that these two definitions of Loève spectral coherences share the same denominator but differ in their numerators.

In general, the Loève spectral coherences $C_{\text{inner}}(f_1, f_2)$ and $C_{\text{outer}}(f_1, f_2)$ are both complex-valued, which means that each one of them will have a magnitude and associated phase of its own. In practice, we find that a quantity called the *two-frequency magnitude-squared coherence* (TF-MSC) is more useful than the spectral coherence itself. With the two spectral coherences of (3.41) and (3.42) at hand, we have two TF-MSCs to consider, namely $|C_{\text{inner}}(f_1, f_2)|^2$ and $|C_{\text{outer}}(f_1, f_2)|^2$, whose respective definitions follow directly from (3.41) and (3.42).

### 3.5.3    Two special cases of the dynamic spectrum $D(t_0, f)$

#### 3.5.3.1    Wigner–Ville distribution

From the defining equation (3.38), we immediately recognize that[9]

$$W(t_0, f) = \int_{-\infty}^{\infty} e^{-j2\pi\tau f} x\left(t_0 + \frac{\tau}{2}\right) x^*\left(t_0 - \frac{\tau}{2}\right) d\tau$$
(3.43)

is the formula for the *Wigner–Ville distribution* of the sample function $x(t)$. In other words, we see that the Wigner–Ville distribution is the *instantaneous estimate* of the *dynamic spectrum* of the nonstationary signal $x(t)$. As such, it is simpler to compute than $D(t_0, f)$ in the classification of signals.

### 3.5.3.2    Cyclic power spectrum

The dynamic spectrum $D(t_0, f)$ also embodies another special case, namely the *cyclic power spectrum* of a sample function $x(t)$ that is known to be *periodic*. Let $T_0$ denote the period of $x(t)$. Then, replacing the time $t_0$ in (3.38) with $T_0 + t$, we may express the *time-varying power spectrum* of $x(t)$ as

$$S_x(t,\ f) = \int_{-\infty}^{\infty} e^{-j2\pi\tau f} \mathbb{E}\left[\left\{x\left(t + T_0 + \frac{\tau}{2}\right)x^*\left(t + T_0 - \frac{\tau}{2}\right)\right\}\right] d\tau$$

$$= \int_{-\infty}^{\infty} e^{-j2\pi\tau f} R_x\left(t + T_0 + \frac{\tau}{2}, t + T_0 - \frac{\tau}{2}\right) d\tau.$$

(3.44)

The new function

$$R_x\left(t + T_0 + \frac{\tau}{2}, t + T_0 - \frac{\tau}{2}\right) = \mathbb{E}\left[x\left(t + T_0 + \frac{\tau}{2}\right)x^*\left(t + T_0 - \frac{\tau}{2}\right)\right] \qquad (3.45)$$

is the *time-varying autocorrelation function* of the signal $x(t)$. The stochastic process, represented by the sample function $x(t)$, is said to be *cyclostationary in the second-order sense* if this autocorrelation sequence is itself periodic with period $T_0$, as shown by

$$R_x\left(t + T_0 + \frac{\tau}{2}, t + T_0 - \frac{\tau}{2}\right) = R_x\left(t + \frac{\tau}{2}, t - \frac{\tau}{2}\right). \qquad (3.46)$$

Under this condition, (3.44) reduces to

$$S_x(t, f) = \int_{-\infty}^{\infty} e^{-j2\pi f\tau} R_x\left(t + \frac{\tau}{2}, t - \frac{\tau}{2}\right) d\tau, \qquad (3.47)$$

which, as expected, is independent of the period $T_0$. Equation (3.47) is recognized as the cyclostationary extension of the well-known *Wiener–Khintchine relation* for stochastic processes.

To be more complete, for a stochastic process to be cyclostationary in the second-order sense, its mean must also be periodic with the same period, $T_0$. When the mean of the stochastic process under study is zero for all time $t$, as assumed here, then this condition is immediately satisfied.

### 3.5.4    Instrumentation for computing Loève spectral correlations

Before proceeding to discuss the cyclostationarity characterization of nonstationary processes in the next section on Fourier theory, we find it instructive to have a diagrammatic instrumentation for computing the Loève spectral correlations using the MTM. To do this, we look to the defining equations (3.2), (3.41), and (3.42), where $t$ in (3.2) denotes discrete time and $f$ in all three equations denotes continuous frequency. Let $x(t)$ denote a time series of length $N$. Then, the inspection of (3.2), (3.41), and (3.42) leads to the *basic instrument* diagrammed in Figure 3.4a. In particular, in accordance with (3.2), the identical functional blocks labeled "multitaper method" in the upper and lower paths of the figure produce the Fourier transforms $X_k(f_1)$ and $X_k(f_2)$ respectively. The designation "basic" is intended to signify that the instrument applies to both kinds of the Loève spectral correlation, depending on how the cross-correlation of the Fourier transforms $X_T(f_1)$ and $X_T(f_2)$ is computed over the set of $K$ Slepian tapers. To be specific, we say that, for the overall output:

- the instrument of Figure 3.4a computes the estimate $\hat{\gamma}_{L,\mathrm{inner}}(f_1, f_2)$ of (3.41) if the cross-correlation is of the first kind, and
- it computes $\hat{\gamma}_{L,\mathrm{outer}}(f_1, f_2)$ of (3.42) if the cross-correlation is of the second kind.

As for Figure 3.4b, it applies to spectral correlations rooted in the classical Fourier framework, which is considered next.

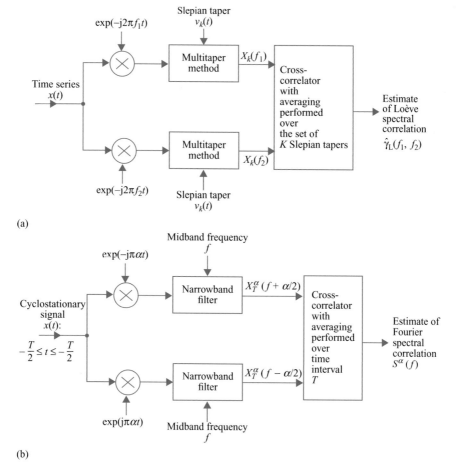

(a)

(b)

**Figure 3.4.** Illustrating the one-to-one correspondences between the Loève and Fourier theories for cyclostationarity. Basic instrument for computing (a) the Loève spectral correlations of a time series $x(t)$ and (b) the Fourier spectral correlations of cyclostationary signal $x(t)$. (Reproduced with permission from S. Haykin, D. J. Thomson and J. Reed, 2009, Spectrum sensing for cognitive radio, *Proceedings of the IEEE*, 97, 849–877.

## 3.6    Cyclostationarity

When the issue of interest is the characterization of digital modulated signals in cognitive-radio applications, for example, we find that there is a large body of literature on the subject, the study of which has been an active area of research for over 50 years. The prominence of interest in cyclostationarity for signal detection and classification[10] is attributed to the works of Gardner (1988, 1994) and the subsequent work of Giannakis and Serpedin (1998) on alternative views and applications of cyclostationarity. For the literature on cyclostationarity, see Note 11 at the end of the chapter.

### 3.6.1    Fourier framework of cyclic statistics

As defined previously, a stochastic process represented by the sample function $x(t)$ is said to be *cyclostationary in the second-order sense* if its time-varying autocorrelation function $R_x(t + \tau/2, t - \tau/2)$ satisfies the periodicity condition of (3.46). Moreover, if the mean of the process is nonzero, it would also have to be time varying with the same period $T_0$. For the present discussion, the mean is assumed to be zero for all time $t$, and the focus of attention, therefore, is confined to second-order statistics.

A cyclostationary process may also be described in terms of its power spectrum, which assumes a periodic form of its own. With interest in this chapter focused on spectral coherence, we now go on to define the inner and outer forms of spectral coherence of a nonstationary process using Fourier theory.

Let $x(t)$ denote a sample function of a complex-valued cyclostationary process with period $T_0$. Using the Fourier series, we may characterize the process by its *cyclic power spectrum of the first kind*, as shown by the Fourier expansion:

$$S_{inner}(t, f) = \sum_{\alpha} s_{inner}^{\alpha}(f) e^{j2\pi\alpha t}, \tag{3.48}$$

where the new parameter $\alpha$, in theory, scans an infinite set of frequencies; that is, $\alpha = n/T_0$, where $n = 0, 1, 2, \ldots$. The power spectrum of (3.48) is *cyclic*, in that it satisfies the condition of periodocity, as shown by

$$S_{inner}(t + T_0, f) = S_{inner}(t, f).$$

The *Fourier coefficients* in (3.48), namely $s_{inner}^{\alpha}(f)$ for varying $\alpha$, are defined by

$$s_{inner}^{\alpha}(f) = \lim_{T \to \infty} \lim_{\Delta t \to 0} \frac{1}{\Delta t} \int_{-\Delta t/2}^{\Delta t/2} \frac{1}{T} \left( X_T(t, f + \alpha) X_T^*(t, f - \alpha) \right) dt. \tag{3.49}$$

The infinitesimally small $\Delta t$ is included in (3.49) to realize the continuous-time nature of the cyclostationary signal $x(t)$ in the limit as $\Delta t$ approaches zero. The *time-varying Fourier transform* of $x(t)$, denoted by $X_T(t, f)$, is defined by

$$X_T(t, f) = \int_{t-T/2}^{t+T/2} x(\xi) e^{-j2\pi f \xi} d\xi. \tag{3.50}$$

Most importantly, for prescribed values of $f$, $s_{\text{inner}}^\alpha(f)$ is the time average of the inner product $X_T(t, f + \alpha/2)X_T^*(t, f - \alpha/2)$; it follows, therefore, that $s_{\text{inner}}^\alpha(f)$ is the *inner spectral correlation* of the cyclostationary signal $x(t)$ for the two frequencies $f_1 = f + \alpha/2$ and $f_2 = f - \alpha/2$.

In light of the rationale presented in Section 3.5, we say that (3.48) and (3.50) provide a *partial description* of the second-order statistics of a complex-valued cyclostationary process. To complete the statistical description of $x(t)$, we need to introduce the *cyclic power spectrum of the second kind*, as shown by

$$S_{\text{outer}}(t, f) = \sum_\alpha s_{\text{outer}}^\alpha(f)e^{j2\pi\alpha t}, \tag{3.51}$$

where $s_{\text{outer}}^\alpha(f)$ is the time average of the outer product $X_T(t, f + \alpha/2)X_T(t, f - \alpha/2)$, which does not involve the use of complex conjugation; (3.49), therefore, defines the *outer spectral correlation* of the signal $x(t)$.

With (3.48) and (3.51) at hand, we may now define the two *Fourier spectral coherences* of a cyclostationary process as follows.

(1) *Fourier spectral coherence of the first kind*:

$$C_{\text{inner}}^\alpha(f) = \frac{s_{\text{inner}}^\alpha(f)}{\left(s^0\left(f + \dfrac{\alpha}{2}\right)s^0\left(f - \dfrac{\alpha}{2}\right)\right)^{1/2}}. \tag{3.52}$$

(2) *Fourier spectral coherence of the second kind*:

$$C_{\text{outer}}^\alpha(f) = \frac{s_{\text{outer}}^\alpha(f)}{\left(s^0\left(f + \dfrac{\alpha}{2}\right)s^0\left(f - \dfrac{\alpha}{2}\right)\right)^{1/2}}. \tag{3.53}$$

Both spectral coherences have the same denominator, where the Fourier coefficient $s^0(f)$ corresponds to $\alpha = 0$; putting $\alpha = 0$ in the expressions for $s_{\text{inner}}^\alpha(f)$ and $s_{\text{outer}}^\alpha(f)$ yields the common formula

$$s^0(f) = \lim_{T\to\infty}\lim_{\Delta t\to 0}\frac{1}{\Delta t}\int_{-\Delta t/2}^{\Delta t/2}\frac{1}{T}|X_T(t, f)|^2\, dt. \tag{3.54}$$

As with the Loève spectral coherences, the Fourier spectral coherences $C_{\text{outer}}^\alpha(f)$ and $C_{\text{inner}}^\alpha(f)$ are both complex valued in general, with each one of them having a magnitude and associated phase of its own.

The use of the Fourier spectral coherences of the first and second kinds in (3.52) and (3.53) can be computationally demanding in practice. To simplify matters, their *cycle frequency-domain profile* (CFDP) versions are often used, as shown by

$$\text{CFDP}_{\text{inner}}(\alpha) = \max_f |C_{\text{inner}}^\alpha(f)|, \tag{3.55}$$

and similarly for the outer spectral coherence $C_{\text{outer}}^\alpha(f)$.

### 3.6.2          Instrumentation for computing the Fourier spectral correlations

The block diagram of Figure 3.4b depicts the instrument for computing the first and second kinds of the Fourier spectral correlations at frequencies $f_1 = f + \alpha/2$ and $f_2 = f - \alpha/2$ in accordance with (3.52) and (3.53). A cyclostationary signal $x(t)$ is applied in parallel to the pair of paths in the figure, both of which use a pair of identical narrowband filters. Both filters have the midband frequency $f$ and bandwidth $\Delta f$, where the $\Delta f$ is small compared with $f$ but large enough compared with the reciprocal of the time $T$ that spans the total duration of the input signal $x(t)$. In any event, the Fourier transform of the input $x(t)$ is shifted due to the multiplying factors $\exp(\pm j\pi\, \alpha t)$, producing the filter outputs $X_T(f + \alpha/2)$ in the upper path and $X_T(f - \alpha/2)$ in the lower path. Depending on how these two filter outputs are processed by the spectral correlator, the overall output produced in Figure 3.4b is $s_{inner}^{\alpha}(f)$ or $s_{outer}^{\alpha}(f)$.

### 3.6.3          Relationship between the Fourier and Loève spectral coherences

Much of the communications literature on cyclostationarity and related topics such as spectral coherence differ from that on multitaper spectral analysis. Nevertheless, these two approaches to cyclostationarity characterization of an input signal are actually related. In particular, examining Figure 3.4a and b, we see that the two basic instruments depicted therein are similar in signal-processing terms, exhibiting the following one-to-one correspondences:

(1) The multiplying factors $e^{-j2\pi f_1 t}$ and $e^{-j2\pi f_2 t}$ for the MTM in Figure 3.4a play similar frequency-shifting roles as the factors $e^{j2\pi\, \alpha t}$ and $e^{-j2\pi\, \alpha t}$ for their Fourier counterparts in Figure 3.4b.
(2) The MTM in Figure 3.4a for a prescribed Slepian taper and the narrowband filter in Figure 3.4b for prescribed mid-band frequency $f$ and parameter $\alpha$ perform similar filtering operations.
(3) Finally, the cross-correlator operates on the MTM outputs $X_k(f_1)$ and $X_k(f_2)$ in Figure 3.4a to produce estimates of the Loève spectral correlations, while the cross-correlator in Figure 3.4b operates on the filter outputs $X_T(f + \alpha/2)$ and $X_T(f - \alpha/2)$ to produce the Fourier spectral correlations with $f_1 = f + \alpha/2$ and $f_2 = f - \alpha/2$.

Naturally, the instruments depicted in Figure 3.4a and b differ from each other in the ways in which their individual components are implemented.

### 3.6.4          Contrasting the two theories on cyclostationarity

The theory of cyclostationarity presented in this section follows the framework formulated in Gardner (1988, 1994). This framework is rooted in the traditional Fourier-transform theory of stationary processes with an important modification: introduction of the parameter $\alpha$ (having the same dimension as frequency) in the statistical characterization of cyclostationary processes. Accordingly, the cyclic spectral features computed from this framework depend on how well the parameter $\alpha$ matches the underlying statistical periodicity of the original signal $x(t)$.

The other theory on cyclostationarity, discussed previously in Section 3.5, follows the framework originally formulated by Thomson (2000). This latter framework combines the following two approaches:

- the Loève transform for dealing with nonstationary processes; and
- the MTM for resolving the bias–variance dilemma through the use of Slepian sequences.

This alternative two-pronged mathematically rigorous framework for the time–frequency analysis of nonstationary processes has a *built-in capability to adapt* to the underlying statistical periodicity of the signal under study. In other words, it is nonparametric and, therefore, *robust*.[12]

Summing up the basic similarities and differences between the Loève and Fourier theories of stochastic processes, we say the following:

- both theories perform similar signal-processing operations on their inputs; but
- the Fourier theory assumes that the stochastic process is cyclostationary, whereas the Loève theory applies to any nonstationary process regardless of whether it is cyclostationary or not.

## 3.7 Harmonic *F*-test for spectral line components

The MTM spectra computed using the theory described in Section 3.3 are not expected to be always good estimates, particularly so when it is known that spectral line components exist in the signal under study. For the MTM, we can apply the *statistical F-test* to check for and estimate existing line components in the spectrum. The *F*-test is a statistical test that assigns a probability value to each of two hypotheses concerning samples taken from a parent population: one hypothesis assumes that line components are present and the other hypothesis assumes that they are not. These samples are assumed to follow a $\chi^2$ distribution, which is the case for an unknown mean and variance sample; this distribution consists of a of sum of squares taken from a Gaussian (normal) population.

### 3.7.1 Brief outline of the *F*-test

Suppose that we have a model described by the equation

$$\mathbf{y} = \mathbf{A}\mathbf{x} + \mathbf{e},$$

which is linear with respect to a $p$-by-1 parameter vector $\mathbf{x}$, and where the $n$-by-$p$ coefficient matrix $\mathbf{A}$ and $n$-by-1 vector $\mathbf{y}$ are known or can be estimated from a given data set. We assume that the error vector $\mathbf{e}$ has independent components that come from a

Gaussian distribution of zero mean and variance $\sigma^2$. Therefore, another way to express our assumed linear model is to write

$$\mathbb{E}[\mathbf{y}] = \mathbf{Ax},$$

where $\mathbb{E}$ is the statistical expectation operator. In order to get the best possible estimate of our parameter vector $\mathbf{x}$ in the *least-squares sense*, we have to address the minimization

$$\min_{\mathbf{x}} \| \mathbf{y} - \mathbf{Ax} \|^2.$$

Using the superscript † to denote the Hermitian transposition of a matrix, we may define the squared error

$$\mathcal{E}(\mathbf{x}) = \mathbf{e}^\dagger \mathbf{e}$$
$$= \| \mathbf{y} - \mathbf{Ax} \|^2,$$

which assumes its minimum value at the well-known *linear least-squares solution*

$$\hat{\mathbf{x}} = \mathbf{A}^+ \mathbf{y},$$

where $\mathbf{A}^+ = (\mathbf{A}^\dagger \mathbf{A})^{-1} \mathbf{A}^\dagger$ is the *pseudo-inverse* of matrix $\mathbf{A}$.

The $F$-test, based on the *F-distribution* (Percival and Walden, 1993), comes about from observing that we can break the observed total variance, $\mathbf{y}^\dagger \mathbf{y}$, of our model into two components, one due to the regression itself, $\| \mathbf{A}\hat{\mathbf{x}} \|^2$, and the other $\| \mathbf{y} - \mathbf{A}\hat{\mathbf{x}} \|^2$ due to residual errors. Each of these components has associated with it a number of DoFs, $v_1$ and $v_2$, respectively. For $N$ complex data points, the total number of DoFs is

$$v_1 + v_2 = 2N.$$

It turns out now that, provided the errors are independent and zero-mean Gaussian random variables, each of the two variance components follows a $\chi^2$ distribution with $v_1$ or $v_2$ DoFs respectively. Their ratio follows the $F(v_1, v_2)$ distribution, on the basis of which we can make hypothesis testing at a desired level of significance.

We test the hypothesis $H_0$: $\mathbf{x} = 0$ against the alternative hypothesis $H_1$: $\mathbf{x} \neq 0$ at a significance level $\alpha$ as follow. If $H_0$ is true, then the ratio

$$F = \frac{\dfrac{\| \mathbf{A}\hat{\mathbf{x}} \|^2}{v_1}}{\dfrac{\| \mathbf{y} - \mathbf{A}\hat{\mathbf{x}} \|^2}{v_2}}$$

follows the $F(v_1, v_2)$ distribution. The value of this distribution for a significance level of $\alpha$ (not to be confused with $\alpha$ in Section 3.6 on cyclostationarity) is found from statistical tables. If the computed ratio is larger than the table value, then our hypothesis is rejected with $100(1 - \alpha)\%$ confidence. This means that at least one of the components of $\mathbf{x}$ is different from zero. In practice, of course, we require a computed $F$ ratio that is much larger than the tabulated one.

The material just covered on linear regression provides us with enough background to set up partial $F$-tests for spectral lines.

## 3.7.2 Point regression single-line *F*-test

Given a time series $\{x(n)\}$ with a single line component at frequency $f_0$, the expected value of the eigencoefficients of this time series, referring back to Section 3.3, is given by

$$\mathbb{E}[X_k(f)] = \mu V_k(f - f_0), \qquad k = 0, 1, \ldots, K - 1, \qquad (3.56)$$

where $K$ is the number of Slepian sequences. With the $\{v_n^{(k)}\}_{n=1}^N$ denoting the $k$th Slepian sequence, in terms of which we may define the $k$th Fourier transform

$$V_k(f) = \sum_{n=0}^{N-1} v_n^{(k)} e^{-j2\pi fn}, \qquad k = 0, 1, \ldots, K - 1. \qquad (3.57)$$

At a given frequency $f$, the $X_k(f)$ correspond to **y** in the linear regression model, the parameter $\mu$ corresponds to **x**, and the $V_k(f-f_0)$ correspond to the matrix **A**. The number of parameters $p = 1$ here, and by setting it up as a least-squares problem we have to minimize the residual error $\|\mathbf{x} - \mathbf{V}\mu\|^2$ with respect to the complex-valued scalar parameter $\mu$, where

$$\mathbf{x} = \begin{bmatrix} X_0(f) \\ X_1(f) \\ \vdots \\ X_{K-1}(f) \end{bmatrix},$$

$$\mu = \mu(f),$$

and

$$\mathbf{v} = \begin{bmatrix} V_0(f - f_0) \\ V_1(f - f_0) \\ \vdots \\ V_{K-1}(f - f_0) \end{bmatrix}.$$

The term *point regression* comes about because we consider each frequency point $f$ as a possible candidate for $f_0$. We set $f_0 = f$ and test the model (3.56) for significance.

The residual error at $f$ can also be written as

$$\mathbb{E}(\mu, f) = \sum_{K=0}^{K-1} |X_k(f) - \mu(f)V_k(0)|^2. \qquad (3.58)$$

Following the least-squares solution based on the pseudo-inverse matrix presented earlier in the section, we may write

$$\hat{\mu}(f) = \frac{\sum_{k=0}^{K-1} V_k^*(0) X_k(f)}{\sum_{k=0}^{K-1} |V_k(0)|^2}.$$

This solution may also be expressed in the form

$$\mu(f) = \sum_{n=0}^{N-1} h_n x(n) e^{-j2\pi fn}, \tag{3.59}$$

where the harmonic data tapers (windows) are defined by

$$h_n = \frac{\sum_{k=0}^{K-1} V_k^*(0) v_n^{(k)}}{\sum_{k=0}^{K-1} |V_k(0)|^2}. \tag{3.60}$$

We can now test the hypothesis $H_0$ that when $f_0 = f$, the model (3.56) is false; that is, the parameter $\mu = 0$, versus the hypothesis $H_1$, where it is not. In other words, the ratio of the energy explained by the assumption of a line component at the given frequency to the residual energy gives us the $F$-ratio as explained in the outline on linear regression model; that is,

$$F(f) = \frac{\frac{1}{v} |\hat{\mu}(f)|^2 \sum_{k=0}^{K-1} |V_k(0)|^2}{\left(\frac{1}{2K-v}\right) e^2(\hat{\mu} - f)}, \tag{3.61}$$

where $v$ is equal to two DoFs (real and imaginary parts of a complex spectral line component). The total number of DoFs, $2K$, comes about from the set of complex data points $\{v_k(f)\}_{k=0}^{K-1}$, from which we may draw relevant information. If the $F$-ratio is large at a certain frequency $f$, then the hypothesis is rejected; that is, a line component does not exist there; otherwise, it does. The location of the maximum of the $F$-ratio provides an estimate of the line frequency that has a resolution within 5–10% of the Cramér–Rao lower bound (CRLB) (Thomson, 1982); the CRLB is discussed in Chapter 6.

The test works well if the lines considered are isolated; that is, if there is only a single line in the band $(f - W, f + W)$. The total number of lines is not important as long as they occur singly. For lines spaced closer than $W$, we may have to use a multiple-line test, which is similar but algebraically a more complicated regression of the eigencoefficients on a matrix of functions $V_k(f - f_i)$ with the simple $F$-test replaced by partial $F$-tests (Thomson, 1982; Drosopoulos and Haykin, 1992). A cautionary note is in order in using multiple-line tests: the CRLBs for line parameter estimation tend to degrade rapidly when the line spacing becomes less than $2/N$.

Note also that the $F$-test is a statistical test. This means that, given a large number of different realizations of a data sequence, highly significant values can sometimes be seen which, in reality, are only sampling fluctuations. A good rule of thumb, as Thomson (1982) points out, is not to get excited by significance levels below $(1 - 1/N)$. Experience also suggests to try the test for a variety of $NW$ values. Line components that disappear from one case to the other are almost certainly sampling fluctuations.

## 3.8 Summary and discussion

### 3.8.1 The MTM for power spectrum estimation

Perception of the environment can be viewed from different perspectives, depending on the application of interest. In this chapter, we focused attention on power-spectrum estimation as the tool for sensing the environment. This perspective is of particular interest in the study of cognitive radio, which is taken up in Chapter 7.

There are two ways of approaching power-spectrum estimation:

(1) *Parametric methods*, which rely on the adoption of a model and the issue becomes one of adjusting the model parameters such that the power spectrum of the model output provides a reliable estimation of the observables from the environment.
(2) *Nonparametric methods*, which are model free; in this case, the observable data are analyzed directly in a methodical way to provide an estimate of the desired power spectrum.

More specifically, we viewed the power-spectrum estimation problem from a finite set of observable data as an *ill-posed inverse problem*. With this viewpoint in mind, the fundamental equation of power-spectrum estimation assumes the form of a *Fredholm integral equation of the first kind;* see Note 2 for detail. To solve this equation, use is made of *prolate spheroidal wave functions* that were discussed early on by Slepian (1965); hence the reference to them in the chapter as *Slepian sequences*. The remarkable property of Slepian sequences is that their Fourier transforms have the *maximal energy concentration in a prescribed bandwidth*. It is this property that befits the use of Slepian sequences as the mathematical basis for defining a corresponding set of *windows* or *tapers* in formulation of the MTM for solving the power-spectrum estimation problem, originally described by Thomson (1982).

Basically, the MTM replaces the bias–variance tradeoff (that has plagued the traditional approach for power-spectrum estimation) with *bias–resolution tradeoff*. Once this tradeoff is taken care of, the variance problem is also solved. Stated in another way, the MTM provides an analytic basis for computing the *best approximation to a desired power spectrum*, which is not possible from the observable data alone.

### 3.8.2 Extensions of the MTM

The MTM also provides the mathematical basis for the following extensions:

• *space–time processing*, and hence the ability for sensing the environment in both frequency and spatial direction; and
• *time–frequency analysis*, which, in turn, provides a rigorous mathematical basis for describing cyclostationarity.

Cyclostationarity means that second-order statistics of the observable data exhibit periodicity across time. As such, cyclostationarity is a basic property of modulated signals, hence the practical interest in the use of cyclostationarity in cognitive radio applications. We will have more to say on this topic on cognitive radio in Chapter 7.

In the latter part of the chapter, we extended the MTM to deal with situations where the power spectrum contains spectral line components; to this end, we introduced the $F$-test based on the $F$-distribution in statistics.

## 3.8.3      Concluding remarks

In the opening paragraph to this chapter, we pointed out that spectrum estimation is of particular interest to cognitive radio, which will be studied in Chapter 7. With interest in this application of cognition growing exponentially, it is apropos that we conclude Chapter 3 with two messages: one about the mathematical framework and the other about the practical requirements.

### 3.8.3.1      Mathematical framework

Insofar as cognitive radio is concerned, spectrum sensing plays the role of perception: the first pillar of cognition. Moreover, by virtue of its physical nature, spectrum sensing requires a thorough understanding of statistical characterization of the radio environment. In light of this requirement, the first message to take from this chapter is summed up as follows:

- First, spectrum estimation using MTM is the only method that provides a reliable and accurate strategy for solving the spectrum-sensing problem.
- Second, given a finite set of observable data, the MTM provides a rigorous mathematical framework for spatio-temporal processing of the observables in a coordinated manner: temporally, by involving the Loève transform to account for nonstationary character of the radio environment, and *spatially*, by incorporating SVD applied to a receiving array of elemental antennas.

Naturally, frequency plays the role of a common variable in both the temporal and spatial processing of the observable data.

### 3.8.3.2      Practical requirement

It is equally important that we pay attention to the computational efficiency of the MTM. Looking ahead to the material covered in Chapter 7 on cognitive radio, our second message is summed up as follows:

In computational terms, practical implementation of the MTM can be accomplished in a matter of tens of microseconds.

Bearing in mind the mathematical and practical considerations described above, the MTM is the method that solves the spectrum-sensing problem in a coordinated manner by accounting for all three variables (frequency, time, and space) that are involved in statistical characterization of the radio environment, and it does so in a computationally efficient manner; see Note 2 of Chapter 7.

## Notes and practical references

1. *Other methods for spectrum sensing in cognitive radio*
   In addition to spectrum estimation, there are several other methods that have been described in the literature for cognitive radio applications. For a comprehensive overview of other methods for spectrum sensing in cognitive radio, the reader is referred to the paper by Zeng *et al.* (2010). However, for reasons described in this chapter and experimental tests highlighted in Note 6 of Chapter 7, spectrum estimation is viewed as the method of choice.

2. *Power-spectrum estimation is an ill-posed inverse problem*
   Consider a time series $\{x(n)\}_{n=0}^{N-1}$, where $n$ denotes discrete time. The *Cramér representation* of this process is written as follows:

$$x(n) = \int_{-1/2}^{1/2} e^{j2\pi fn} \, dX(f),\qquad(3.62)$$

where $dX(f)$ is the *generalized Fourier transform*. With $x(n)$ assumed to have zero mean, it follows that $dX(f)$ will also have zero mean. For its second-order statistic, we write

$$\mathbb{E}[dX(f)dX^*(v)] = S(f)\delta(f-v)\,df\,dv,\qquad(3.63)$$

where $\delta(f)$ is the *Dirac delta function* in the frequency domain and $S(f)$ is the *power spectrum* of $x(n)$. Note that, with the mainlobe of the power-spectrum estimator occupying a passband of width $2W$, the frequency in (3.62) has been normalized with respect to $2W$, centered on $f=0$; hence the limits of integration in that equation.

   Next, by definition, the ordinary *discrete-time Fourier transform* of the time series $\{x(n)\}_{n=0}^{N-1}$ is given by

$$X_N(f) = \sum_{n=0}^{N-1} x(n)e^{-j2\pi fn}.\qquad(3.64)$$

Substituting (3.62) into (3.64), we write

$$X_N(f) = \int_{-1/2}^{1/2} \sum_{n=0}^{N-1} x(n)\,e^{-j2\pi(f-v)n} dX(v),\qquad(3.65)$$

where we have interchanged the order of summation and integration. Now, we introduce the definition

$$K_N(f) = \sum_{n=1}^{N} e^{-j2\pi fn},\qquad(3.66)$$

which is called the *Dirichlet kernel*. Summing the series in (3.66), we may express this kernel in the equivalent form

$$K_N(f) = \frac{\sin(N\pi f)}{\sin(\pi f)},\qquad(3.67)$$

where $K_N(0) = N$. Thus, referring back to (3.65), we may redefine this equation in terms of the Dirichlet kernel as the convolution integral

$$X_N(f) = \int_{-1/2}^{1/2} K_N(f-v)dX(v). \tag{3.68}$$

This integral equation, defining a linear input–output relationship with $K_N(f-v)$ as the weigting function is referred to as the *fundamental equation of spectral analysis*. In the literature, it is viewed as an example of the *Fredholm equation of the first kind* (Morse and Feshback, 1953).

From the perspective of power-spectrum estimation, we are really interested in the relationship between the expected value of $|X_N(f)|^2$ and the power spectrum $S(f)$. To this end, using (3.63) and (3.68), we obtain the desired relationship, after some straightforward algebraic manipulations, as shown by

$$\mathbb{E}[|X_N(f)|^2] = \int_{-1/2}^{1/2} K_N^2(f-v)S(f)dv. \tag{3.69}$$

Note that, as the size of the data set $N$ is allowed to approach infinity, the kernel $K_N(f)$ approaches the Dirac delta function $\delta(f)$, and with it (3.69) reduces to

$$S(f) = \lim_{N\to\infty} \frac{1}{N}\mathbb{E}[|X_N(f)|^2], \tag{3.70}$$

which is the mathematical definition of the power spectrum. In the general setting of (3.69), $S(f)$ is the unknown function and $|X_N(f)|^2$ is the known function given a finite set of observations $\{x(n)\}_{n=0}^{N-1}$. Hence, we may make the following important statement:

To recover the power spectrum $S(f)$, from the linear input–output relationship described in the fundamental equation of spectral analysis, we have an inverse problem that is ill-posed because, given a finite set of observations, we cannot uniquely determine the power spectrum $S(f)$ defined over an infinite number of points in the frequency domain.

The lack of uniqueness, emphasized in this statement, violates one of the three conditions due to Hadamard for the well-posedness of an inverse problem: existence, uniqueness, and continuity. The issue of an inverse problem being ill-posed was discussed previously in Section 2.14. So, indeed, we may say that power-spectrum estimation is an *ill-posed inverse problem*.

3. *Slepian sequences*

   Detailed information on Slepian sequences is given in Slepian (1978). A method for computing such sequences, for large data length, is given in the appendix of the paper by Thomson (1982).

4. *Regularization property of the MTM*

   Percival and Walden (1993), building on the fundamental equation of spectral analysis discussed in Note 2 of this chapter, show that the generic formula of (3.69) in the note above takes the following form:

$$\mathbb{E}[|X_k(f)|^2] = \int_{f-\frac{1}{2}}^{f+\frac{1}{2}} \left( \frac{|V_k(f-v)|^2}{\lambda_k} \right) S(v)dv, \tag{3.71}$$

where $X_k(f)$ is the Fourier transform of the time series $\{x(n)\}_{n=0}^{N-1}$, $V_k(f)$ is the Fourier transform of the $k$th Slepian sequence $\{v_k\}_{k=0}^{K-1}$ and $\lambda_k$ is the associated eigen-coefficient. $S(f)$ is the unknown power spectrum to be estimated. Examining the *convolution integral* of (3.71), we may view $|V_k(f)|^2$ as a *smoothing function*. Most importantly, this smoother is a good one because its smoothing property is confined entirely to the main lobe of the spectrum that occupies the frequency interval $[-1/2, 1/2]$ and thus avoids sidelobe leakage. This smoothing operation represents a form of regularization that stabilizes an ill-posed inverse problem (Haykin, 2009). We may therefore conclude that regularization is mathematically built into the MTM.

5. An *inconsistent estimate* is one where the variance of the estimate does not decrease with sample size.

6. *Distinction between the Loève spectrum and bispectrum*

   Care should be exercised in distinguishing the Loève spectrum $\gamma_L(f_1, f_2)$ from the bispectrum $B(f_1, f_2)$. Both are functions of two frequencies, $f_1$ and $f_2$, but the Loève spectrum $\gamma_L(f_1, f_2)$ is a *second*-moment description of a possibly nonstationary process; in contrast, the bispectrum describes the *third*-moments of a stationary process and has an implicit third frequency $f_3 = f_1 + f_2$.

7. For most complex-valued signals, the expectation $\mathbb{E}[x(t_1)x(t_2)]$, and therefore $\mathbb{E}[X(f_1)X(f_2)]$, is zero. For communication signals, however, this expectation is often not zero.

8. *Mooers' approach to complex-valued stochastic processes*

   In a terminological context, there is confusion in how second-order moments of complex-valued stochastic processes are defined in the literature:

   • Thomson (1982) and Picinbone (1996) use the terms "forward" and "reversed" to distinguish, for example, the second-order moments $\mathbb{E}[X_k(f_1)X_k^*(f_2)]$ and $\mathbb{E}[X_k(f_1)X_k(f_2)]$ respectively.

   • Mooers (1973) uses the terms "inner" and "outer," borrowed from the notions of inner products and outer products, to distinguish between these two second-order moments.

   • In the cyclostationarity literature on communication signals, the terms "spectral correlation" and "conjugate spectral correlation" are used to refer to the expectations $\mathbb{E}[X(f_1)X^*(f_2)]$ and $\mathbb{E}[X(f_1)X(f_2)]$ respectively. This terminology is misleading: if $\mathbb{E}[X(f_1)X^*(f_2)]$ stands for spectral correlation, then the expression for conjugate spectral correlation would be $\mathbb{E}[X^*(f_1)X(f_2)]$, which is not the intention.

   As stated in the text, in this chapter we follow Mooers' terminology.

9. *Cautionary note on the use of (3.43)*

   The naive implementation of the Wigner–Ville distribution, as defined in this equation using a finite sample size, may result in bias and sampling properties that are worse than the periodogram.

10. *Cyclostationarity and IEEE 802.22*

    Research interest in cyclostationarity for signal detection and classification has experienced resurgence with the emergence of cognitive radio, resulting a signal-detection technique in the draft form of the IEEE 802.22 standard for cognitive radio applications.

11. The literature on cyclostationarity includes the book by Hurd and Miamee (2007) and the bibliography due to Serpedin *et al.* (2005) that lists over 1500 papers on the subject.

12. *More on Fourier-based cyclostationarity*

The Fourier-based approach to cyclostationarity may also acquire an adaptive capability of its own. In many spectrum-sensing applications based on this approach, the Fourier spectral coherence of the first and second kinds, defined in (3.52) and (3.53), are computed over the entire spectral domains of interest, and the *actual* cycle frequencies (i.e. statistical periodicity of the signal) may thus be accurately estimated. According to Spooner (personal communication, 2008), applicability of the Fourier-based cyclostationarity approach to spectrum sensing can be extended to a wide range of situations, from completely blind (no prior knowledge of periodicity) to highly targeted ones (known periodicities with possible errors).

# 4 Bayesian filtering for state estimation of the environment

Perception of the environment, viewed as a problem in spectrum estimation as discussed in Chapter 3, is most appropriate for applications where spectrum sensing of the environment is crucial to the application at hand; cognitive radio is one such important application. A distinctive aspect of spectrum estimation is the fact that it works *directly* on environmental measurements (i.e. observables). However, in many other environments encountered in the study of cognitive dynamic systems, perception of the environment boils down to state estimation, which is the focus of attention in this chapter.

We defer a formal definition of the state to Section 4.4, where the issue of state estimation is taken up. For now, it suffices to say that estimating the *state* of a physical environment is compounded by two practical issues:

(1) The state of the environment is *hidden* from the observer, with information about the state being available only indirectly through dependence of the observables (measurements) on the state.
(2) Evolution of the state across time and measurements on the environment are both corrupted by the unavoidable presence of physical *uncertainties* in the environment.

To tackle perception problems of the kind just described, the first step is to formulate a *probabilistic model* that accounts for the underlying physics of the environment. Logically, the model consists of a pair of equations:

- a *system equation*,[1] which accounts for evolution of the environmental state across time, and
- a *measurement equation*, which describes dependence of the measurements on the state.

These two equations constitute the well-known *state-space model* of the environment, on which the state-estimation problem (i.e. perception of the environment) is based. In particular, we look to the system equation for a value of the state at some prescribed time.

To solve this problem, we look to the *Bayesian framework*[2] as a general formalism of the state-space model, with the unknown state (assumed to be multidimensional) and associated uncertainties all being treated as random vectors. Naturally, in this formalism, the measurement equation characterizes the information available to the perceptor of a cognitive dynamic system under study. However, this is only one side of the story as to how the environment is being perceived by the perceptor. To characterize how the perceptor actually makes *inferences* about the state of the environment, which is the other side of the story, the perceptor looks to the system equation for exploiting dependence of the observables on the state. In a way, the two sides of the story briefly described here sum up how the visual brain perceives its surrounding environment. By the same token, the story applies equally well to cognitive radar as an important engineering example of cognitive dynamic systems.

Given a probabilistic state-space model of the environment, the aim of this chapter is to develop algorithms for solving the state-estimation problem in an *iterative on-line manner*. The emphasis on iterative computation is for the following practical reason:

The updated estimate of the state builds on its old value; hence, its iterative computation is ideally suited for the use of a computer.

The optimal solution to this state-estimation problem is to be found in the *Bayesian filter*. When, however, the state-space model is nonlinear, which is frequently the case in practice, the Bayesian filter defies computational feasibility. In situations of this kind, we have to be content with an *approximation* to the Bayesian filter. Given this practical reality, we would like to develop an algorithm that is the "best" approximation to the Bayesian filter in some sense. This development is indeed an important aim of this chapter.

With the state-space model being probabilistic, however, it is apropos that we begin the discussion with a brief review of relevant concepts and ideas in probability theory.

## 4.1        Probability, conditional probability, and Bayes' rule

Let $A$ denote an *event* that describes the outcome of a probabilistic experiment or subset of such outcomes in a sample space denoted by $S$. Henceforth, we will use the symbol $\mathbb{P}$ to signify the "operator" for the probability of an event enclosed inside a pair of square brackets. The probability of event $A$, denoted by $\mathbb{P}[A]$, is governed by three *axioms of probability*, formulated in the context of set theory.[3]

Axiom I    *Nonnegativity.* This states that *the probability of event A is a nonnegative number bounded by unity*, as shown by

$$0 \le \mathbb{P}[A] \le 1 \quad \text{for any event } A.$$

Axiom II    *Additivity.* This states that *if A and B are two disjoint events, then the probability of their union satisfies the equality*

$$\mathbb{P}[A \cup B] = \mathbb{P}[A] + \mathbb{P}[B],$$

where the symbol $\cup$ denotes "union." In general, if the sample space has $N$ elements and $A_1, A_2, \ldots, A_N$ is a sequence of disjoint events, then the probability of the union of these $N$ events satisfies the equality

$$\mathbb{P}[A_1 \cup A_2 \cup \cdots A_N] = \mathbb{P}[A_1] + \mathbb{P}[A_2] + \cdots \mathbb{P}[A_N],$$

which is intuitively satisfying.

Axiom III *Normalization.* This states that *the probability of the entire sample space S is equal to unity,* as shown by

$$\mathbb{P}[S] = 1.$$

These three axioms provide an implicit definition of probability.

## 4.1.1 Conditional probability

When an experiment is performed and we only obtain *partial information* on the outcome of the experiment, we may *reason* about that particular outcome by invoking the notion of conditional probability. Stated the other way round:

Conditional probability provides the premise for *probabilistic reasoning* or *inference.*

To be specific, suppose we perform an experiment that involves a pair of events $A$ and $B$. Let $\mathbb{P}[A|B]$ denote the probability of event $A$ given that event $B$ has occurred. The probability $\mathbb{P}[A|B]$ is called the *conditional probability of A given B.* Assuming that $B$ has nonzero probability, the conditional probability $\mathbb{P}[A|B]$ is formally defined by

$$\mathbb{P}[A|B] = \frac{\mathbb{P}[A \cap B]}{\mathbb{P}[B]}, \tag{4.1}$$

where $\mathbb{P}[A \cap B]$ is the joint probability of events $A$ and $B$ and $\mathbb{P}[B]$ is nonzero. The proof of (4.1) follows directly from the intuitively satisfying relationship

$$\mathbb{P}[A \cap B] = \mathbb{P}[A|B]\mathbb{P}[B].$$

Most importantly, for a fixed event $B$, the conditional probability $\mathbb{P}[A|B]$ is a legitimate probability law, in that it obeys all three axioms of probability in its own way. In a sense, the conditional probability $\mathbb{P}[A|B]$ captures the *partial information that the occurrence of event B provides about event A.* We may, therefore, view the conditional probability $\mathbb{P}[A|B]$ as a probability law concentrated on event $B$.

## 4.1.2 Bayes' rule

Suppose we are confronted with a situation where the conditional probability $\mathbb{P}[A|B]$ and the individual probabilities $\mathbb{P}[A]$ and $\mathbb{P}[B]$ are all easily determined directly, but it is the conditional probability $\mathbb{P}[B|A]$ that is desired. To deal with this situation, we first rewrite (4.1) in the form

$$\mathbb{P}[A \cap B] = \mathbb{P}[A|B]\mathbb{P}[B].$$

Clearly, we may equally write

$$\mathbb{P}[A \cap B] = \mathbb{P}[B|A]\mathbb{P}[A].$$

The left-hand sides of these two relations are identical; therefore, we have

$$\mathbb{P}[A|B]\mathbb{P}[B] = \mathbb{P}[B|A]\mathbb{P}[A].$$

Provided that $\mathbb{P}[A]$ is nonzero, we may determine the desired conditional probability $\mathbb{P}[B|A]$ by using the relation

$$\mathbb{P}[B\,|\,A] = \frac{\mathbb{P}[A\,|\,B]\mathbb{P}[B]}{\mathbb{P}[A]}. \qquad (4.2)$$

This relation is known as *Bayes' rule*.

As simple as it looks, Bayes' rule provides the correct language for describing *inference*; its formulation, however, cannot be done without making assumptions about the probabilities on the right-hand side of (4.2).

## 4.2    Bayesian inference and importance of the posterior

The brief review material just presented on probabilistic models sets the stage to study the role of probability theory in *probabilistic reasoning* based on the Bayesian view of inference.[4]

To proceed with the discussion, consider Figure 4.1. This depicts two finite-dimensional spaces: a *parameter space* and an *observation space*, with the parameter space being hidden from the observer that only has access to the observation space. A parameter vector $\boldsymbol{\theta}$, drawn from the parameter space $\Theta$, is mapped probabilistically by the *probabilistic transition mechanism* (e.g. telephone channel) onto the observation space $\chi$, producing the observation vector $\mathbf{x}$. The vector $\mathbf{x}$ is itself the sample value of a random vector $\mathbf{X}$. Given the probabilistic scenario depicted in Figure 4.1, we may identify two different operations that are the dual of each other.[5] Specifically, we have:

(1) *Probabilistic modeling.* The aim of this operation is to formulate the conditional probability density function $p_{\mathbf{X}|\Theta}(\mathbf{x}|\boldsymbol{\theta})$, which provides an adequate description of the underlying physical behavior of the observation space.

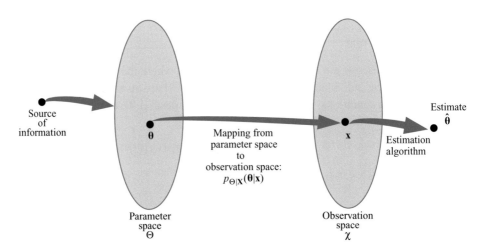

**Figure 4.1.** Schematic model for the parameter-estimation problem.

(2) *Statistical analysis.* The aim of this second operation is the *inverse of probabilistic modeling*, for which we need the alternative conditional probability density function $p_{\Theta|\mathbf{X}}(\theta|\mathbf{x})$.

In a fundamental sense, statistical analysis is more profound than probabilistic modeling. We may justify this assertion by viewing the unknown parameter vector $\theta$ as the *cause* for the physical behavior of the probabilistic transition mechanism, and viewing the observation vector $\mathbf{x}$ as the *effect*. In essence, statistical analysis solves an *inverse problem* by retrieving the causes (i.e. the parameter vector $\theta$) from the effects (i.e. the observation vector $\mathbf{x}$). Indeed, we may go on to say that *probabilistic modeling* helps us to characterize the *future behavior* of $\mathbf{x}$ conditioned on $\theta$; on the other hand, statistical analysis permits us to make *inference* about the unknown $\theta$ given $\mathbf{x}$, which makes it particularly important in probabilistic reasoning.

To formulate the conditional probability density function $p_{\mathbf{X}|\Theta}(\mathbf{x}|\theta)$, we recast Bayes' rule of (4.2) in its *continuous version* by writing

$$p_{\Theta|\mathbf{X}}(\theta|\mathbf{x}) = \frac{p_{\mathbf{X}|\Theta}(\mathbf{x}|\theta)p_{\Theta}(\theta)}{p_{\mathbf{X}}(\mathbf{x})}. \tag{4.3}$$

The denominator of the right-hand side of (4.3) is itself defined in terms of the numerator through integration, as shown by

$$\begin{aligned} p_{\mathbf{X}}(\mathbf{x}) &= \int_{\Theta} p_{\mathbf{X}|\Theta}(\mathbf{x}|\theta)p_{\Theta}(\theta)\,d\theta \\ &= \int_{\Theta} p_{\mathbf{X},\Theta}(\mathbf{x},\theta)\,d\theta \end{aligned} \tag{4.4}$$

where $p_{\mathbf{X},\Theta}(\mathbf{x}, \theta)$ is the joint probability density function of $\mathbf{x}$ and $\theta$. In effect, the left-hand side of (4.4) is obtained by integrating out the dependence of the joint probability density function $p_{\mathbf{X},\Theta}(\mathbf{x}, \theta)$ on the parameter vector $\theta$. Using probabilistic terminology, $p_{\mathbf{X}}(\mathbf{x})$ is said to be a *marginal density* of the joint probability density function $p_{\mathbf{X},\Theta}(\mathbf{x}, \theta)$. The inversion formula of (4.3) is sometimes referred to as the *principle of inverse probability*.

In light of this principle, we may now introduce four distinct distributions:

(1) *Observation distribution.* This stands for the conditional probability density function $p_{\mathbf{X}|\Theta}(\mathbf{x}|\theta)$ pertaining to the "observation" vector $\mathbf{x}$ given the parameter vector $\theta$.
(2) *Prior distribution.* This stands for the probability density function $p_{\Theta}(\theta)$ pertaining to the parameter vector $\theta$ "prior" to receiving the observation vector $\mathbf{x}$.
(3) *Posterior distribution.* This stands for the conditional probability density function $p_{\mathbf{X}|\Theta}(\mathbf{x}|\theta)$ pertaining to the unknown parameter vector $\theta$ "after" receiving the observation vector $\mathbf{x}$.
(4) *Evidence.* This stands for the probability density function $p_{\mathbf{X}}(\mathbf{x})$ referring to the "information" contained in the observation vector $\mathbf{x}$ solely for statistical analysis.

The posterior distribution $p_{\Theta|\mathbf{X}}(\theta|\mathbf{x})$ is central to *Bayesian inference*. In particular, we may view it as the updating of information available on the parameter vector $\theta$ in light of the information about $\theta$ that is contained in the observation vector $\mathbf{x}$, while the prior distribution $p_{\Theta}(\theta)$ is the information available on $\theta$ prior to receiving the observation vector $\mathbf{x}$.

## 4.2.1    Likelihood

The inversion aspect of statistical analysis manifests itself in the notion of the *likelihood function* or just simply *likelihood*.[6] In a formal sense, the likelihood, denoted by $l(\boldsymbol{\theta}|\mathbf{x})$, is just the prior density $p_{\mathbf{X}|\Theta}(\mathbf{x}|\boldsymbol{\theta})$, but reformulated in a different order, as shown by

$$l(\boldsymbol{\theta}|\mathbf{x}) = p_{\mathbf{X}|\Theta}(\mathbf{x}|\boldsymbol{\theta}). \tag{4.5}$$

The important point to note here is that the likelihood and the observation distribution are both governed by exactly the same function that involves the parameter vector $\boldsymbol{\theta}$ and the observation vector $\mathbf{x}$. There is, however, a subtle difference between them in probabilistic interpretation: the likelihood function $l(\boldsymbol{\theta}|\mathbf{x})$ is treated as a function of the unknown parameter vector $\boldsymbol{\theta}$ given $\mathbf{x}$, whereas the observation density $p_{\mathbf{X}|\Theta}(\mathbf{x}|\boldsymbol{\theta})$ is treated as a function of the observation vector $\mathbf{x}$ given $\boldsymbol{\theta}$. However, note that, unlike $p_{\mathbf{X}|\Theta}(\mathbf{x}|\boldsymbol{\theta})$, the likelihood $l(\boldsymbol{\theta}|\mathbf{x})$ is *not* a distribution; rather, it is simply a function of the parameter vector $\boldsymbol{\theta}$ for some given $\mathbf{x}$,

In light of the terminologies just introduced, namely the posterior, prior, likelihood, and evidence, we may now use (4.3) to express the underlying rule for Bayesian statistical analysis (i.e. inference) in words as follows:

$$\text{posterior} = \frac{\text{likelihood} \times \text{prior}}{\text{evidence}}.$$

## 4.2.2    The likelihood principle

For convenience of presentation, henceforth we will let

$$\pi(\boldsymbol{\theta}) = p_{\Theta}(\boldsymbol{\theta}). \tag{4.6}$$

Then, recognizing that the evidence defined in (4.4) plays merely the role of a *normalizing function* that is independent of $\boldsymbol{\theta}$, we may now sum up the principle of inverse probability succinctly as follows:

The Bayesian statistical model is essentially made up of two components: the likelihood $l(\boldsymbol{\theta}|\mathbf{x})$ and the prior $\pi(\boldsymbol{\theta})$, where $\boldsymbol{\theta}$ is an unknown parameter vector and $\mathbf{x}$ is the observation vector.

To elaborate on the significance of the defining equation (4.6), consider the two likelihood functions $l(\boldsymbol{\theta}|\mathbf{x}_1)$ and $l(\boldsymbol{\theta}|\mathbf{x}_2)$ on parameter vector $\boldsymbol{\theta}$. If, for a prescribed prior $\pi(\boldsymbol{\theta})$, these two likelihood functions are scaled versions of each other, then the corresponding posterior densities of $\boldsymbol{\theta}$ are essentially identical, the validity of which is a straightforward consequence of Bayes' rule. In light of this result, we may now formulate the so-called *likelihood principle*[7] as follows:

If $\mathbf{x}_1$ and $\mathbf{x}_2$ are two observation vectors depending on an unknown parameter vector $\boldsymbol{\theta}$, such that

$$l(\boldsymbol{\theta}|\mathbf{x}_1) = cl(\boldsymbol{\theta}|\mathbf{x}_2) \qquad \text{for all } \boldsymbol{\theta},$$

where $c$ is a scaling factor, then these two observation vectors lead to an identical inference on $\boldsymbol{\theta}$ for any prescribed prior $p_{\boldsymbol{\theta}}(\boldsymbol{\theta})$.

### 4.2.3  Sufficient statistic

Consider, next, a model parameterized by the vector $\boldsymbol{\theta}$ given the observation vector $\mathbf{x}$. In statistical terms, the model is described by the posterior density $p_{\Theta|\mathbf{X}}(\boldsymbol{\theta}|\mathbf{x})$. In this context, we may now introduce a function $\mathbf{t}(\mathbf{x})$, which is said to be a *sufficient statistic* if the probability density function of the parameter vector $\boldsymbol{\theta}$ given $\mathbf{t}(\mathbf{x})$ satisfies the condition

$$p_{\Theta|\mathbf{X}}(\boldsymbol{\theta}|\mathbf{x}) = p_{\Theta|\mathrm{T}(\mathbf{x})}(\boldsymbol{\theta}|\mathbf{t}(\mathbf{x})). \qquad (4.7)$$

This condition imposed on $\mathbf{t}(\mathbf{x})$, for it to be a sufficient statistic, appears intuitively appealing, as evidenced by the following statement:

The function $\mathbf{t}(\mathbf{x})$ provides a sufficient summary of the whole information about the unknown parameter vector $\boldsymbol{\theta}$, which is contained in the observation vector $\mathbf{x}$.

We may thus view the notion of sufficient statistic as a tool for "data reduction," the use of which results in considerable simplification in statistical analysis.[8]

## 4.3  Parameter estimation and hypothesis testing: the MAP rule

### 4.3.1  Parameter estimation

As pointed out previously, the posterior density $p_{\Theta|\mathbf{X}}(\boldsymbol{\theta}|\mathbf{x})$ is central to the formulation of a Bayesian probabilistic model, where $\boldsymbol{\theta}$ is an unknown parameter vector and $\mathbf{x}$ is the observation vector. It is logical therefore, that we use this probability density function for parameter estimation. Accordingly, we define the *maximum a-posteriori probability (MAP) estimate* of the parameter vector $\boldsymbol{\theta}$, assumed to be continuous, as follows:[9]

$$\begin{aligned}
\hat{\boldsymbol{\theta}}_{\mathrm{MAP}} &= \arg\max_{\boldsymbol{\theta}} \, p_{\Theta|\mathbf{X}}(\boldsymbol{\theta}\,|\,\mathbf{x}) \\
&= \arg\max_{\boldsymbol{\theta}} \, l(\boldsymbol{\theta}\,|\,\mathbf{x})\pi(\boldsymbol{\theta}),
\end{aligned} \qquad (4.8)$$

where $l(\boldsymbol{\theta}|\mathbf{x})$ is the likelihood defined in (4.5) and $\pi(\boldsymbol{\theta})$ is the prior defined in (4.6). Note that, in formulating (4.8), we have ignored the evidence $p_{\mathbf{X}}(\mathbf{x})$; this omission is justified, since the evidence is merely a normalizing function that is independent of the parameter vector $\boldsymbol{\theta}$; as such, it does not affect the maximization in (4.8).

Clearly, to compute the estimate $\hat{\boldsymbol{\theta}}_{\mathrm{MAP}}$, we require availability of the prior $\pi(\boldsymbol{\theta})$. In words, the right-hand side of (4.8) reads as follows:

Given the observation vector $\mathbf{x}$, the estimate $\hat{\boldsymbol{\theta}}_{\mathrm{MAP}}$ is that particular value of the parameter vector $\boldsymbol{\theta}$ in the argument of the posterior density $p_{\Theta|\mathbf{X}}(\boldsymbol{\theta}|\mathbf{x})$ for which the posterior density attains its maximum value.

We may generalize this statement by going on to say that the posterior $p_{\Theta|\mathbf{X}}(\boldsymbol{\theta}|\mathbf{x})$ *contains all the information about the unknown multidimensional parameter vector $\boldsymbol{\theta}$ given the observation vector $\mathbf{x}$*. Recognition of this fact leads us to make the follow-up important statement:

The MAP estimate $\hat{\boldsymbol{\theta}}_{\mathrm{MAP}}$ of the unknown parameter vector $\boldsymbol{\theta}$ is the globally optimal solution to the parameter-estimation problem, in the sense that there is no other estimator that can achieve a better estimate.

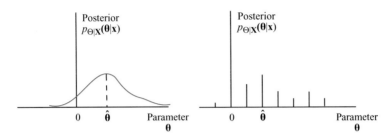

**Figure 4.2.** Illustrating the MAP rule for two cases of parameter $\theta$: (a) continuous parameter; (b) discrete parameter.

In referring to $\hat{\theta}_{\text{MAP}}$ as the MAP estimate, we have made a slight change in our terminology: we have, in effect, referred to $p_{\Theta|\mathbf{X}}(\theta|\mathbf{x})$ as the *a-posteriori probability density* rather than the *posterior density* of $\theta$ for the case of continuous $\theta$. We have made this minor change so as to conform to the MAP terminology that is well and truly embedded in the literature on Bayesian statistical theory.

The statement just made on the MAP rule applies to the case when the unknown parameter vector $\theta$ is *continuous*. For the case when $\theta$ is discrete, taking a *finite* number of possible values, we modify the statement on the MAP rule to embody the following important property (Bertsekas and Tsitsiklis, 2008):

> The MAP rule maximizes the probability of correct decision for any observation vector $\mathbf{x}$, because it chooses $\hat{\theta}$ to be the most likely estimate. (4.9)

Stated in other words, the MAP rule minimizes the probability of making an incorrect decision for each observation vector $\mathbf{x}$, with the overall probability of error being averaged over all $\mathbf{x}$. Figure 4.2 illustrates the MAP rule for a one-dimensional scenario with parts (a) and (b) of the figure referring to the continuous and discrete cases respectively.

The discussion presented in this subsection has focused on parameter estimation. In the next subsection, we study hypothesis testing, which is another way of referring to decision-making.

## 4.3.2    Hypothesis testing

Suppose that the *posterior-density* computer for the MAP rule has been derived, as depicted in Figure 4.3, which sets the stage for addressing the *hypothesis-testing problem*. In a general setting, there are $M$ possible hypotheses to consider, in which case we speak of *M-ary hypothesis testing*, with $M$ being typically a small integer depending on the problem of interest. Binary-hypothesis testing is a special case with $M = 2$.

Let the $M$ hypotheses be denoted by the set $\{H_m\}_{m=1}^M$. Then, following the MAP terminology introduced in Figure 4.3, we say that the $m$th hypothesis refers to the *event* described by the random variable $\Theta = \theta_m$, where the subscript $m$ takes the value $1, 2, \ldots, M$.

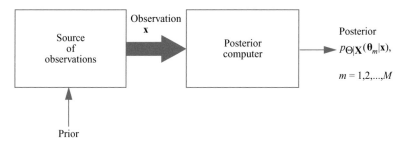

**Figure 4.3.** Block diagram for illustrating the MAP rule.

In practice, we typically find that the observation vector **x** is discrete. Then, to solve the *M*-ary hypothesis-testing problem using the MAP rule, we proceed as follows:

- Given an observation vector **x**, we use Bayes' rule to find the *posterior probabilities*

$$\mathbb{P}[\Theta = \boldsymbol{\theta}_m \,|\, \mathbf{X} = \mathbf{x}] = p_{\Theta|\mathbf{X}}(\boldsymbol{\theta}_m \,|\, \mathbf{x}), \qquad m = 1, 2, \ldots, M, \qquad (4.10)$$

  where **X** is a random vector and $p_{\Theta|\mathbf{X}}(\boldsymbol{\theta}_m \,|\, \mathbf{x})$ is a discrete probability density function.
- Next, we use the posterior computer in Figure 4.3 to select the particular hypothesis *m* for which the posterior probability in (4.10) is the largest.

When there is a tie in the decision-making with two or more hypotheses attaining the same largest posterior probability, we break the tie by arbitrarily choosing one of them; here, it is assumed that all the *M* hypotheses are *equiprobable*.

Most importantly, the result of applying the MAP rule to the *M*-ary hypothesis-testing problem is *optimal*, in the sense that the probability of correct decision is maximized over all *M* possible decisions.

To compute the probability of a correct decision for a specific observation vector **x**, let $h_{\mathrm{MAP}}(\mathbf{x})$ denote the hypothesis selected by the MAP rule as configured in Figure 4.3. Then, according to (4.10), the *probability of correct decision* is described by the expression

$$\mathbb{P}[\Theta = h_{\mathrm{MAP}}(\mathbf{x}) \,|\, \mathbf{X} = \mathbf{x}].$$

Moreover, let $S_m$ denote the set of all **x** in the observation space, for which the MAP rule selects hypothesis $H_m$. Hence, we may define the *overall probability of correct decision* for the *M*-ary hypothesis-testing problem as follows:

$$\mathbb{P}[\Theta = h_{\mathrm{MAP}}(\mathbf{x})] = \sum_{m=1}^{M} \mathbb{P}[\Theta = \boldsymbol{\theta}_m, \mathbf{x} \in S_m]. \qquad (4.11)$$

Correspondingly, the *overall probability of error* is described by the expression

$$\sum_{m=1}^{M} \mathbb{P}[\Theta \neq \boldsymbol{\theta}_m, \mathbf{x} \in S_m].$$

The overall probabilities of correct and erroneous decisions, as described herein, add up to unity, which is exactly how it should be. In effect, having calculated the overall probability of correct decision using (4.11), the overall probability of error is obtained simply by subtracting the result of (4.11) from unity.

### 4.3.3    Summarizing remarks on Bayesian inference

The main purpose of Bayesian statistical inference may be summed up as follows:

Given the observations of a random phenomenon, Bayesian inference provides a coherent methodology for deriving the underlying probability distribution of that phenomenon.

To be more specific, the probability distribution referred to in this statement is the a-posteriori probability density function or the *posterior* for short. The posterior is defined as the product of the likelihood and the prior, except for a normalizing factor called the evidence. For an observation vector denoted by $\mathbf{x}$ and a related unknown parameter vector $\boldsymbol{\theta}$, the likelihood is written as $l(\boldsymbol{\theta}|\mathbf{x})$, which, in reality, is the probability density function of $\mathbf{x}$, conditioned on $\boldsymbol{\theta}$; in this context, it is noteworthy that the Bayesian approach is the only one that allows for conditioning on the observations. As for the prior, denoted by $\pi(\boldsymbol{\theta})$, it provides for information known about the parameter vector $\boldsymbol{\theta}$ prior to collection of the observations. Thus, ignoring the evidence, the posterior, denoted by $p(\boldsymbol{\theta}|\mathbf{x})$, is defined by the product $l(\boldsymbol{\theta}|\mathbf{x})\pi(\boldsymbol{\theta})$.

What makes the posterior $p(\boldsymbol{\theta}|\mathbf{x})$ so very important is the fact that it contains all the information that we need to know, including the prior about the unknown parameter vector $\boldsymbol{\theta}$ given the observation vector $\mathbf{x}$.

Moreover, when it comes to *decision-making*, we look to the posterior as the distribution to formulate the *MAP rule*. Basically, the MAP rule maximizes the probability of correct decision for any observation vector $\mathbf{x}$ by virtue of the fact that it chooses the parameter vector $\hat{\boldsymbol{\theta}}$, for which the posterior attains its maximum value. As such, when considering the issues of parameter estimation and hypothesis testing, the MAP rule is optimal in that we cannot do better.

For parameter estimation, the MAP rule assumes that the unknown parameter vector $\boldsymbol{\theta}$ is continuous. When dealing with hypothesis testing, the MAP rule is still applicable, except that this time the posterior is reformulated as a discrete probability density function compatible with the number of hypotheses under test.

Our primary interest in Bayesian inference in this chapter is motivated by the second viewpoint on how to deal with perception in the perception–action cycle that was discussed in Section 2.11. With that viewpoint in mind, our objective is to estimate the environmental state of a dynamic system, given a set of observations received from the environment. With the environmental state being "hidden" from the perceptor, the only course open to us is to formulate a state-space model for the system, which is done in the next section. Such a model, in turn, sets the stage for deriving the Bayesian filter that is the best that we can do for state estimation, at least in a conceptual sense.

Another important point of interest is that state estimation is an example of statistical analysis. As such, in light of the discussion presented in Section 4.2 on Bayesian inference, state estimation is an inverse problem. Moreover, it is *ill-posed* on account of unavoidable physical uncertainties in the environment.

## 4.4       State-space models

The material on Bayesian inference presented in Sections 4.2 and 4.3 provides the right background for *Bayesian filtering*, which is aimed at estimating the *state* of a dynamic system. To this end, we begin the discussion with the following formal definition:

The state of a dynamic system is defined as the minimal amount of information about the effects of past inputs applied to the system, such that it is sufficient to completely describe the future behavior of the system.

Typically, the state is *not* measurable directly, as it is *hidden* from the perceptor. Rather, in an indirect manner, the state makes its effect on the environment (outside world) to be estimatable through a set of *observables*. As such, characterization of the dynamic system is described by a *state-space model*, which embodies a pair of equations:[10]

(1) *System equation (model)*, which, formulated as a *first-order Markov chain*, describes the evolution of the state as a function of time, as shown by

$$\mathbf{x}_{n+1} = \mathbf{a}_n(\mathbf{x}_n, \boldsymbol{\omega}_n), \tag{4.12}$$

where $n$ denotes discrete time, the vector $\mathbf{x}_n$ denotes the current value of the state, and $\mathbf{x}_{n+1}$ denotes its immediate future value; the vector $\boldsymbol{\omega}_n$ denotes *system noise* and $\mathbf{a}_n(.,.)$ is a vectorial function of its two arguments, representing transition from state $\mathbf{x}_n$ to state $\mathbf{x}_{n+1}$. A cautionary note in the context of terminology is in order: in Sections 4.2 and 4.3, $\mathbf{x}$ was used to denote an observation vector. In Section 4.4, on the other hand, $\mathbf{x}_n$ is used to denote the state of a dynamic system at discrete time $n$.

(2) *Measurement equation (model)*, which is formulated as

$$\mathbf{y}_n = \mathbf{b}_n(\mathbf{x}_n, \mathbf{v}_n), \tag{4.13}$$

where the vector $\mathbf{y}_n$ denotes a set of measurements (observables), the vector $\mathbf{v}_n$ denotes *measurement noise*, and $\mathbf{b}_n(.,.)$ denotes another vectorial function.

The subscript $n$ in both $\mathbf{a}_n$ and $\mathbf{b}_n$ is included to cover situations where these two functions are *time varying*. For the state-space model to be of practical value, it must, however, relate to the underlying physics of the system under study. Stated in a practical way: to formulate the state-space model of a dynamic system, we need to understand the underlying physics of the system; it is only in this way that we get meaningful results.

Figure 4.4 depicts a signal-flow graph representation of the state-space model defined by (4.12) and (4.13); and Figure 4.5 depicts its evolution across time as a *first-order Markov chain*. In their own individual ways, both of these figures emphasize the hidden nature of the state, viewed from the perceptor. Most importantly, however, adoption of the state-space model, described in (4.12) and (4.13), offers certain attributes:

• mathematical and notational convenience;
• a close relationship of the mode to physical reality; and
• a meaningful basis of accounting for the statistical behavior of the dynamic system.

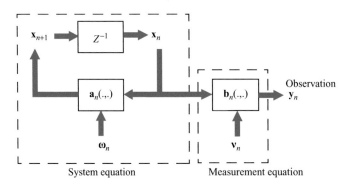

**Figure 4.4.** Generic state-space model of a time-varying, nonlinear dynamic system, where $Z^{-1}$ denotes a block of time-unit delays.

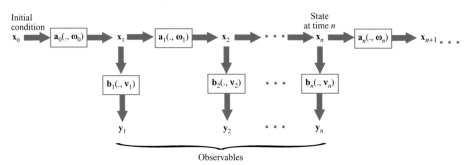

**Figure 4.5.** Evolution of the state across time, viewed as a first-order Markov chain.

Henceforth, the following assumptions are made:

(1) The initial state $\mathbf{x}_0$ is uncorrelated with the system noise $\boldsymbol{\omega}_n$ for all $n$.
(2) The two sources of noise, $\boldsymbol{\omega}_n$ and $\mathbf{v}_n$, are statistically independent of each other, which means that

$$\mathbb{E}[\boldsymbol{\omega}_n \mathbf{v}_k^\mathsf{T}] = \mathbf{0} \qquad \text{for all } n \text{ and } k. \tag{4.14}$$

This equation is a sufficient condition for independence when $\boldsymbol{\omega}_n$ and $\mathbf{v}_n$ are *jointly Gaussian*.

## 4.4.1    Sequential state-estimation problem

Although, indeed, the state is hidden from the perceptor, the environment does provide information about the state through *measurements (observables)*, which prompts us to make the following statement:

Given a record of measurements, consisting of $\mathbf{y}_1, \mathbf{y}_2, \ldots, \mathbf{y}_n$, the requirement is to compute an estimate of the unknown state $\mathbf{x}_k$ that is optimal in some statistical sense, with the estimation being performed in a sequential manner.

In a way, this statement embodies two systems:

- The unknown dynamic system, whose observable $\mathbf{y}_n$ is a function of the hidden state.
- The sequential state-estimator or filter, which exploits information about the state that is contained in the observables.

We may, therefore, view state estimation as an *encoding–decoding* problem in the following sense:

The measurements represent an "encoded" version of the state, and the state estimate produced by the filter represents a "decoded" version of the measurements.

As a corollary to this statement, we may say state estimation is an *inverse problem*, reconfirming what we said in "Summarizing remarks on Bayesian inference" in Section 4.3.

The state-estimation problem is commonly referred to as *prediction* if $k > n$, *filtering* if $k = n$, and *smoothing* if $k < n$. Typically, a smoother is statistically more accurate than both the predictor and filter, as it uses more observables, past and present. On the other hand, both prediction and filtering can be performed in *real time*, whereas smoothing cannot.

## 4.4.2   Hierarchy of state-space models

The mathematical difficulty of solving the state-estimation problem is highly dependent on how good the state-space model is a descriptor of the dynamic system under study. In general, we may identify the following hierarchy of models in increasing complexity:

(1) *Linear, Gaussian model*
   In this model, which is the simplest of state-space models, (4.12) and (4.13) respectively reduce to

$$\mathbf{x}_{n+1} = \mathbf{A}_{n+1,n}\mathbf{x}_n + \boldsymbol{\omega}_n \tag{4.15}$$

   and

$$\mathbf{y}_n = \mathbf{B}_n\mathbf{x}_n + \mathbf{v}_n. \tag{4.16}$$

   where $\mathbf{A}_{n+1,n}$ is called the *transition matrix* and $\mathbf{B}_n$ is called the *measurement matrix*. The system noise $\boldsymbol{\omega}_n$ and measurement noise $\mathbf{v}_n$ are both additive and assumed to be *statistically independent zero-mean Gaussian processes*, whose covariance matrices are respectively denoted by $\mathbf{Q}_{\omega,n}$ and $\mathbf{Q}_{v,n}$. The state-space model defined by (4.15) and (4.16) is indeed the model that was used by Kalman to derive his classic recursive filter that is mathematically elegant and devoid of any approximation (Kalman, 1960).
(2) *Linear, non-Gaussian model*
   In this second model, we still use (4.15) and (4.16), but the system noise $\boldsymbol{\omega}_n$ and measurement noise $\mathbf{v}_n$ are now assumed to be additive, statistically independent, *non-Gaussian* processes. The non-Gaussianity of these two processes is, therefore, the only source of mathematical difficulty. In situations of this kind, we may extend the application of Kalman filter theory by using the *Gaussian-sum approximation*, summarized as follows:

   Any probability density function $p(\mathbf{x})$, describing a multidimensional non-Gaussian vector represented by the sample value $\mathbf{x}$, can be approximated as closely as desired by the Gaussian-sum formula

$$p(\mathbf{x}) \approx \sum_{i=1}^{N} \alpha_i \mathcal{N}(\mathbf{x}; \bar{\mathbf{x}}_i, \boldsymbol{\Sigma}_i) \tag{4.17}$$

   for some integer $N$ and positive scalers $\alpha_i$ with $\sum_{i=1}^{N} \alpha_i = 1$.

The commonly used term $\mathcal{N}(\mathbf{x};\overline{\mathbf{x}}_i, \Sigma_i)$ stands for the Gaussian (normal) density function of $\mathbf{x}$ with mean $\overline{\mathbf{x}}_i$ and covariance matrix $\Sigma_i$ for $i = 1, 2, ..., N$. The Gaussian-sum on the right-hand side of (4.17) converges uniformly to the given probability density function $p(\mathbf{x})$ as the number of terms $N$ increases and the elemental covariance matrix $\Sigma_i$ approaches zero for all $i$ (Anderson and Moore, 1979). Note, however, that the terms in a Gaussian-sum model tend to grow exponentially over the course of time, which may, therefore, require the use of a pruning algorithm in computing (4.17).

(3) *Nonlinear Gaussian model*

The third model in the hierarchy of state-space models of increasing complexity is formulated as

$$\mathbf{x}_{n+1} = \mathbf{a}_n(\mathbf{x}_n) + \boldsymbol{\omega}_n, \tag{4.18}$$

$$\mathbf{y}_n = \mathbf{b}_n(\mathbf{x}_n) + \mathbf{v}_n, \tag{4.19}$$

where the dynamic noise $\boldsymbol{\omega}_n$ and measurement noise $\mathbf{v}_n$ are both assumed to be additive and Gaussian. Although this twofold assumption simplifies the state-estimation procedure, it is here, however, where we start to experience serious mathematical difficulty in solving the sequential state-estimation problem due to the nonlinear vectorial functions $\mathbf{a}_n(\cdot)$ and $\mathbf{b}_n(\cdot)$.

(4) *Nonlinear, non-Gaussian models*

This last class of state-space models is described by (4.12) and (4.13), where both the system model and the measurement model are nonlinear, and the system noise $\boldsymbol{\omega}_n$ and measurement noise $\mathbf{v}_n$ are not only non-Gaussian but also nonadditive. In this kind of scenario, the solution narrows down basically to particle filters as the only viable way that we presently know for solving the sequential state-estimation problem; more will be said on particle filters later in the next section under the heading "Indirect numerical approximation of the posterior."

## 4.5     The Bayesian filter

The adoption of a *Bayesian filter* to solve the state estimation of a dynamic system, be it linear or nonlinear, is motivated by the fact that it provides a *general unifying framework for sequential state estimation*, at least in a conceptual sense. In any event, from a theoretical perspective, the state estimation may be viewed as an expanded form of parameter estimation under the Bayesian paradigm; hence the name "Bayesian filter": instead of estimating an unknown parameter vector, the new issue of interest is that of estimating the hidden state of a dynamic state given a sequence of measurements.

Naturally, probability theory is central to the Bayesian approach to state estimation. To simplify the presentation, henceforth we will do two things:

- drop the subscript from the symbol denoting a probability density function and
- use the term "distribution" to refer to a probability density function.

Moreover, referring back to the system model of (4.12) and the measurement model of (4.13), henceforth we also use the following notations:

$\mathbf{Y}_n$ — *sequence of observations*, denoting $\{\mathbf{y}_i\}_{i=1}^n$;

$p(\mathbf{x}_n|\mathbf{Y}_{n-1})$ — *predictive distribution* of the state $\mathbf{x}_n$ at the current time $n$, given the entire sequence of past observations up to and including $\mathbf{y}_{n-1}$;

$p(\mathbf{x}_n|\mathbf{Y}_n)$ — *posterior distribution* of the current state $\mathbf{x}_n$, given the entire sequence of observations up to and including the current time $n$; this distribution is commonly referred to as simply the "posterior";

$p(\mathbf{x}_n|\mathbf{x}_{n-1})$ — *transition-state distribution* of the current state $\mathbf{x}_n$ given the immediate past state $\mathbf{x}_{n-1}$; this distribution is simply referred to as the "prior";

$l(\mathbf{y}_n|\mathbf{x}_n)$ — *likelihood function* or simply *likelihood* of the current observation $\mathbf{y}_n$ given the current state $\mathbf{x}_n$.

For derivation of the Bayesian filter, the only assumption that we will make is that the evolution of the state is *Markovian*, which embodies the combination of two conditions:

(1) Given the sequence of states, $\mathbf{x}_0, \mathbf{x}_1, \ldots, \mathbf{x}_{n-1}, \mathbf{x}_n$, the current state $\mathbf{x}_n$ depends only on the immediate past state $\mathbf{x}_{n-1}$ through the state transition distribution $p(\mathbf{x}_n|\mathbf{x}_{n-1})$. The *initial state* $\mathbf{x}_0$ is distributed according to

$$p(\mathbf{x}_0|\mathbf{y}_0) = p(\mathbf{x}_0).$$

(2) The observations $\mathbf{y}_1, \mathbf{y}_2, \ldots, \mathbf{y}_n$ are conditionally dependent on the corresponding states $\mathbf{x}_1, \mathbf{x}_2, \ldots, \mathbf{x}_n$; this assumption implies that the conditional joint likelihood of the observations (i.e., the joint distribution of all the observations conditioned upon all the states up to and including time $n$) factors as

$$l(\mathbf{y}_1, \mathbf{y}_2, \ldots, \mathbf{y}_n|\mathbf{x}_1, \mathbf{x}_2, \ldots \mathbf{x}_n) = \prod_{i=1}^n l(\mathbf{y}_i|\mathbf{x}_i). \tag{4.20}$$

The posterior $p(\mathbf{x}_n|\mathbf{Y}_n)$ plays a key role in Bayesian analysis, in that it embodies the entire information that we have about the state $\mathbf{x}_n$ at time $n$ *after* having received the *entire* observation sequence $\mathbf{Y}_n$; hence, the posterior contains all the information necessary for state estimation. What we are saying here is merely a reassertion of statements made in Section 4.3.

Suppose, for example, we wish to determine the *filtered estimate* of the state $\mathbf{x}_n$, optimized in the minimum mean-squared error (MMSE) sense; the desired solution is

$$\hat{\mathbf{x}}_{n|n} = \mathbb{E}[\mathbf{x}_n|\mathbf{Y}_n]$$
$$= \int \mathbf{x}_n p(\mathbf{x}_n|\mathbf{Y}_n)\, d\mathbf{x}_n. \tag{4.21}$$

Correspondingly, for a measurement of accuracy of the filtered estimate $\hat{\mathbf{x}}_{n|n}$, we compute the *covariance matrix*

$$\mathbf{P}_{n|n} = \mathbb{E}[(\mathbf{x}_n - \hat{\mathbf{x}}_{n|n})(\mathbf{x}_n - \hat{\mathbf{x}}_{n|n})^T]$$
$$= \int (\mathbf{x}_n - \hat{\mathbf{x}}_{n|n})(\mathbf{x}_n - \hat{\mathbf{x}}_{n|n})^T p(\mathbf{x}_n|\mathbf{Y}_n)\, d\mathbf{x}_n \tag{4.22}$$

With *computational efficiency* being a compelling practical factor, particularly with the computer as the machine to do the computation, there is a strong desire to compute the filtered estimate $\hat{\mathbf{x}}_{n|n}$ and related parameters in a *recursive* (iterative) manner. To explain, suppose that we have the posterior distribution of the state $\mathbf{x}_{n-1}$ at time $n-1$, denoted by $p(\mathbf{x}_{n-1}|\mathbf{Y}_{n-1})$. Then, computation of the updated value of the posterior of the state at time $n$ is governed by two basic time steps:

(1) *Time update.* The first update involves *computing the predictive distribution* of $\mathbf{x}_n$ given the observations sequence $\mathbf{Y}_{n-1}$, as shown by

$$\underbrace{p(\mathbf{x}_n\mid\mathbf{Y}_{n-1})}_{\substack{\text{predictive}\\\text{distribution}}} = \int \underbrace{p(\mathbf{x}_n\mid\mathbf{x}_{n-1})}_{\text{prior}}\ \underbrace{p(\mathbf{x}_{n-1}\mid\mathbf{Y}_{n-1})}_{\text{old posterior}}\ d\mathbf{x}_{n-1} \tag{4.23}$$

where the integration is performed over the state space, embodying $\mathbf{x}_{n-1}$. Equation (4.23) is justified by the basic laws of probability theory: multiplication of the old posterior $p(\mathbf{x}_{n-1}|\mathbf{Y}_{n-1})$ by the prior $p(\mathbf{x}_n|\mathbf{x}_{n-1})$ results in a joint distribution of the old state $\mathbf{x}_{n-1}$ and the current state $\mathbf{x}_n$, conditioned on $\mathbf{Y}_{n-1}$; that is, $p(\mathbf{x}_n, \mathbf{x}_{n-1}|\mathbf{Y}_{n-1})$. Then, integrating this conditional joint distribution with respect to $\mathbf{x}_{n-1}$ yields the desired predictive distribution $p(\mathbf{x}_n|\mathbf{Y}_{n-1})$. The old posterior, $p(\mathbf{x}_{n-1}|\mathbf{Y}_{n-1})$, is obviously available from the previous filtering recursion at time $n-1$. As for the prior $p(\mathbf{x}_n|\mathbf{x}_{n-1})$, we proceed as follows:
  (i) At the initialization stage, we use two measurements to find an initial value for the prior.
  (ii) Once the initialization stage is done, we use the state equation in the state-space model to find the prior; from then on, the prior takes the form of a predictive distribution, which it is.
(2) *Measurement update.* This second update exploits information about the current state $\mathbf{x}_n$ that is contained in the new observation $\mathbf{y}_n$, so as to update the old posterior $p(\mathbf{x}_{n-1}|\mathbf{Y}_{n-1})$. In particular, applying Bayes' rule to the predictive distribution $p(\mathbf{x}_n|\mathbf{Y}_{n-1})$ yields

$$\underbrace{p(\mathbf{x}_n|\mathbf{Y}_n)}_{\substack{\text{updated}\\\text{posterior}}} = \frac{1}{Z_n}\underbrace{p(\mathbf{x}_n|\mathbf{Y}_{n-1})}_{\substack{\text{predictive}\\\text{distribution}}}\underbrace{l(\mathbf{y}_n|\mathbf{x}_n)}_{\text{likelihood}}. \tag{4.24}$$

The normalizing constant in (4.24) is defined by

$$Z_n = p(\mathbf{y}_n|\mathbf{Y}_{n-1})$$

$$= \int l(\mathbf{y}_n|\mathbf{x}_n)p(\mathbf{x}_n|\mathbf{Y}_{n-1})\,d\mathbf{x}_n. \tag{4.25}$$

It is commonly referred to as the *partition function*; its inclusion in (4.24) ensures that the total volume under the multidimensional surface of the posterior distribution $p(\mathbf{x}_n|\mathbf{Y}_n)$ is unity, as it should be. The sequence of partition functions $\{Z_i\}_{i=1}^{n}$ produces the *log-likelihood* of the corresponding sequence of observations $\{\mathbf{y}_i\}_{i=1}^{N}$, as shown by

$$\log(p(\mathbf{y}_1,\mathbf{y}_2,...,\mathbf{y}_n)) = \sum_{i=1}^{n}\log(Z_i). \tag{4.26}$$

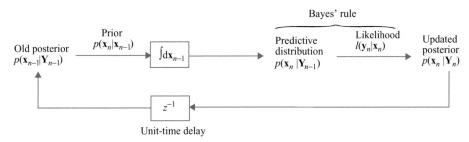

**Figure 4.6.** Block diagram describing computational recursion of the Bayesian filter, with its updated posterior $p(\mathbf{x}_n | \mathbf{Y}_n)$ as the output of interest.

Equations (4.22)–(4.25) are all consequences of the Markovian assumption described previously.

The time update and measurement update are both carried out at every time step throughout the computation of the Bayesian filter. In effect, they constitute a *computational recursion* of the filter, as depicted in Figure 4.6; the factor $Z_n$ has been left out in the figure for convenience of presentation.

### 4.5.1 Optimality of the Bayesian filter

The Bayesian filter of Figure 4.6 is *optimal*, with two important properties:

(1) The filter operates in a *recursive* manner by *propagating* the posterior $p(\mathbf{x}_n | \mathbf{Y}_n)$ from one recursion to the next.
(2) Knowledge about the state $\mathbf{x}_n$, extracted from the entire observations process $\mathbf{Y}_n$ by the filter, is completely contained in the posterior $p(\mathbf{x}_n | \mathbf{Y}_n)$, which is the "best" that can be achieved, at least in a conceptual sense.

With the posterior as the focus of analytic attention, we now lay down the groundwork for our filtering objective. To be specific, consider an arbitrary function of the state $\mathbf{x}_n$, denoted by $h(\mathbf{x}_n)$. In a filtering context, we are interested in the *on-line estimation of signal characteristics* of the function $h(\mathbf{x}_n)$. These characteristics are embodied in the *Bayes estimator*, defined by the ensemble average of the function $h(\mathbf{x}_n)$, namely

$$\bar{h}_n = \mathbb{E}_p[h(\mathbf{x}_n)]$$
$$= \int \underbrace{h(\mathbf{x}_n)}_{\text{arbitrary function}} \underbrace{p(\mathbf{x}_n | \mathbf{Y}_n)}_{\text{posterior}} \mathrm{d}\mathbf{x}_n \tag{4.27}$$

where $\mathbb{E}_p$ is the expectation operator with respect to the posterior $p(\mathbf{x}_n | \mathbf{Y}_n)$ that pertains to a dynamic system, be it linear or nonlinear. Equation (4.27) includes (4.21) for the filtered estimate of the state and (4.22) for the covariance matrix of the estimate as two special cases, illustrating the general unifying framework of the Bayesian model. For (4.21) we simply have

$$h(\mathbf{x}_n) = \mathbf{x}_n$$

and for (4.22) we have

$$\mathbf{h}(\mathbf{x}_n) = (\mathbf{x}_n - \hat{\mathbf{x}}_{n|n})(\mathbf{x}_n - \hat{\mathbf{x}}_{n|n})^{\mathrm{T}},$$

where the arbitrary function now assumes the form of a vectorial function $\mathbf{h}(.)$.

## 4.5.2    Approximation of the Bayesian filter

For the special case of a dynamic system described by the linear, Gaussian model of (4.15) and (4.16), the recursive solution of (4.27) is realized *exactly* through the celebrated Kalman filter. However, when the dynamic system is nonlinear or non-Gaussian or both, then the product distribution constituting the integrand of (4.27) is no longer Gaussian, which makes computation of the Bayes estimator $\hat{h}_n$ a difficult proposition. In situations of this latter kind, we have no option but to abandon the notion of optimality in the Bayesian sense and seek an *approximate* estimator that is computationally feasible, yet related to the Bayesian filter in some manner. Henceforth, we confine the discussion to a nonlinear model operating under the Gaussian assumption, which may be justified in practice by invoking the *central limit theorem* of probability theory.[11]

Under the Gaussian assumption, we are now ready to state our *nonlinear filtering objective* formally:

Given the entire sequence of observations $\mathbf{Y}_n$, at time $n$, pertaining to the nonlinear state-space model of (4.18) and (4.19), derive an approximate realization of the Bayes estimator $\bar{h}(\mathbf{x}_n)$, defined in (4.27), that is subject to two practical requirements:

(1) computational plausibility and
(2) recursive implementability.

To find suboptimal solutions of the nonlinear-filtering problem by approximating the Bayesian filter, we may follow one of two routes, depending on the way in which the approximation is made.

(1) *Direct numerical approximation of the posterior.* The rationale behind this direct approach to nonlinear filtering is summed up as follows:

In general, it is easier to approximate the posterior $p(\mathbf{x}_n|\mathbf{Y}_n)$ directly and in a local sense than it is to approximate the nonlinear function characterizing the system model of the filter.

To be specific, the posterior $p(\mathbf{x}_n|\mathbf{Y}_n)$ is approximated *locally* around the point $\mathbf{x}_n = \hat{\mathbf{x}}_{n|n}$, where $\hat{\mathbf{x}}_{n|n}$ is the *filtered estimate* of the state $\mathbf{x}_n$, given all the observables up to and including time $n$; the emphasis on locality makes the design of the filter computationally simple and fast to execute. The objective of the approximation is to facilitate the subsequent application of *Kalman filter theory*, for which both linearity and Gaussianity hold.

(2) *Indirect numerical approximation of the posterior.* The rationale behind this second approach to nonlinear filtering is summed up as follows:

The posterior distribution $p(\mathbf{x}_n|\mathbf{Y}_n)$ is approximated indirectly and in a global sense through the use of Monte Carlo simulations, so as to make the Bayesian framework for nonlinear filtering computationally tractable.

*Particle filters*[12] constitute a popular example of this second approach to nonlinear filtering. To be more specific, particle filters rely on a technique called the *sequential Monte Carlo (SMC) method*, which uses a set of randomly chosen samples, referred to as *particles*, with associated weights to approximate the posterior distribution $p(\mathbf{x}_n|\mathbf{Y}_n)$. The sampling process is called *importance sampling*, which is based on the premise that in certain situations it is more appropriate to sample from a *proposal* or *instrumental distribution*; then, a "change-of-measure" formula is applied to the filtering procedure to account for the fact that the proposal distribution is naturally different from the target distribution (Cappé *et al.*, 2005). Now, as the number of particles used in the simulation becomes larger, the Monte Carlo computation of the posterior distribution becomes more accurate, which, of course, is the desired objective. However, the increased number of particles makes the use of the SMC method computationally more expensive. In other words, computational cost is traded for improved nonlinear filtering accuracy.

From this brief discussion, it is apparent that the locally direct approach to approximate Bayesian filtering builds on the well-established Kalman filter theory. On the other hand, the globally indirect approach charts a path of its own by departing from Kalman filter theory. Generally speaking, the globally indirect approach to nonlinear filtering is more demanding in computational terms than the locally direct approach, particularly when the nonlinear filtering problem is difficult.

## 4.6   Extended Kalman filter

The EKF is the simplest locally direct approximation to the Bayesian filter, under the Gaussian assumption. The basic idea behind derivation of the EKF is summed up as follows (Maybeck, 1982):

Linearize the nonlinear equation (i.e., system or measurement equation) about the most recent estimate of the state; once the new state estimate is computed, a new and better reference state trajectory is incorporated into the next step of the state-estimation process.

To elaborate, the filtered estimate of the state, $\hat{\mathbf{x}}_{n|n}$, is used in linearization of the system equation; and the predicted estimate of the state, $\hat{\mathbf{x}}_{n|n-1}$, is used in linearization of the measurement equation. In both cases, linearization is accomplished by using the *Taylor series* and retaining only first-order terms in the resulting expansion of the system or measurement equation.

Thus, referring to the state equation (4.18), we write

$$\mathbf{A}_{n+1|n} = \left.\frac{\partial \mathbf{a}_n(\mathbf{x})}{\partial \mathbf{x}}\right|_{\mathbf{x}=\hat{\mathbf{x}}_{n|n}}. \tag{4.28}$$

Similarly, referring to the measurement equation (4.19), we write

$$\mathbf{B}_n = \left.\frac{\partial \mathbf{b}_n(\mathbf{x})}{\partial \mathbf{x}}\right|_{\mathbf{x}=\hat{\mathbf{x}}_{n|n-1}}. \tag{4.29}$$

Once the transition matrix $\mathbf{A}_{n+1|n}$ and measurement matrix $\mathbf{B}_n$ have been constructed, they are respectively used in first-order Taylor series approximations of the nonlinear functions $\mathbf{a}_n(\mathbf{x}_n)$, first, and $\mathbf{b}_n(\mathbf{x}_n)$ around the estimates $\hat{\mathbf{x}}_{n|n}$, obtaining

$$\mathbf{a}_n(\mathbf{x}_n) \approx \mathbf{a}_n(\hat{\mathbf{x}}_{n-1}) + \mathbf{A}_{n+1|n}(\mathbf{x}_n - \hat{\mathbf{x}}_{n|n}) \tag{4.30}$$

and then

$$\mathbf{b}_n(\mathbf{x}_n) \approx \mathbf{b}_n(\hat{\mathbf{x}}_{n-1}) + \mathbf{B}_n(\mathbf{x}_n - \hat{\mathbf{x}}_{n|n-1}). \tag{4.31}$$

With this pair of linearized approximations of the nonlinear functions $\mathbf{a}_n(\mathbf{x}_n)$ and $\mathbf{b}_n(\mathbf{x}_n)$ at hand and assuming that the system noise $\boldsymbol{\omega}_n$ and measurement noise $\mathbf{v}_n$ are both Gaussian, the stage is set for the application of Kalman filter theory, yielding the EKF summary presented in Table 4.1 (Maybeck, 1982; Haykin, 2009).

**Table 4.1.** Summary of the EKF.

*Input process*:
   Observations $= \{\mathbf{y}_1, \mathbf{y}_2, \dots, \mathbf{y}_n\}$

*Known parameters*:
   Nonlinear state-transition vectorial function $= \mathbf{a}_n(\mathbf{x}_n)$
   Nonlinear measurement vectorial function $= \mathbf{b}_n(\mathbf{x}_n)$
   Correlation matrix of process noise vector $= \mathbf{Q}_{\omega,n}$
   Correlation matrix of measurement noise vector $= \mathbf{Q}_{v,n}$

*Computation*: $n = 1, 2, 3, \dots$

$$\mathbf{G}_n = \mathbf{P}_{n,n-1}\mathbf{B}_n^T[\mathbf{B}_n\mathbf{P}_{n,n-1}\mathbf{B}_n^T + \mathbf{Q}_{v,n}]^{-1}$$
$$\alpha_n = \mathbf{y}_n - \mathbf{b}_n(\hat{\mathbf{x}}_{n|n-1})$$
$$\hat{\mathbf{x}}_{n|n} = \hat{\mathbf{x}}_{n|n-1} + \mathbf{G}_n\alpha_n$$
$$\hat{\mathbf{x}}_{n+1|n} = \mathbf{a}_n(\hat{\mathbf{x}}_{n|n})$$
$$\mathbf{P}_{n|n} = \mathbf{P}_{n|n-1} - \mathbf{G}_n\mathbf{B}_n\mathbf{P}_{n|n-1}$$
$$\mathbf{P}_{n+1|n} = \mathbf{A}_{n+1|n}\mathbf{P}_{n|n}\mathbf{A}_{n+1|n}^T\mathbf{Q}_{\omega,n}$$

*Notes*:
   1. The linearized matrices $\mathbf{A}_{n+1,n}$ and $\mathbf{B}_n$ are computed from their nonlinear counterparts $\mathbf{a}_n(\mathbf{x}_n)$ and $\mathbf{b}_n(\mathbf{x}_n)$ using (4.28) and (4.29) respectively.
   2. The values $\mathbf{a}_n(\hat{\mathbf{x}}_{n|n})$ and $\mathbf{b}_n(\hat{\mathbf{x}}_{n|n-1})$ are obtained by substituting the filtered state estimate $\hat{\mathbf{x}}_{n|n}$ and the predicted state estimate $\hat{\mathbf{x}}_{n|n-1}$ for the state $\mathbf{x}_n$ in the nonlinear vectorial functions $\mathbf{a}_n(\mathbf{x}_n)$ and $\mathbf{b}_n(\mathbf{x}_n)$ respectively, as in (4.30) and (4.31).
   3. Examining the order of iterations in this table, we now see the reason for evaluating $\mathbf{A}_{n+1,n}$ and $\mathbf{B}_n$ in the manner described in (4.28) and (4.29).

*Initial conditions*:
$$\hat{\mathbf{x}}_{1|0} = \mathbb{E}[\mathbf{x}_1]$$
$$\mathbf{P}_{1,0} = \mathbb{E}[(\mathbf{x}_1 - \mathbb{E}[\mathbf{x}_1])(\mathbf{x}_1 - \mathbb{E}[\mathbf{x}_1])^T] = \Pi_0,$$

where $\Pi_0 = \delta^{-1}\mathbf{I}$, $\delta$ is a small positive constant, and $\mathbf{I}$ is the identity matrix.

## 4.6.1    Summarizing remarks on the extended Kalman filter

The EKF is attractive for nonlinear state estimation for two compelling reasons:

(1) It builds on the framework of Kalman filter theory in a principled way.
(2) It is relatively simple to understand and, therefore, straightforward to put into practical use, for which it has established a long track record over the past four decades in diverse applications.

However, it has two fundamental drawbacks that tend to limit its practical usefulness:

(1) For the EKF to function satisfactorily, the nonlinearity of the state-space model, embodying (4.18) and (4.19), has to be of a *mild* sort, so as to justify the use of the first-order Taylor series expansion, upon which its derivation is built.
(2) Its derivation requires knowledge of first-order partial derivatives (i.e. Jacobians) of the state-space model of the nonlinear dynamic system under study; however, for many practical applications, the computation of Jacobians is undesirable or simply not feasible because the state-space model may contain nondifferentiable terms.

There is one other matter of theoretical interest that is noteworthy: the traditional approach adopted in the literature for deriving the EKF, exemplified by that described in this section, does not teach us the relationship between the EKF and Bayesian filter. It is in Elliott and Haykin (2010) where we see that *the EKF is indeed an approximate form of the Bayesian filter under the Gaussian assumption.* The theory discussed therein for establishing this relationship is based on the well-known *Zakai equation* in nonlinear filter theory.

## 4.7    Cubature Kalman filters

For a more powerful approximation of the Bayesian filter, we look to a new nonlinear filter, named the CKF (Arasaratnam and Haykin, 2009). Derivation of this new filter exploits the fact that, under the Gaussian assumption, approximation of the Bayesian filter reduces to computing multidimensional integrals of a special form whose integrand is described, in words, as follows:

$$(\text{nonlinear function}) \times (\text{Gaussian function}) \tag{4.32}$$

Specifically, given an arbitrary nonlinear function $\mathbf{f}(\mathbf{x})$ of the vector $\mathbf{x} \in R_M$ and using a Gaussian function, we consider an integral of the form

$$h(\mathbf{f}) = \int_{R_M} \underbrace{\mathbf{f}(\mathbf{x})}_{\substack{\text{Arbitrary} \\ \text{function}}} \underbrace{\exp(-\mathbf{x}^T\mathbf{x})}_{\substack{\text{Gaussian} \\ \text{function}}} \, d\mathbf{x}, \tag{4.33}$$

which is, of course, defined in the Cartesian coordinate system.

For numerical approximation of the nonlinear function $h(\mathbf{f})$ in (4.33), we will use a *cubature rule* based on *monomials* (Stroud, 1971; Cools, 1997). In mathematics, the term "monomials" refers to a product of powers of variables.[13] A monomial-based cubature rule satisfies our numerical approximation needs for the following reasons:

- the provision of a reasonably accurate approximation;
- the requirement of a small number of function evaluations in the approximation; and
- the relative ease of extending the approximation to high dimensions.

Moreover, we will focus the approximation on a *third-degree spherical radial cubature rule*. The motivation behind considering spherical–radial integration is the fact that it befits the integral described in (4.33), as shown in what follows. As for the choice of a third-degree cubature rule, the reader is referred to Note 14 at the end of the chapter.

To proceed then, we will construct the cubature rule in such a way that the numerical approximation of (4.33) is built on a *set of weighted cubature points that is an even symmetric set*. In so doing, the complexity in solving a set of nonlinear equations for a set of desired weights and cubature points is reduced markedly. Before going into details about the cubature rule, we introduce a number of notations and definitions.

- Using $\mathcal{D}$ to denote the region of integration in (4.33), we say that a weighting function $w(\mathbf{x})$ defined on the region $\mathcal{D}$ is *fully symmetric* if the following two conditions hold:
  - (1) $\mathbf{x} \in \mathcal{D}$ implies $\mathbf{y} \in \mathcal{D}$, where $\mathbf{y}$ is any point obtainable from $\mathbf{x}$ by permutations and by changes of sign of the coordinates of $\mathbf{x}$.
  - (2) $w(\mathbf{x}) = w(\mathbf{y})$ on the region $\mathcal{D}$.
- In a fully symmetric region, we call a point $\mathbf{u}$ as a *generator* if $\mathbf{u} = (u_1, u_2, \ldots, u_r, 0, \ldots 0) \in R_M$, where $u_i \geq u_{i+i} > 0$ for $i = 1, 2, \ldots, (r-1)$.
- We use the notation $[u_1, u_2, \ldots, u_r]$ to represent the complete set of points that can be obtained by permutating and changing the signs of the generator $\mathbf{u}$ in all possible ways. For the sake of brevity, we suppress the $(n-r)$ zero nodes in the notation. For example, we write $[1] = R_2$ to represent the following permuted set of points:

$$\left\{ \begin{pmatrix} 1 \\ 0 \end{pmatrix}, \begin{pmatrix} 0 \\ 1 \end{pmatrix}, \begin{pmatrix} -1 \\ 0 \end{pmatrix}, \begin{pmatrix} 0 \\ -1 \end{pmatrix} \right\}.$$

- We use the notation $[u_1, u_2, \ldots, u_r]_i$ to denote the $i$th point from the generator $\mathbf{u}$.

### 4.7.1     Converting to spherical–radial integration

The key step in this conversion is a change of variables from the Cartesian vector $\mathbf{x} \in R_M$ to a spherical–radial one defined by radius r and direction vector $\mathbf{z}$, as outlined here:

Let $\mathbf{x} = r\mathbf{z}$ with $\mathbf{z}^\mathsf{T}\mathbf{z} = 1$, so that $\mathbf{x}^\mathsf{T}\mathbf{x} = r^2$ for $r \in [0, \infty)$.

Then, the integral of (4.33) can be rewritten in a spherical–radial coordinate system as shown by the double integral

$$h(\mathbf{f}) = \int_0^\infty \int_{\mathcal{U}_M} \mathbf{f}(r\mathbf{z}) r^{M-1} \exp(-r^2) \, d\sigma(\mathbf{z}) \, dr, \tag{4.34}$$

where $\mathcal{U}_M$ is the region defined by $\mathcal{U}_M = \{z; z^T z = 1\}$ and $\sigma(.)$ is a *spherical surface measure* on $\mathcal{U}_M$ in an integral defined by

$$S(r) = \int_{\mathcal{U}_M} \mathbf{f}(r\mathbf{z}) \, d\sigma(\mathbf{z}). \tag{4.35}$$

The integral of (4.35) is computed numerically by the spherical rule. Then, having computed $S(r)$ and using (4.34), the radial integral

$$h = \int_0^\infty S(r) r^{M-1} \exp(-r^2) \, dr \tag{4.36}$$

is computed numerically by using the Gaussian quadrature; and with it, computation of (4.33) will have been accomplished. Both of these two rules are described in the following, in that order.

## 4.7.2 Spherical rule

We first derive a *third-degree spherical rule* that takes the form

$$\int_{\mathcal{U}_M} \mathbf{f}(\mathbf{z}) \, d\sigma(\mathbf{z}) \approx w \sum_{i=1}^{2M} \mathbf{f}[u]_i. \tag{4.37}$$

The rule in (4.37) entails from the generator $[u]$ a total of $2M$ *cubature points* symmetrically positioned on a unit hypersphere (ellipsoid) centered on the origin; the cubature points are located at the intersection of an $M$-dimensional hypersphere and its axes. To find the unknown parameters $u$ and $w$, it suffices to consider *monomials* $\mathbf{f}(\mathbf{z}) = 1$ and $\mathbf{f}(\mathbf{z}) = z_1^2$ due to the fully symmetric generators, as shown here:

$$\mathbf{f}(\mathbf{z}) = 1: \qquad 2Mw = \int_{\mathcal{U}_M} d\sigma(\mathbf{z}) = A_M, \tag{4.38}$$

$$\mathbf{f}(\mathbf{z}) = z_1^2: \qquad 2wu^2 = \int_{\mathcal{U}_M} z_1^2 \, d\sigma(\mathbf{z}) = \frac{A_M}{M}, \tag{4.39}$$

where $M$ is the dimension of vector $\mathbf{x}$ and the surface area of the unit hypersphere is defined by

$$A_M = \frac{2\sqrt{\pi^M}}{\Gamma(M/2)},$$

where

$$\Gamma(M) = \int_0^\infty x^{M-1} \exp(-x) \, dx$$

is the *Gamma function*. Given the $A_M$ just defined, solving (4.38) and (4.39) for $w$ and $u$ yields

$$w = \frac{A_M}{2M},$$

$$u^2 = 1.$$

### 4.7.3    Radial rule

For the radial rule, we propose to use a Gaussian quadrature, which is known to be the most efficient numerical method to compute an integral in a single dimension. An $m$-point *Gaussian quadrature* is exact up to polynomials of degree $(2M - 1)$ and constructed as

$$\int_D f(x)w(x)\,dx \approx \sum_{i=1}^{m} w_i f(x_i), \tag{4.40}$$

where $m = 2M$, $w(x)$ denotes a weighting function, and $w_i = w(x_i)$ (Press and Teukolsky, 1990). The $x_i$ and the $w_i$ are respectively the *quadrature points* and associated *weights* to be determined. Comparison of the integrals in (4.36) and (4.40) yields the weighting function and the region of integration to be $w(x) = x^{M-1}\exp(-x^2)$ and $[0, \infty)$, respectively. Thus, using a final change of variables, $t = x^2$, we obtain the desired radial integral

$$\int_0^\infty f(x)x^{M-1}\exp(-x^2)\,dx = \frac{1}{2}\int_0^\infty \tilde f(t)t^{(M/2)-1}\exp(-t)\,dt, \tag{4.41}$$

where $\tilde f(t) = f(\sqrt{t})$. The integral on the right-hand side of (4.41) is now in the form of the well-known *generalized Gauss–Laguerre formula* (Stroud, 1966; Press and Teukolsky, 1990).

A first-degree Gauss–Laguerre rule is exact for $f(x) = 1, x$. Correspondingly, the rule is exact for $f(x) = 1, x^2$; it is not exact for odd-degree polynomials, such as that in $f(x) = x, x^3$. Fortunately, however, when the radial rule is combined with the spherical rule to compute the integral (4.33), the resulting spherical–radial rule vanishes for all odd-degree polynomials; the reason for this nice result is that the spherical rule vanishes by virtue of the symmetry for any odd-degree polynomial; see (4.34). Hence, the spherical–radial rule for computing (4.33) is exact for all odd-degree polynomials. Following this argument, for a spherical–radial rule to be exact for all third-degree polynomials in $\mathbf{x} \in R_M$, it suffices to consider the first-degree generalized Gauss–Laguerre rule, which entails the use of a single point and a single weight. We may thus write

$$\int_0^\infty f_i(x)x^{M-1}\exp(-x^2)\,dx \approx w_1 f_i(x_1), \qquad i = 0,1,$$

where

$$w_1 = \frac{1}{2}\Gamma\left(\frac{M}{2}\right),$$

and

$$x_1 = \sqrt{M/2}.$$

### 4.7.4    Spherical–radial rule

In this final subsection, we describe two useful results that are used: the first is used to combine the spherical and radial rules and the second is used to extend the spherical–radial rule for a Gaussian weighted integral. The respective results are presented as two propositions (Arasaratnam and Haykin, 2009):

**Proposition 4.1.** *Let the radial integral be computed numerically by an $m_r$-point Gaussian quadrature rule:*

$$\int_0^\infty f(r) r^{M-1} \exp(-r^2)\, dr = \sum_{i=1}^{m_r} a_i f(r_i).$$

*Let the spherical integral be computed numerically by an $m_s$-point spherical rule:*

$$\int_{\mathcal{U}_M} \mathbf{f}(r\mathbf{s})\, d\sigma(\mathbf{s}) = \sum_{j=1}^{m_s} b_j \mathbf{f}(r\mathbf{s}_j).$$

*Then, an $(m_s \times m_r)$-point spherical–radial cubature rule is approximately given by the double summation*

$$\int_{R_M} \mathbf{f}(\mathbf{x}) \exp(-\mathbf{x}^T\mathbf{x})\, d\mathbf{x} \approx \sum_{j=1}^{m_s} \sum_{i=1}^{m_r} a_i b_j \mathbf{f}(r_i\mathbf{s}_j).$$

**Proposition 4.2.** *Let two weighting functions be denoted by $w_1(\mathbf{x}) = \exp(-\mathbf{x}^T\mathbf{x})$ and $w_2(\mathbf{x}) = \mathcal{N}(\mathbf{x}; \boldsymbol{\mu}, \boldsymbol{\Sigma})$, where, for a given vector $\mathbf{x}$, the $\mathcal{N}(\mathbf{x}; \boldsymbol{\mu}, \boldsymbol{\Sigma})$ denotes the Gaussian distribution of $\mathbf{x}$ with mean $\boldsymbol{\mu}$ and covariance matrix $\boldsymbol{\Sigma}$. Then, for every square-root matrix $\boldsymbol{\Sigma}^{1/2}$ such that $\boldsymbol{\Sigma}^{1/2}\boldsymbol{\Sigma}^{T/2} = \boldsymbol{\Sigma}$, we have*

$$\int_{R_M} \mathbf{f}(\mathbf{x}) w_2(\mathbf{x})\, d\mathbf{x} = \frac{1}{\sqrt{\pi^M}} \int_{R_M} \mathbf{f}(\sqrt{2\boldsymbol{\Sigma}}\mathbf{x} + \boldsymbol{\mu}) w_1(\mathbf{x})\, d\mathbf{x},$$

where the superscript T/2 signifies transpose of the square root of $\boldsymbol{\Sigma}$.

For the third-degree spherical–radial rule, $m_r = 1$ and $m_s = 2M$. Accordingly, we only require an *even* set of $2M$ cubature points. Moreover, the rule is exact for integrands that can be written as a linear combination of polynomials of degree up to three and all odd-degree polynomials. Invoking Propositions 4.1 and 4.2, we may now extend this third-degree spherical–radial rule to numerically compute the *standard Gaussian-weighted integral*

$$h_N(\mathbf{f}) = \int_{R_M} \mathbf{f}(\mathbf{x}) \mathcal{N}(\mathbf{x}; \mathbf{0}, \mathbf{I})\, d\mathbf{x}$$

$$\approx \sum_{i=1}^{m} w_i \mathbf{f}(\boldsymbol{\xi}_i),$$

(4.42)

where $m = 2M$ with $M$ being dimensionality of the state-space,

$$\boldsymbol{\xi}_i = \sqrt{\frac{m}{2}}[1]_i,$$

and

$$w_i = \frac{1}{m}, \qquad i = 1, 2, ..., m = 2M.$$

In effect, the $\boldsymbol{\xi}_i$ are the *cubature-point representations* of the M-dimensional vector $\mathbf{x}$. That is, the cubature-point set $\{\boldsymbol{\xi}_i, w_i\}_{i=1}^{2M}$ is the desired set for approximating the standard Gaussian-weighted integral of (4.42).

## 4.7.5    Derivation of the CKF

The formula of (4.42) is the *cubature rule* we have been seeking for *numerical approximation* of the moment integral of (4.33). Indeed, the cubature rule is central to the computation of all the moment integrals embodied in the Bayesian framework for nonlinear filtering. As with the EKF, we assume that the dynamic noise $\boldsymbol{\omega}_n$ and measurement noise $\mathbf{v}_n$ are jointly Gaussian. Under this Gaussian assumption, we may now proceed to approximate the Bayesian filter using the cubature rule applied to the traditional pair of updates:

(1) *Time update.* Suppose that the prior $p(\mathbf{x}_{n-1}|\mathbf{Y}_{n-1})$ is approximated by a Gaussian distribution whose mean is the filtered state-estimate $\hat{\mathbf{x}}_{n-1|n-1}$ and its covariance matrix is equal to the filtered-error covariance matrix $\mathbf{P}_{n-1|n-1}$. Then, using the formula for the Bayes estimator, we may express the predicted estimate of the state $\mathbf{x} \in R_M$ as shown by

$$
\hat{\mathbf{x}}_{n|n-1} = \mathbb{E}[\mathbf{x}_n \mid \mathbf{Y}_{n-1}]
$$
$$
= \int_{R_M} \underbrace{\mathbf{a}(\mathbf{x}_{n-1})}_{\substack{\text{Nonlinear}\\\text{state-}\\\text{transition}\\\text{function}}} \underbrace{\mathcal{N}(\mathbf{x}_{n-1};\hat{\mathbf{x}}_{n-1|n-1},\mathbf{P}_{n-1|n-1})}_{\text{Gaussian distribution}} \, d\mathbf{x}_{n-1},
\tag{4.43}
$$

where we have used knowledge of the system model of (4.18) and the fact that the system noise $\boldsymbol{\omega}_{n-1}$ is uncorrelated with the sequence of observations $\mathbf{Y}_{n-1}$; in (4.43), we have assumed that the vectorial function $\mathbf{a}(.)$ is time invariant. Similarly, we obtain the prediction-error covariance matrix

$$
\mathbf{P}_{n|n-1} = \int_{R_M} \mathbf{a}(\mathbf{x}_{n-1})\mathbf{a}^{\mathrm{T}}(\mathbf{x}_{n-1})\mathcal{N}(\mathbf{x}_{n-1};\hat{\mathbf{x}}_{n-1|n-1},\mathbf{P}_{n-1|n-1})
$$
$$
\times d\mathbf{x}_{n-1} - \hat{\mathbf{x}}_{n-n-1}\hat{\mathbf{x}}_{n|n-1}^{\mathrm{T}} + \mathbf{Q}_{\boldsymbol{\omega}_{n-1}}.
\tag{4.44}
$$

The pair of (4.43) and (4.44) address time-update aspects of approximating the Bayesian filter.

(2) *Measurement update.* Next, to find a formula for the measurement update, suppose that the joint distribution of the state $\mathbf{x}_n$ and the observation $\mathbf{y}_n$, conditioned on the sequence $\mathbf{Y}_{n-1}$, is also Gaussian as shown by

$$
\mathcal{N}\left(\underbrace{\begin{bmatrix} \mathbf{x}_n \\ \mathbf{y}_n \end{bmatrix}}_{\substack{\text{Joint}\\\text{variables}}};\underbrace{\begin{bmatrix} \hat{\mathbf{x}}_{n|n-1} \\ \hat{\mathbf{y}}_{n|n-1} \end{bmatrix}}_{\substack{\text{Joint}\\\text{mean}}},\underbrace{\begin{bmatrix} \mathbf{P}_{n|n-1} & \mathbf{P}_{xy,n|n-1} \\ \mathbf{P}_{yx,n|n-1} & \mathbf{P}_{yy,n|n-1} \end{bmatrix}}_{\substack{\text{Joint}\\\text{covariance matrix}}}\right),
\tag{4.45}
$$

where $\hat{\mathbf{x}}_{n|n-1}$ is the predicted state-estimate defined in (4.43) and $\hat{\mathbf{y}}_{n|n-1}$ is the *predicted estimate of the observation* $\mathbf{y}_n$ given the sequence $\mathbf{Y}_{n-1}$, as shown by

$$
\hat{\mathbf{y}}_{n|n-1} = \int_{R_M} \underbrace{\mathbf{b}(\mathbf{x}_n)}_{\substack{\text{Nonlinear}\\\text{measurement}\\\text{function}}} \underbrace{\mathcal{N}(\mathbf{x}_n;\hat{\mathbf{x}}_{n|n-1},\mathbf{P}_{n|n-1})}_{\text{Gaussian function}} \, d\mathbf{x}_n.
\tag{4.46}
$$

The prediction-error covariance matrix of $\hat{\mathbf{y}}_{n|n-1}$ is defined by

$$P_{yy,n|n-1} = \int_{R_M} \underbrace{\mathbf{b}(\mathbf{x}_n)\mathbf{b}^T(\mathbf{x}_n)}_{\substack{\text{Outer product} \\ \text{of} \\ \text{nonlinear} \\ \text{measurement} \\ \text{function} \\ \text{with itself}}} \underbrace{\mathcal{N}(\mathbf{x}_n; \hat{\mathbf{x}}_{n|n-1}, \mathbf{P}_{n|n-1})}_{\text{Gaussian function}} \, d\mathbf{x}_n - \underbrace{\hat{\mathbf{y}}_{n|n-1}\hat{\mathbf{y}}^T_{n|n-1}}_{\substack{\text{Outer product} \\ \text{of the} \\ \text{estimate } \hat{\mathbf{y}}_{n|n-1} \\ \text{with itself}}} + \underbrace{\mathbf{Q}_{v,n}}_{\substack{\text{Covariance} \\ \text{matrix of} \\ \text{measurement} \\ \text{noise}}} \quad (4.47)$$

Lastly, the cross-covariance matrix of the state $\mathbf{x}_n$ and the observation $\mathbf{y}_n$ is given by

$$\mathbf{P}_{xy,n|n-1} = \mathbf{P}_{yx,n|n-1}$$

$$= \int_{R_M} \underbrace{\mathbf{x}_n \mathbf{b}^T(\mathbf{x}_n)}_{\substack{\text{Outer} \\ \text{product} \\ \text{of } \mathbf{x}_n \text{ with} \\ \mathbf{b}(\mathbf{x}_n)}} \underbrace{\mathcal{N}(\mathbf{x}_n; \hat{\mathbf{x}}_{n|n-1}, \mathbf{P}_{n|n-1})}_{\text{Gaussian function}} \, d\mathbf{x}_n - \underbrace{\hat{\mathbf{x}}_{n|n-1}\hat{\mathbf{y}}^T_{n|n-1}}_{\substack{\text{Outer product} \\ \text{of the} \\ \text{estimates} \\ \hat{\mathbf{x}}_{n|n-1} \text{ and } \hat{\mathbf{y}}_{n|n-1}}}. \quad (4.48)$$

The four integral formulas, described in (4.43), (4.44), (4.47), and (4.48) address other aspects of approximating the Bayesian filter. However, as different as these four formulas are, their integrands have a common form similar to that described in (4.32): the product of a nonlinear function and a corresponding Gaussian function of known mean and covariance matrix. To be more precise, all these four integrals lend themselves to approximation by means of the cubature rule, using the cubature-point set $\{\xi_i, w_i\}_{i=1}^{2M_x}$ that follows (4.42) for the normalized case of standard Gaussian-weighted integral.

Most importantly, once the cubature rule is done with, recursive computation of the filtered estimate of the state builds on *linear Kalman filter theory* by proceeding along the following three steps:

(1) *Kalman gain.* The computation of this is accomplished by using the formula

$$\mathbf{G}_n = \mathbf{P}_{xy,n|n-1}\mathbf{P}^{-1}_{yy,n|n-1} \quad (4.49)$$

where $\mathbf{P}^{-1}_{yy,n|n-1}$ is the inverse of the covariance matrix $\mathbf{P}_{yy,n|n-1}$ defined in (4.47), and $\mathbf{P}_{xy,n|n-1}$ is the cross-covariance matrix defined in (4.48).

(2) *Updating of the filtered state estimation.* Upon receiving the new observation $\mathbf{y}_n$, the filtered estimate of the state $\mathbf{x}_n$ is computed in accordance with the *predictor–corrector formula*:

$$\underbrace{\hat{\mathbf{x}}_{n|n}}_{\substack{\text{Updated} \\ \text{filtered-} \\ \text{state} \\ \text{estimate}}} = \underbrace{\hat{\mathbf{x}}_{n|n-1}}_{\substack{\text{Predicted} \\ \text{state} \\ \text{estimate}}} + \underbrace{\mathbf{G}_n}_{\substack{\text{Kalman} \\ \text{gain}}} \underbrace{(\mathbf{y}_n - \hat{\mathbf{y}}_{n|n-1})}_{\substack{\text{Innovations} \\ \text{process}}}, \quad (4.50)$$

where the predicted estimates $\hat{\mathbf{x}}_{n|n-1}$ and $\hat{\mathbf{y}}_{n|n-1}$ are respectively defined in (4.43) and (4.46), and $\mathbf{G}_n$ is defined in (4.49).

(3) *Filtered-error covariance matrix.* The filtered-state estimate of (4.50) is associated with the filtering-error covariance matrix, which is computed in accordance with the formula:

$$\mathbf{P}_{n|n} = \mathbf{P}_{n|n-1} - \mathbf{G}_n \mathbf{P}_{yy,n|n-1} \mathbf{G}_n^{\mathrm{T}}, \tag{4.51}$$

where the matrices $\mathbf{P}_{n|n-1}$, $\mathbf{P}_{yy,n|n-1}$, and $\mathbf{G}_n$ are respectively defined in (4.44), (4.47), and (4.49).

Finally, the posterior distribution of the state $\mathbf{x}_n$ is computed as the Gaussian distribution

$$p(\mathbf{x}_n|\mathbf{Y}_n) = \mathcal{N}(\mathbf{x}_n; \hat{\mathbf{x}}_{n|n}, \mathbf{P}_{n|n}), \tag{4.52}$$

where the mean $\hat{\mathbf{x}}_{n|n}$ is defined by (4.50) and the covariance matrix $\mathbf{P}_{n|n}$ is defined by (4.51).

Thus, having started the computation with the old posterior $p(\mathbf{x}_{n-1}|\mathbf{Y}_{n-1})$ under the time update, the computational recursion of the CKF has moved forward systematically through the measurement update by propagating the old posterior $p(\mathbf{x}_{n-1}|\mathbf{Y}_{n-1})$ as in the Bayesian filter, culminating in the computation of the updated posterior distribution $p(\mathbf{x}_n|\mathbf{Y}_n)$; and the recursion is then repeated as required. The $n$th recursion is defined by (4.43) progressing to (4.52), in that order.

## 4.7.6    Properties of the CKF

This new nonlinear filter, rooted in Bayesian estimation, has some important properties, summarized in what follows:

*Property 1*  Unlike the EKF, the CKF is a *derivative-free on-line sequential-state estimator*, relying on integration from one iteration to the next for its operation; hence, the CKF has a built-in *noise-smoothing* capability.

*Property 2*  Approximations of the moment integrals in the Bayesian filter are all *linear* in the number of function evaluations.

*Property 3*  As with the EKF, computational complexity of the CKF grows as $M^3$, where $M$ is dimensionality of the state.

*Property 4*  Regularization is naturally built into the CKF by virtue of the fact that the *prior* in Bayesian filtering is known to play the role of a *stabilizer*.

*Property 5*  The CKF *inherits well-known properties of the linear Kalman filter*, including square-root filtering that can be employed for improved accuracy and reliability.

*Property 6*  The CKF provides a method of choice for approximating the Bayesian filter under the Gaussian assumption in a nonlinear setting.

*Property 7*  By virtue of its uniquely characteristic even set of weighted cubature points, the CKF has the built-in signal-processing power to ease the curse-of-dimensionality problem in an additive Gaussian noise environment, but in its present form the curse cannot be mitigated.

The curse-of-dimensionality problem refers to the exponential growth of computational cost with increasing dimensionality of the state-space, measurement space, or both. More will be said on Property 7 in the next section, where the CKF is contrasted with another nonlinear filter called the unscented Kalman filter (UKF).

### 4.7.7    Summarizing remarks on the CKF

The CKF is a new nonlinear filter that approximates the optimal Bayesian filter by exploiting the third-degree cubature rule,[14] which is well suited for the numerical approximation of integrals whose integrand is of the form described in (4.33). With this approximation in place, the stage is set for building on Kalman filter theory, hence the name "the cubature Kalman filter." Perhaps the simplest way to describe the CKF is to say that it is the best-known approximation to the optimal Bayesian filter under the Gaussian assumption in a nonlinear setting. Simply put, the CKF is an important addition to the "kit of tools for nonlinear filtering."

## 4.8    On the relationship between the cubature and unscented Kalman filters

### 4.8.1    Unscented Kalman filter

Julier and coworkers (Julier *et al.*, 2000; Julier and Uhlmann, 2004) described a new nonlinear filter, named the UKF. The UKF is another direct localized method of approximating the Bayesian filter, but, unlike the CKF, it uses a completely different set of deterministic weighted points for the approximation.

To elaborate on the UKF, consider an $M$-dimensional random vector $\mathbf{x}$ that has a symmetric prior distribution $\pi(\mathbf{x})$ with mean $\boldsymbol{\mu}$ and covariance matrix $\boldsymbol{\Sigma}$, for which the Gaussian distribution is a special case. Given the dimensionality of the state-space $M$, an *odd* set of sample points, numbering $(2M + 1)$ and a corresponding set of weights, identified by $\{\chi_i, w_i\}_{i=0}^{2M}$, are chosen to satisfy the following pair of *moment-matching conditions*:

$$\sum_{i=0}^{2M} w_i \chi_i = \boldsymbol{\mu},$$    (4.53)

$$\sum_{i=0}^{2M} w_i (\chi_i - \boldsymbol{\mu})(\chi_i - \boldsymbol{\mu})^{\mathrm{T}} = \boldsymbol{\Sigma}.$$    (4.54)

Among the many candidate sets of weighted points that can satisfy (4.53) and (4.54), the following symmetrically distributed point-set is chosen for the UKF:

$$\chi_0 = \boldsymbol{\mu}, \qquad w_0 = \frac{\kappa}{M + \kappa},$$    (4.55)

$$\chi_i = \boldsymbol{\mu} + [(M + \kappa)\boldsymbol{\Sigma}]_i^{1/2}, \qquad w_i = \frac{1}{2(M + \kappa)},$$    (4.56)

$$\chi_{M+i} = \boldsymbol{\mu} - [(M + \kappa)\boldsymbol{\Sigma}]_i^{1/2}, \qquad w_i = \frac{1}{2(M + \kappa)},$$    (4.57)

where $i = 1, 2, \ldots, M$ and the $i$th column of covariance matrix $\boldsymbol{\Sigma}$ is denoted by $\boldsymbol{\Sigma}_i$. The symmetric set of $2M$ weighted points, centered around the zero weighted point $\{\chi_0, w_0\}$, is called the *sigma-point set*. The new parameter $\kappa$ is picked so as to *scale* the distribution of the $2M$ sigma points from the prior mean $\chi_0 = \mu$; hence the reference to $\kappa$ as the *scaling parameter*. Furthermore, to capture the kurtosis of the prior distribution $\pi(\mathbf{x})$ closely enough, it is recommended to set

$$\kappa = 3 - M. \tag{4.58}$$

The rationale behind this choice is to preserve exactly the moments in the approximation to the fifth order in the simple one-dimensional case of Gaussian distribution; for details, see Appendices I in Julier and coworkers (Julier *et al.*, 2000; Julier and Uhlmann, 2004).

With the so-called *unscented transformation*, described by (4.53) to (4.58), the stage is set to formulate a procedure for computing the posterior statistics of a new vector $\mathbf{y} \in R_{M_y}$, which is related to the vector $\mathbf{x}$ by the nonlinear transformation

$$\mathbf{y} = \mathbf{b}(\mathbf{x}). \tag{4.59}$$

Given the odd set of $(2M + 1)$ weighted sigma-points, the unscented transformation is now extended to approximate the mean and covariance matrix of the vector $\mathbf{y}$ by the following set of *projected sigma points* in the $R_{M_y}$ space:

$$\mathbb{E}[\mathbf{y}] = \int_{R_M} \mathbf{b}(\mathbf{x})\pi(\mathbf{x})\, d\mathbf{x}$$
$$= \sum_{i=0}^{2M} w_i \mathcal{Y}_i \tag{4.60}$$

$$\mathrm{cov}[\mathbf{y}] = \int_{R_M} (\mathbf{b}(\mathbf{x}) - \mathbb{E}[\mathbf{y}])\,(\mathbf{b}(\mathbf{x}) - \mathbb{E}[\mathbf{y}])^{\mathrm{T}}\, \pi(\mathbf{x})\, d\mathbf{x}$$
$$= \sum_{i=0}^{2M} w_i (\mathcal{Y}_i - \mathbb{E}[\mathbf{y}])\,(\mathcal{Y}_i - \mathbb{E}[\mathbf{y}])^{\mathrm{T}}, \tag{4.61}$$

where $\mathcal{Y}_i$ is a column vector, defined by

$$\mathcal{Y}_i = \mathbf{b}(\chi_i), \qquad i = 0, 1, \ldots, 2M. \tag{4.62}$$

Provided that the sigma-point set captures the first $p$th-order moments of the prior distribution $\pi(\mathbf{x})$, then the sigma-point transformation to approximate the mean and covariance matrix of $\mathbf{y}$, presented in (4.60) and (4.61), is correct for a $p$th-order nonlinearity. This statement is proved by comparing the Taylor-series expansion of the nonlinear function $\mathbf{b}(\mathbf{x})$ with the statistics computed by the unscented transformation.

With the unscented transformation in place, as just described, we have in effect *linearized* the nonlinear relationship of (4.59) in a way that preserves second-order information about $\mathbf{x}$ that is contained in $\mathbf{y}$. So, finally, with additive system noise $\omega_n$ to the vectorial nonlinear evolution of $\mathbf{x}_n$ over time $n$ in accordance with the system equation (4.18) and additive measurement noise $\mathbf{v}_n$ to the vectorial nonlinear dependence of $\mathbf{y}_n$ on $\mathbf{x}_n$ in accordance with the measurement equation (4.19), where $n$ denotes discrete time, we are now in a position to invoke linear Kalman filter theory. In so doing, we will have developed a recursive formulation of the UKF, so named by virtue of building on the combined use of unscented transformation and relevant parts of the Kalman filter.

## 4.8.2    On the relationship between UKF and CKF

From this exposition of UKF just presented and that of CKF presented in Section 4.7, we readily see that the UKF and CKF share a common property:

To approximate the Bayesian filter, they both use a weighted set of symmetric points.

To illustrate this statement, Figure 4.7a and b, based on a two-dimensional space, show distributions of the weighted sigma-point set for the UKF and the corresponding weighted cubature-point set for the CKF respectively. The two sets of weighted points are signified by the location and heights of the stems shown therein. This figure clearly illustrates that these two sets of weighted points are quite different from each other.

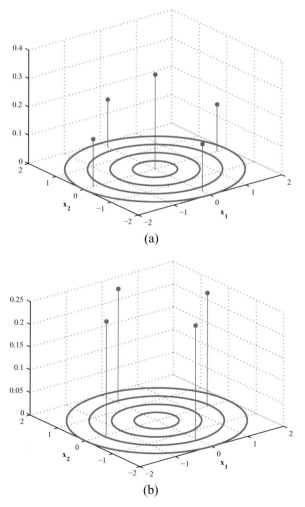

(a)

(b)

**Figure 4.7.** Two kinds of distributions of point sets in two-dimensional space: (a) sigma point-set for the UKF; (b) third-degree spherical–radial cubature point-set for the CKF.

To elaborate more deeply on the fundamental differences that distinguish these two nonlinear approximations of the Bayesian filter from each other, the following three important points are to be borne in mind.

### 4.8.2.1     Theoretical considerations

Derivation of the UKF proceeds from an "unscented" transformation perspective, which is heuristic in some of the algorithmic steps taken in formulating it; the evidence for this point is to be found in (4.55)–(4.57), where the scaling parameter $\kappa$ is introduced. In a related context, in the original proposition of UKF described by Julier *et al.* (2000) right up to relatively recent work on the UKF reported by Sarkka (2007), the scaling parameter $\kappa$ is emphasized as an essential ingredient in computing the sigma-point set.

On the other hand, derivation of the CKF is *mathematically rigorous* from beginning to end. Specifically, the third-degree cubature rule is used to numerically approximate Gaussian-weighted integrals, as described in Section 4.7. Although, indeed, the idea of using the cubature rule for constructing such weighted integrals has been known in the mathematics literature for four decades, there has been no reference made to it in the nonlinear filtering literature, except by Arasaratnam and Haykin (2009).

### 4.8.2.2     Geometric considerations

In general, for an $M$-dimensional state space there are $(2M + 1)$ sigma points that characterize the UKF. One sigma point is located at the origin of a $2M$-dimensional ellipsoid and the remaining $2M$ sigma points are symmetrically distributed on the surface of the ellipsoid. Occurrence of the sigma point at the origin of the ellipsoid is attributed to the scaling parameter $\kappa$; see (4.55).

In direct contrast, for the same $M$-dimensional state space we have an even set of $2M$ cubature points to characterize the CKF. These $2M$ cubature points are also distributed symmetrically on the surface of a $2M$-dimensional ellipsoid. With an even number of cubature points, the CKF distinguishes itself from the UKF in not having a cubature point at the origin of the ellipsoid.

Most importantly, locations of the points and associated weights that characterize these two nonlinear filters are entirely different. Naturally, this fundamental geometric difference has serious consequences. More precisely, in light of the two weighted-point distributions portrayed in Figure 4.7 for the simple case of a two-dimensional state space, we are emboldened to make the following statement:

Given an $M$-dimensional state space, the presence of a weighted sigma-point at the origin of the $2M$-dimensional ellipsoid, characterizing the UKF, has the effect of weakening the approximating power of the UKF, compared with the CKF that has no cubature point at the origin of its own, and albeit different, $2M$-dimensional ellipsoid structure.

### 4.8.2.3     Curse-of-dimensionality problem

Weakness of the UKF is attributed to an *odd* set of weighted sigma-points resulting from the use of scaling parameter $\kappa$; which is directly responsible for its *inability* to *match* the signal-processing power of the CKF characterized by an *even* set of weighted cubature points. This weakness shows up when they are both confronted with a difficult nonlinear filtering problem, where dimensionality of the state space, measurement space, or both is high enough for the filtering computational complexity to become unmanageable due to the curse-of-dimensionality problem, discussed in the next section. Suffice it to say, the experimental results presented therein demonstrate superiority of the continuous-discrete version of the CKF over its UKF counterpart.

### 4.8.3  Summarizing remarks

The UKF, as originally described by Julier and coworkers (Julier *et al.*, 2000; Julier and Uhlmann, 2004), is fundamentally different from the CKF first described by Arasaratnam and Haykin (2009).

The fundamental difference between these two nonlinear filters can be traced to the introduction of the scaling parameter $\kappa$ in the UKF and the emphasis placed on its use not only by the originators of this filter, but also subsequently by many other investigators in the past 10 years.

What is truly surprising is that if the scaling parameter $\kappa$ is set equal to zero in the defining equation (4.55) (i.e. the state-space dimensionality $M = 3$ in (4.58)), then the weight $w_0$ is reduced to zero, and with it the weighted sigma-point-set of the UKF collapses precisely onto the weighted cubature-point set of the CKF. What is all the more surprising is another fact: this observation based on setting $\kappa = 0$ was rarely pointed out in the nonlinear filtering literature; and, by the same token, the cubature rule, basic to the rigorous mathematical derivation of the CKF, was ignored by the nonlinear filtering community for four decades!

## 4.9  The curse of dimensionality

In solving sequential state-estimation problems, it is very important that we pay careful attention to the following statement:

Computational complexity of the state-estimation problem grows exponentially with increasing dimensionality of the state-space model.

This challenging computational issue is encountered whenever we are faced with solving large-scale nonlinear filtering problems, where high dimensionality applies to the state space, measurement space, or both. More than likely, problems of this kind will become more common as we go on to tackle problems rooted in the study of a *system of systems*.

To explain the origin of the computational problem just stated, we begin by reminding ourselves that, in mathematical terms, the hidden state of a *physical environment* may naturally evolve across time in accordance with the *stochastic differential equation*

$$\frac{\mathrm{d}}{\mathrm{d}t}\mathbf{x}(t) = \mathbf{a}(\mathbf{x}(t-1)) + \boldsymbol{\omega}(t), \tag{4.63}$$

where $t$ denotes *continuous time*. However, with the usual emphasis on digital processing of the observables, given the abundant availability of computers nowadays, the measurement equation is formulated in *discrete time*. In reality, therefore, the state-space model is *hybrid* in nature, involving the use of continuous time in the system equation and discrete time in the measurement equation. To proceed with the state-estimation problem in practice, the usual first step, therefore, is to *discretize* the system equation, so as to set the stage for solving the problem on a computer; the original hybrid model is thereby converted into the fully discrete model, as formulated in (4.18) and (4.19). In fact, it is because of this *analog-to-digital conversion process* that the state dimensionality becomes a problem in the first place. Simply put, if we wish to work in the discrete-time domain, which in today's world is the order of the day, then we may have a serious computational issue to deal with.

To elaborate further on this issue, consider Figure 4.8a, b, and c, corresponding to the dimensionality $M = 1$, 2, and 3 respectively. The figure clearly shows that growth

(a)

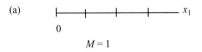

$M = 1$

(b)

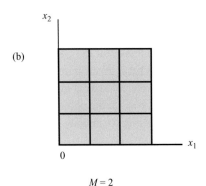

$M = 2$

(c)

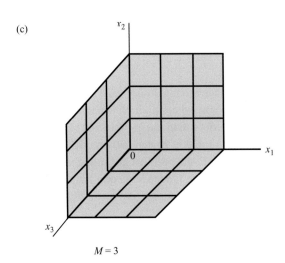

$M = 3$

**Figure 4.8.** Illustrating the curse-of-dimensionality problem as the dimension $M$ assumes the sequence of values 1, 2, 3.

in the number of "resolution cells" in a grid follows a linear law in Figure 4.8a, square law in Figure 4.8b, and cubic law in Figure 4.8c; hence the statement on computational complexity made at the beginning of this section.

The problem being described here is commonly referred to as *the curse-of-dimensionality problem*. The terminology was coined by Bellman in his classic book *Adaptive Control Processes*. Quoting from the book (Bellman, 1961: 94):

In view of all that we have said in the foregoing sections, the many obstacles we appear to have surmounted, what casts the pall over our victory celebration? It is the curse-of-dimensionality, a malediction that has plagued the scientist from earliest days.

## 4.9.1 Case study on the curse-of-dimensionality problem

Arasaratnam *et al.* (2010) studied the case of a difficult radar tracking problem to compare the continuous-discrete (CD) versions of three nonlinear filters: EKF, UKF, and CKF; correspondingly, these three filters are referred to as CD-EKF, CD-UKF, and CD-CKF respectively.

The study focused on a typical air-traffic control scenario, where the objective is to track the trajectory of an aircraft that executes a maneuver at (nearly) constant speed and turn rate in the horizontal plane. Specifically, the motion in the horizontal plane and the motion in the vertical plane are considered to be decoupled from each other. In the language used in aviation, this kind of motion is commonly referred to as a *coordinated turn* (Bar-Shalom *et al.*, 2001). The radar problem of tracking coordinated turns is considered to be challenging for the following reasons:

- The problem is nonlinear in both the system and measurement models (equations).
- The dimensionality of the state space is seven. This is accounted for by position and velocity in each of the three Cartesian coordinates plus the coordinated turn; this dimensionality is high for a radar tracking application.
- There is control over the degree of nonlinearity in the measurement space via the coordinated-turn rate parameter; as this parameter is increased, the aircraft maneuvers more quickly, which, in turn, makes the task of tracking the aircraft more difficult.

Simply put, this seven-dimensional state-estimation problem may well be the highest dimensional problem in the radar tracking literature.

The conclusion drawn from results of the experiments performed under this case study is summarized as follows:

Among the three approximate Bayesian filters, the CD-CKF is the most accurate and reliable, followed by the CD-UKF, and then the CD-EKF.

This ordering of the three approximate Bayesian filters for radar tracking applies equally well to nonlinear state-space models of the discrete kind. Therefore, we are emboldened to make the following statement:

Under the Gaussian assumption, a CKF is a method of choice for challenging radar tracking problems.

As a corollary to this statement, we may reiterate Property 7 in Section 4.7, by saying that although the CKF does not solve the curse-of-dimensionality problem, it does ease it under the Gaussian assumption to a greater extent than is feasible with other approximate Bayesian filters.

## 4.10    Recurrent multilayer perceptrons: an application for state estimation

For an interesting application of nonlinear sequential state estimation on a topic that is highly relevant to cognitive dynamic systems, we look to the supervised training of *RMLP*. This class of neural networks was discussed in Chapter 2. As discussed therein, an RMLP has two distinguishing features:

(1) The network consists of multiple layers of neurons that operate in parallel; hence the usefulness of the network as a memory.
(2) The network involves the use of feedback from one layer to another, thereby enhancing the computational power of the network.

To describe how a nonlinear sequential state-estimator can be used to train an RMLP in a supervised manner, consider such a network with $W$ synaptic weights and $p$ output nodes. With $n$ denoting a time step in the supervised training of the network, let the vector $\mathbf{w}_n$ denote the entire set of synaptic weights in the network, computed at time $n$. For example, we may construct the vector $\mathbf{w}_n$ by stacking the weights associated with neuron 1 in the first hidden layer on top of each other, followed by those of neuron 2, and then carry on in this manner until we have accounted for all the neurons in the first hidden layer, and then do the same for the second and any other hidden layer in the network, until all the weights in the network have been accounted for in the vector $\mathbf{w}_n$ in the orderly fashion just described.

With sequential state estimation in mind, the *state-space model* of the network under training is defined by the following pair of equations:

(1) *System equation*, which is described by

$$\mathbf{w}_{n+1} = \mathbf{w}_n + \boldsymbol{\omega}_n. \tag{4.64}$$

The dynamic noise $\boldsymbol{\omega}_n$ is a white Gaussian noise of zero mean and covariance matrix $\mathbf{Q}_\omega$; the noise is purposely included in the system equation to *anneal* the supervised training of the network over time. In the early stages of the training, the covariance matrix $\mathbf{Q}_\omega$ is large to encourage the supervised learning algorithm to escape local minima, and then it is gradually reduced to some finite but small value.

(2) *Measurement equation*, which is described by

$$\mathbf{d}_n = \mathbf{b}(\mathbf{w}_n, \mathbf{v}_n, \mathbf{u}_n) + \mathbf{v}_n, \tag{4.65}$$

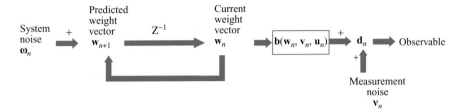

**Figure 4.9.** Nonlinear state-space model depicting the underlying dynamics of a recurrent neural network undergoing supervised training; $Z^{-1}$ denotes a bank of unit-time delays.

where the new entires are as follows:

- $\mathbf{d}_n$ is the *observable*.
- $\mathbf{v}_n$ is the vector representing the *recurrent node activities* inside the network, with its elements listed in an order consistent with those of the weight vector $\mathbf{w}_n$.
- $\mathbf{u}_n$ is the vector denoting the input signal applied to the network; that is, $\mathbf{u}_n$ serves the purpose of a *driving force*.
- $\mathbf{v}_n$ is the vector denoting *measurement noise* corrupting the vector $\mathbf{d}_n$; it is assumed to be a *multivariate white noise process of zero mean and diagonal covariance matrix* $\mathbf{R}_n$. The source of this noise is attributed to the way in which $\mathbf{d}_n$ is actually obtained.

The *vector-valued measurement function* $\mathbf{b}(.,.,.)$ in (4.65) accounts for the overall *nonlinearity* of the RMLP from the input to the output layer; it is the only source of nonlinearity in the state-space model of the recurrent network. This model is described in Figure 4.9.

Insofar as the notion of *state* is concerned, there are two different contexts in which this notion naturally features in the supervised training of the recurrent network:

(1) *Externally adjustable state*, which manifests itself in adjustments applied to the network's weights through supervised training; hence the inclusion of the weight vector $\mathbf{w}_n$ in the state-space model described by both (4.64) and (4.65).
(2) *Internally adjustable state*, which is represented by the vector of recurrent node activities $\mathbf{v}_n$; these activities are outside the scope of the presently configured supervised training process, and it is for this reason that the vector $\mathbf{v}_n$ is included only in the measurement model of (4.65). The externally applied driving force (input vector) $\mathbf{u}_n$, the dynamic noise $\boldsymbol{\omega}_n$, and the global feedback around the MLP account for the evolution of $\mathbf{v}_n$ over time $n$.

## 4.10.1   Description of the supervised training framework using the EKF

Given the training sample $\{\mathbf{u}_n, \mathbf{d}_n\}_{n=1}^{N}$, the issue of interest is how to undertake the supervised training of an RMLP by means of a sequential state-estimator. Since the RMLP is nonlinear by virtue of the nonlinear measurement model of (4.65), the sequential state-estimator would have to be correspondingly nonlinear. With this requirement in mind, we begin the discussion by considering how, for example, the EKF, summarized in Table 4.1, can be used to fulfill this role.

To adapt Table 4.1 for our present needs, we first recognize the following pair of equations, based on the state-space model of (4.64) and (4.65) of the RMLP:

(1) The *innovation*, defined by

$$\alpha_n = \mathbf{d}_n - \mathbf{b}(\hat{\mathbf{w}}_{n-1}, \mathbf{v}_n, \mathbf{u}_n) \tag{4.66}$$

where the desired (target) response $\mathbf{d}_n$ plays the role of the "observable" for the EKF.

(2) The *weight update*, defined by

$$\hat{\mathbf{w}}_{n|n} = \hat{\mathbf{w}}_{n|n-1} + \mathbf{G}_n\alpha_n, \tag{4.67}$$

where $\hat{\mathbf{w}}_{n|n-1}$ is the *predicted (old)* estimate of the RMLP's weight vector $\mathbf{w}$ at time $n$ given the desired response up to and including time $n-1$, and $\hat{\mathbf{w}}_{n|n}$ is the filtered (updated) estimate (4.67) of $\mathbf{w}$. The matrix $\mathbf{G}_n$ is the *Kalman gain*, which is an integral part of the EKF algorithm.

Accordingly, we may formulate Table 4.2, describing the steps involved in supervised training of the RMLP, using the EKF algorithm.

**Table 4.2.** Summary of the EKF algorithm for supervised training of the RMLP

---

*Training sample*
$$\mathcal{T} = \{\mathbf{u}_n, \mathbf{d}_n\}_{n=1}^{N},$$
where $\mathbf{u}_n$ is the input vector applied to the RMLP and $\mathbf{d}_n$ is the corresponding desired response.

*RMLP and Kalman filter: parameters and variables*

| | |
|---|---|
| $\mathbf{b}(.,.,.)$ | vector-valued measurement function |
| $\mathbf{B}_n$ | linearized measurement matrix at time-step $n$ |
| $\mathbf{w}_n$ | weight vector at time-step $n$ |
| $\hat{\mathbf{w}}_{n|n-1}$ | predicted estimate of the weight vector |
| $\hat{\mathbf{w}}_{n|n}$ | filtered estimate of the weight vector |
| $\mathbf{v}_n$ | vector of recurrent node activities in the RMLP |
| $\mathbf{y}_n$ | output vector of the RMLP produced in response to the input vector $\mathbf{u}_n$ |
| $\mathbf{Q}_\omega$ | covariance matrix of the dynamic noise $\omega_n$ |
| $\mathbf{Q}_\mathbf{v}$ | covariance matrix of the measurement noise $\mathbf{v}_n$ |
| $\mathbf{G}_n$ | Kalman gain |
| $\mathbf{P}_{n|n-1}$ | prediction-error covariance matrix |
| $\mathbf{P}_{n|n}$ | filtering-error covariance matrix |

*Computation*
For $n = 1, 2, \ldots$, compute the following:

$$\mathbf{G}_n = \mathbf{P}_{n|n-1}\mathbf{B}_n^{\mathrm{T}}[\mathbf{B}_n\mathbf{P}_{n|n-1}\mathbf{B}_n^{\mathrm{T}} + \mathbf{Q}_{\mathbf{v},n}]^{-1}$$

$$\alpha_n = \mathbf{d}_n - \mathbf{b}_n(\hat{\mathbf{w}}_{n|n-1}, \mathbf{v}_n, \mathbf{u}_n)$$

$$\hat{\mathbf{w}}_{n|n} = \hat{\mathbf{w}}_{n|n-1} + \mathbf{G}_n\alpha_n$$

$$\hat{\mathbf{w}}_{n+1|n} = \hat{\mathbf{w}}_{n|n}$$

$$\mathbf{P}_{n|n} = \mathbf{P}_{n|n-1} - \mathbf{G}_n\mathbf{B}_n\mathbf{P}_{n|n-1}$$

$$\mathbf{P}_{n+1|n} = \mathbf{P}_{n|n} + \mathbf{Q}_{\omega,n}$$

*Initialization*
$$\hat{\mathbf{w}}_{1|0} = \mathbb{E}(\mathbf{w}_1)$$

$\mathbf{P}_{1|0} = \delta^{-1}\mathbf{I}$, where $\delta$ is a small positive constant and $\mathbf{I}$ is the identity matrix

---

Examining the underlying operation of the RMLP, we find that the term $\mathbf{b}(\hat{\mathbf{w}}_{n|n-1}, \mathbf{v}_n, \mathbf{u}_n)$ is the *actual output vector* $\mathbf{y}_n$ produced by the RMLP with its "old" weight vector $\hat{\mathbf{w}}_{n|n-1}$ and internal state $\mathbf{v}_n$ in response to the input vector $\mathbf{u}_n$. Therefore, we may rewrite the combination of (4.66) and (4.67) as a single equation:

$$\hat{\mathbf{w}}_{n|n} = \hat{\mathbf{w}}_{n|n-1} + \mathbf{G}_n(\mathbf{d}_n - \mathbf{y}_n). \tag{4.68}$$

On the basis of this insightful equation, we may now depict the supervised training of the RMLP as the combination of two *mutually coupled components, forming a closed-loop feedback system*, as that shown in Figure 4.10. Specifically:

(1) Figure 4.10a depicts the supervised learning process, viewed partly from the *network's perspective*. With the weight vector set at its old (predicted) value $\hat{\mathbf{w}}_{n|n-1}$, the RMLP computes the output vector $\mathbf{y}_n$ in response to the input vector $\mathbf{u}_n$. Thus, the RMLP supplies the EKF with $\mathbf{y}_n$ as the predicted estimate of the observable, namely $\hat{\mathbf{d}}_{n|n-1}$.
(2) Figure 4.10b depicts the EKF in its role as the *facilitator* of the training process. Supplied with $\hat{\mathbf{d}}_{n|n-1} = \mathbf{y}_n$, the EKF updates the old estimate of the weight vector by operating on the current desired response $\mathbf{d}_n$. The filtered estimate of the weight vector, namely $\hat{\mathbf{w}}_{n|n}$, is thus computed in accordance with (4.67). The $\hat{\mathbf{w}}_{n|n}$ so computed is supplied by the EKF to the RMLP via a *bank of unit-time delays*.

With the transition matrix being equal to the identity matrix, as evidenced by (4.64), we may set $\hat{\mathbf{w}}_{n+1|n}$ equal to $\hat{\mathbf{w}}_{n|n}$ for the next iteration. This equality permits the supervised training cycle to be repeated, until the training process is terminated.

Note that, in the supervised training framework of Figure 4.10, the training sample $\mathcal{T}$ is split between the RMLP and EKF: the input vector $\mathbf{u}_n$ is applied to the RMLP as the

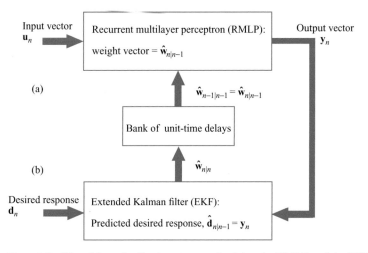

**Figure 4.10.** Closed-loop feedback system, embodying the RMLP and the EKF. (a) The RMLP, with weight vector $\hat{\mathbf{w}}_{n|n-1}$, operates on the input vector $\mathbf{u}_n$ to produce the output vector $\mathbf{y}_n$. (b) The EKF, with the prediction $\mathbf{d}_{n|n-1} = \mathbf{y}_n$, operates on the desired response $\mathbf{d}_n$ to produce the filtered weight vector $\hat{\mathbf{w}}_{n|n} = \hat{\mathbf{w}}_{n+1|n}$, preparing the feedback system for the next iteration.

excitation, and the desired response $\mathbf{d}_n$ is applied to the EKF as the "observable of the hidden weight (state) vector $\mathbf{w}_n$."

The *predictor–corrector property* is an intrinsic property of the Kalman filter, its variants, and extensions. In light of this property, examination of the block diagram of Figure 4.10 leads us to make the following insightful statement:

The recurrent neural network, undergoing training, performs the role of the predictor; and the EKF, providing the supervision, performs the role of the corrector.

Thus, whereas in traditional applications of the Kalman filter for sequential state estimation the roles of predictor and corrector are embodied in the Kalman filter itself, in supervised-training applications these two roles are split between the recurrent neural network and the EKF. Such a split of responsibilities in supervised learning is in perfect accord with the way in which the input and desired response elements of the training sample $\mathcal{T}$ are split in Figure 4.10.

### 4.10.2     The EKF algorithm

For us to be able to apply the EKF algorithm as the facilitator of the supervised learning task, we have to linearize the measurement equation (4.65) by retaining first-order terms in the Taylor-series expansion of the nonlinear part of the equation. With $\mathbf{b}(\mathbf{w}_n, \mathbf{v}_n, \mathbf{u}_n)$ as the only source of nonlinearity, we may *approximate* (4.65) as

$$\mathbf{d}_n = \mathbf{B}_n \mathbf{w}_n + \mathbf{v}_n, \tag{4.69}$$

where $\mathbf{B}_n$ is the *p*-by-$W$ *measurement matrix of the linearized model*, with $p$ denoting the dimensionality of the nonlinear vectorial function $\mathbf{b}(.,.,.)$; that is, the number of output nodes in the network. The linearization process involves computing the partial derivatives of the $p$ outputs of the RMLP with respect to its $W$ weights, obtaining the matrix

$$\mathbf{B} = \begin{bmatrix} \dfrac{\partial b_1}{\partial w_1} & \dfrac{\partial b_1}{\partial w_2} & \cdots & \dfrac{\partial b_1}{\partial w_W} \\[2ex] \dfrac{\partial b_2}{\partial w_1} & \dfrac{\partial b_2}{\partial w_2} & \cdots & \dfrac{\partial b_2}{\partial w_W} \\[2ex] \vdots & \vdots & & \vdots \\[2ex] \dfrac{\partial b_p}{\partial w_1} & \dfrac{\partial b_p}{\partial w_2} & \cdots & \dfrac{\partial b_p}{\partial w_W} \end{bmatrix}, \tag{4.70}$$

where the vector $\mathbf{v}_n$ in $\mathbf{b}(\mathbf{w}, \mathbf{v}_n, \mathbf{u}_n)$ is maintained constant at some value; reference to the time index $n$ has been omitted to simplify the presentation. The $b_i$ in (4.70) denotes the $i$th element of the vectorial function $\mathbf{b}(\mathbf{w}_n, \mathbf{v}_n, \mathbf{u}_n)$. The partial derivatives on the right-hand side of the equation are evaluated at $\mathbf{w}_n = \hat{\mathbf{w}}_{n|n-1}$, where $\hat{\mathbf{w}}_{n|n-1}$ is the prediction of the weight vector $\mathbf{w}_n$ at time $n$ given the desired response up to and including time $n - 1$.

The state-evolution equation (4.64) is linear to begin with; therefore, it is unaffected by the linearization of the measurement equation. Thus the *linearized state-space model* of the recurrent network permits the application of the EKF summarized in Table 4.1.

### 4.10.3   Decoupled EKF

Computational requirements of the EKF are dominated by the need to *store* and *update* the filter error covariance matrix $\mathbf{P}_{n|n}$ at each time step $n$. For a recurrent neural network containing $p$ output nodes and $W$ weights, the computational complexity of the EKF is $O(pW^2)$ and its storage requirement is $O(W^2)$. For large $W$, these requirements may be highly demanding. In such situations, we may look to the *decoupled extended Kalman filter (DEKF)* as a practical remedy for proper management of computational resources (Puskorius and Feldkamp, 2001).

The basic idea behind the DEKF is to *ignore* the interactions between the estimates of certain weights in the recurrent neural network. In so doing, a controllable number of zeros is introduced into the covariance matrix $\mathbf{P}_{n|n}$. More specifically, if the weights in the network are decoupled in such a way that we create *mutually exclusive weight groups*, then the covariance matrix $\mathbf{P}_{n|n}$ is structured into a *block-diagonal form*, as illustrated in Figure 4.11.

Let $g$ denote the designated number of disjoint weight groups created in the manner just described. For $i = 1, 2, \ldots, g$, correspondingly let $\hat{\mathbf{w}}_{n|n}^{(i)}$ be the filtered weight vector for group $i$, $\mathbf{P}_{n|n}^{(i)}$ be the subset of the filtered error covariance matrix for group $i$, $\mathbf{G}_{n}^{(i)}$ be the Kalman gain matrix for group $i$, and so on for the other entries in the DEKF. Let $W_i$

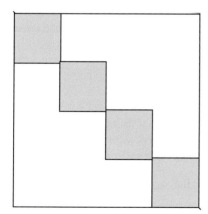

**Figure 4.11.** Block-diagonal representation of the filtered error covariance matrix $\mathbf{P}_{n|n}^{(i)}$ pertaining to the DEKF. The shaded parts of the square represent nonzero values of $\mathbf{P}_{n|n}^{(i)}$, where $i = 1, 2, 3, 4$ for the example illustrated in the figure. As we make the number of disjoint weight groups $g$ larger, more zeros are created in the covariance matrix $\mathbf{P}_{n|n}$; in other words, the matrix $\mathbf{P}_{n|n}$ becomes more sparse; the computational burden is therefore reduced, but the numerical accuracy of the state estimation becomes degraded.

denote the number of weights in group $i$. The concatenation of the filtered weight vectors $\hat{\mathbf{w}}_{n|n}^{(i)}$ forms the overall filtered weight vector $\hat{\mathbf{w}}_{n|n}$; similar remarks apply to $\mathbf{P}_{n|n}^{(i)}$ and $\mathbf{G}_n^{(i)}$ and the other entries in DEKF. In light of these new notations, we may now rewrite the DEKF algorithm for the $i$th weight group as follows:

$$\mathbf{G}_n^{(i)} = \mathbf{P}_{n|n-1}^{(i)}(\mathbf{B}_n^{(i)})^{\mathrm{T}} \left[ \sum_{j=1}^{g} \mathbf{B}_n^{(j)} \mathbf{P}_{n|n-1}^{(j)} (\mathbf{B}_n^{(j)})^{\mathrm{T}} + \mathbf{Q}_{\mathbf{v},n}^{(i)} \right]^{-1}$$

$$\boldsymbol{\alpha}_n^{(i)} = \mathbf{d}_n^{(i)} - \mathbf{b}_n^{(i)} \left( \hat{\mathbf{w}}_{n|n-1}^{(i)}, \mathbf{v}_n^{(i)}, \mathbf{u}_n^{(i)} \right)$$

$$\hat{\mathbf{w}}_{n|n}^{(i)} = \hat{\mathbf{w}}_{n|n-1}^{(i)} + \mathbf{G}_n^{(i)} \boldsymbol{\alpha}_n^{(i)}$$

$$\hat{\mathbf{w}}_{n+1|n}^{(i)} = \hat{\mathbf{w}}_{n|n}^{(i)}$$

$$\mathbf{P}_{n|n}^{(i)} = \mathbf{P}_{n|n-1}^{(i)} - \mathbf{G}_n^{(i)} \mathbf{B}_n^{(i)} \mathbf{P}_{n|n-1}^{(i)}$$

$$\mathbf{P}_{n+1|n}^{(i)} = \mathbf{P}_{n|n}^{(i)} + \mathbf{Q}_{\boldsymbol{\omega},n}^{(i)}.$$

Initialization of the DEKF algorithm proceeds in the same manner described previously in Table 4.1 for the EKF algorithm.

Computational requirements of the DEKF now assume the following orders:

computational complexity        $O\left( p^2 W + p \sum_{i=1}^{g} W_i^2 \right)$,

storage requirement        $O\left( \sum_{i=1}^{g} W_i^2 \right)$.

Thus, depending on the size of $g$ (i.e. the number of disjoint weight groups), the computational requirements of the DEKF can be significantly smaller than those of the EKF.

## 4.10.4    Summarizing remarks on the EKF

An attractive feature of using the EKF as the sequential state estimator for the supervised training of a recurrent neural network is that its basic algorithmic structure (and, therefore, its implementation) is relatively simple, as evidenced by the summary presented in Table 4.1. However, it suffers from two practical limitations:

(1) The EKF requires linearization of the recurrent neural network's vector-valued measurement function $\mathbf{b}(\mathbf{w}_n, \mathbf{v}_n, \mathbf{u}_n)$.
(2) Depending on the size of the weight vector $\mathbf{w}$ (i.e. dimensionality of the state space), we may have to resort to the use of a DEKF to reduce the computational complexity and storage requirement, thereby sacrificing computational accuracy.

We may bypass both of these limitations by using a derivative-free nonlinear sequential state estimator, as discussed next.

## 4.10.5    Supervised training of neural networks using the CKF

In Section 4.7 we discussed the CKF, the formulation of which rests on applying the *cubature rule*. Insofar as supervised training of a recurrent neural network is concerned, the CKF has some unique properties, as summarized here:

(1) Compared with the EKF, the CKF is a more numerically *accurate* approximator of the Bayesian filter, in that it completely preserves second-order information about the state that is contained in the observations.

(2) The CKF is *derivative free*; hence, there is no need for linearizing the measurement matrix of the recurrent neural network.

(3) Last, but by no means least, the cubature rule is used to approximate the time-update integral that embodies the posterior distribution and all the other integral formulas involved in the formulation of the Bayesian filter operating in a Gaussian environment; as a practical rule, integration is preferred over differentiation because of its "smoothing" property.

## 4.10.6    Adaptivity considerations

An interesting property of a recurrent neural network (e.g. RMLP), observed after the network has been trained in a supervised manner, is the *emergence of an adaptive behavior*. This phenomenon occurs despite the fact that the synaptic weights in the network have been fixed. The root of this adaptive behavior may be traced to a fundamental theorem, which is stated as follows (Lo and Yu, 1995):

Consider a recurrent neural network embedded in a stochastic environment with relatively small variability in its statistical behavior. Provided that the underlying probability distribution of the environment is fully represented in the supervised-training sample supplied to the network, it is possible for the network to adapt to the relatively small statistical variations in the environment without any further on-line adjustments being made to the synaptic weights of the network.

This fundamental theorem is valid *only* for recurrent networks. We say so because the dynamic state of a recurrent network actually acts as a "short-term memory" that carries an estimate or statistic of the uncertain environment for adaptation in which the network is embodied.

This adaptive behavior has been referred to differently in the literature. Lo and Bassu (2001) refer to it as *accommodative learning*. In another paper published in the same year (Younger *et al.*, 2001), it is referred to as *meta-learning*, meaning "learning how to learn." Hereafter, we will refer to this adaptive behavior as meta-learning as well.

Regardless of how this adaptive behavior is termed, it is not expected that it will work as effectively as a truly adaptive neural network, where provision is made for automatic on-line weight adjustments if the environment exhibits large statistical variability. This observation has been confirmed experimentally by Lo and Bassu (2001), where comparative performance evaluations were made between a recurrent neural network with meta-learning and an adaptive neural network with long-term and short-term memories; the comparative evaluations were performed in the context of system identification.

Nevertheless, the meta-learning capability of recurrent neural networks should be viewed as a desirable property in control and signal-processing applications, particularly where on-line adjustments of synaptic weights are not practically feasible or they are too costly to perform.

## 4.11    Summary and discussion

### 4.11.1    Optimal Bayesian filter

In this chapter we discussed another important basic issue in the study of cognitive dynamic systems:

- state-space modeling of the environment in which the system is operating and
- Bayesian filtering for estimating the hidden state of the system given a set of state-dependent measurements (observables) supplied by the environment.

The motivation for studying the Bayesian filter for state estimation is that it is *optimal*, at least in a conceptual sense. Unfortunately, except for the special case of a linear dynamic system with Gaussian noise processes and a couple of other special cases, the Bayesian filter is not computationally feasible. Under the assumptions of linearity and Gaussianity, the Bayesian filter reduces to the celebrated Kalman filter.

### 4.11.2    Extended Kalman filter

When the dynamic system is nonlinear and/or non-Gaussian, as in often the case in practice, we have to be content with an approximation of the Bayesian filter. In this chapter we focused on nonlinear filtering for sequential state estimation under the Gaussian assumption. The Gaussian assumption may be justified on the following grounds:

(1) From a mathematical perspective, Gaussian processes are simple and mathematically easy to handle.
(2) Noise processes encountered in many real-world problems may be modeled as Gaussian due to the central limit theorem of probability theory.

Under the Gaussian assumption, we first described the EKF, which is simple to derive and relatively straightforward to implement. Indeed, it is for these two reasons that the EKF is widely used even to this day. However, a major limitation of the Kalman filter is that its application is restricted to nonlinear dynamic systems, where the nonlinearity is of a "mild" sort.

### 4.11.3    Cubature Kalman filters

Practical limitations of the EKF can be overcome by using the CKF, the rigorous derivation of which is based on the third-degree cubature rule in mathematics. The distinguishing characteristic of this new nonlinear filter is its *even* set of weighted cubature points. It is this unique characteristic that gives the filter the signal-processing power to solve nonlinear sequential state-estimation problems, where dimensionality of the state-space model is well beyond the reach of the EKF and UKF. Most importantly, the CKF has many other highly desirable properties, summarized in Section 4.7, which lead us to conclude that the CKF is a method of choice for highly nonlinear state-estimation problems under the Gaussian assumption; this assertion will be well illustrated in Chapter 6 on cognitive radar.

## Notes and practical references

1. The system equation is also referred to as the process equation in the literature.
2. *Bayesian framework for vision*
   In the book entitled *Perception as Bayesian Inference*, edited by Knill and Richards (1996), strong arguments are presented for the Bayesian framework as a general formalism for human perception. As stressed in that book, the Bayesian approach has been successfully applied not only in computer vision (Grenander, 1976–1981; Geman and Geman, 1984; Yuille and Clark, 1993), but also in the modeling of human visual perception (Bennett *et al.*, 1989; Kersten, 1990). The first part of the Knill–Richards book addresses different issues under Bayesian frameworks, followed by the second part devoted to implications and applications of the Bayesian framework.
   Indeed, the Bayesian framework has developed strong roots in human perception and neural coding, so much so that we are becoming emboldened to speak of "The Bayesian brain"; see the paper by Knill and Pouget (2004) and the book co-edited by Doya *et al.* (2007).
3. For an elegant treatment of probability theory on the basis of set theory, see the book entitled *Introduction to Probability*, by Bertsekas and Tsitsiklis (2008).
4. *Historical note on Bayesian*
   Calculations based on Bayes' rule, presented previously as (4.2), are referred to as "Bayesian." In actual fact, Bayes provided a continuous version of the rule; see (4.3). In a historical context, it is also of interest to note that the full generality of (4.3) was not actually perceived by Bayes; rather, the task of generalization was left to Laplace some years later.
5. *The Bayesian paradigm*
   It is because of the duality property that the Bayesian paradigm is sometimes referred to as a *principle of duality*; see Robert (2007: 8). Robert's book presents a detailed and readable treatment of the Bayesian paradigm. For a more advanced treatment of the subject, see Bernardo and Smith (1998).

6. *Origin of the likelihood function*

In a paper published in 1912, Fisher moved away from the Bayesian approach (Fisher, 1912). Then, in a classic paper published in 1922, he introduced the notion of likelihood (Fisher, 1922).

7. *The likelihood principle*

In Appendix B of their book, Bernardo and Smith (1998) show that many non-Bayesian inference procedures do not lead to identical inferences when applied to the proportional likelihoods considered under the likelihood principle.

8. For detailed discussion of the sufficient statistic, see Bernardo and Smith (1998: 247–255).

9. *Maximum likelihood estimation*

In another approach to parameter estimation, known as *maximum likelihood estimation*, the parameter vector $\boldsymbol{\theta}$ is estimated using the formula

$$\theta_{\mathrm{ML}} = \arg\max_{\boldsymbol{\theta}} l(\boldsymbol{\theta}|\mathbf{x}).$$

That is, the *maximum likelihood (ML) estimate* $\hat{\boldsymbol{\theta}}_{\mathrm{ML}}$ is that value of the parameter vector $\boldsymbol{\theta}$ that maximizes the conditional distribution $f_{\mathbf{x}|\Theta}(\mathbf{x}|\boldsymbol{\theta})$, reformulated as $l(\boldsymbol{\theta}|\mathbf{x})$. The ML estimator differs from the MAP estimator in that it ignores the prior $\pi(\boldsymbol{\theta})$; it may, therefore, be said that ML lies at the fringe of the Bayesian approach.

Maximizations of the likelihood function may lead to more than one global maximum, with the result that the procedure used to perform the maximization may *diverge*. To overcome this difficulty, maximum likelihood has to be *stabilized*. This could be done by incorporating prior information on the parameter space, exemplified by the distribution $\pi(\boldsymbol{\theta})$, into the solution, which brings us back to the Bayesian approach and, therefore, (4.8).

10. *Terminology*

In formulating the state-space model described in (4.12) and (4.13), we have started to depart from the terminology in previous chapters by using a subscript for discrete time. This has been done merely to simplify mathematical formalism of the state-space model.

11. *Central limit theorem*

To describe the central limit theorem in probability theory, let the set $\{X_i\}_{i=1}^{N}$ denote a sequence of *independently* and *identically distributed* (iid) random variables with common mean $\mu$ and common variance $\sigma^2$. Define the normalized random variable

$$Y_N = \frac{1}{\sigma N^{1/2}}\left(\sum_{i=1}^{N} X_i - N\mu\right),$$

for which we readily find that the mean

$$\mathbb{E}[Y_N] = 0$$

and the variance

$$\mathrm{var}[Y_N] = 1.$$

The stage is now set for stating the *central limit theorem* in terms of $Y_N$ as follows:

The probability density function of the normalized variable $Y_N$ converges to that of the standard Gaussian variable with zero mean and unit variance; that is,

$$p_Y(y) = \int_{-\infty}^{y} e^{-x^2/2} \, dx$$

in the sense that, for every $y$, we have the limiting probabilistic condition

$$\lim_{N \to \infty} \mathbb{P}[Y_N \le y] = p_Y(y).$$

In words, the central limit theorem asserts that as the number of iid variables $X_i$, $i = 1, 2, \ldots, N$, approaches infinity, the distribution of the normalized random variable $Y_N$ assumes the standard Gaussian distribution.

12. *Particle filters*

The formulation of particle filters is rooted in Monte Carlo simulations. As such, ordinarily, we find that they are:

- simple to code and, therefore, enjoy a broad range of applications;
- computationally demanding and, therefore, unsuitable for large-scale, on-line processing applications.

For a book on Monte Carlo statistical methods, see Robert and Casella (2004). For books on particle filters, see Cappé *et al.* (2005), and Ristic *et al.* (2004).

13. *Monomials*

As mentioned in the text, the term "monomials" is used to refer to a product of powers of variables. Consider, for example, a single variable denoted by $x$; for this simple example, the monomial is 1 or $x^d$, where $d$ is a positive integer. Suppose next that there are several variables, say $x_1$, $x_2$, and $x_3$. In this more general example, each variable is exponentiated individually, such that the monomial is of the product form $x_1^{d_1} x_2^{d_2} x_3^{d_3}$, where the individual powers $d_1$, $d_2$, and $d_3$ are all non-negative integers.

14. *Third-degree cubature rule*

The third-degree spheroidal–radial cubature rule was chosen in Arasaratnam and Haykin (2009) for approximation of the Bayesian filter for a number of practical reasons:

(1) This rule entails the use of an *even* set of weights in the approximation procedure, all of which are *positive*. On the other hand, if we were to adopt fifth-degree or any odd higher degree cubature rule, we would find that some of the weights in the approximation assume negative values. Unfortunately, the appearance of negative weights in the approximation may result in numerical *instability*, which is obviously undesirable.

(2) It may well be that with negative weights resulting from using the fifth-order or higher order cubature rule, it is difficult, if not impossible, to formulate a square-root solution for the CKF that would solve the numerical instability problem.

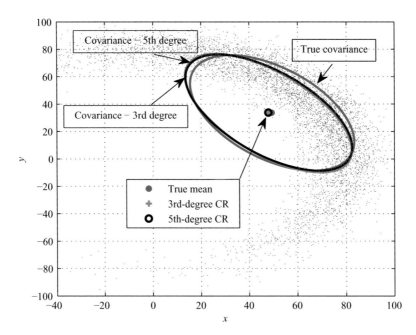

**Figure 4.12.** First two-order statistics of a nonlinearly transformed Gaussian random variable computed by the third- and fifth-degree cubature rules.

(3) Last, but by no means least, experimental results comparing the third-degree with the fifth-degree cubature rule, presented in Figure 4.12, appear to show that estimation accuracy of the CKF based on the fifth-degree rule is marginally improved over the third-degree rule; moreover, the marginal improvement in accuracy was attained at the expense of significantly increased cost in computational complexity.

# 5 Dynamic programming for action in the environment

The previous two chapters of the book have focused on the basic issue: perception of the environment from two different viewpoints. Specifically, Chapter 3 focused on perception viewed as power-spectrum estimation, which is applicable to applications such as cognitive radio, where spectrum sensing of the radio environment is of paramount importance. In contrast, Chapter 4 focused on Bayesian filtering for state estimation, which is applicable to another class of applications exemplified by cognitive radar, where estimating the parameters of a target in a radar environment is the issue of interest. Although those two chapters address entirely different topics, they also have a common perspective: they both pertain to the perceptor of a cognitive dynamic system, the basic function of which is to account for perception in the perception–action cycle. In this chapter, we move on to the action part of this cycle, which is naturally the responsibility of the actuator of the cognitive dynamic system.

Just as the Bayesian framework provides the general formalism for the perceptor to perceive the environment by estimating the hidden state of the environment, Bellman's dynamic programming provides the general formalism for the actuator to act in the environment through control or decision-making.

*Dynamic programming* is a technique that deals with situations where decisions are made in stages (i.e. different time steps), with the outcome of each decision being *predictable* to some extent before the next decision is made. A key aspect of such situations is that decisions cannot be made in isolation. Rather, the desire for a low cost at the present is balanced against the undesirability of a high cost in the future. This is a *credit-assignment problem*, because credit or blame must be assigned to each one in a set of decisions. For *optimal planning*, therefore, it is necessary to have an *efficient* tradeoff between immediate and future costs. Such a tradeoff is indeed captured by the formalism of dynamic programming. In particular, dynamic programming addresses the following fundamental question:

How can a decision-maker learn to improve its long-term performance when the attainment of this improvement may require having to sacrifice short-term performance?

Bellman's dynamic programming provides the mathematical basis for an optimal solution to this fundamental problem in an elegant and principled manner.

In the art of mathematical-model building, the challenge is to strike the right balance between two entities, one practical and the other theoretical. The two entities are respectively

- the realistic description of a given problem and
- the power of analytic and computational methods to apply to the problem.

In this context, an issue of particular concern in dynamic programming is that of a *decision-maker* having to operate in a stochastic environment. To address this issue, we build our *model* around *Markov decision processes*. Given the initial state of a dynamic system, a Markov decision process provides the mathematical basis for choosing a sequence of decisions that will maximize the *returns* from a multistage decision-making process. What we have just described here is the essence of Bellman's dynamic programming. Therefore, it is befitting that we begin the study of dynamic programming with a discussion of this new mathematical tool.

## 5.1     Markov decision processes

As already mentioned, it is the actuator of a cognitive dynamic system that is responsible for decision-making. Henceforth, we simplify matters by using the two terms decision-maker and dynamic system interchangeably.

Consider, then, a decision-maker that interacts with its environment in the manner illustrated in Figure 5.1. The decision-maker operates in accordance with a *finite-discrete-time Markovian decision process*, characterized as follows:

- The environment evolves *probabilistically*, occupying a finite set of *discrete states*. However, the state does *not* contain past statistics, even though those statistics could be useful to the learning process.
- For each environmental state there is a *finite set of possible actions* that may be taken by the decision-maker.
- Every time the decision-maker acts in the environment, a certain *cost* is incurred.
- States are observed, actions are taken, and costs are all incurred at *discrete times*.

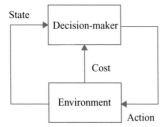

**Figure 5.1.** Block diagram of a decision-maker interacting with its environment.

So, with the notion of state being of profound importance in dynamic programming, we introduce the following definition:

The state of the environment is a summary of the entire past experience gained by a decision-maker from its interactions with the environment, such that the information necessary for the decision-maker to predict future behavior of the environment is contained in that summary.

The state at time step $n$ is denoted by the random variable $X_n$ and the actual state at time step $n$ is denoted by $i_n$; in other words, $i_n$ is a *sample value* of $X_n$. The finite set of possible states is denoted by $\mathcal{H}$. A surprising aspect of dynamic programming is that its applicability depends very little on the nature of the state. Unlike nonlinear filtering for state estimation discussed in Chapter 4, we may, therefore, proceed without any assumption on the structure of the state space. (This statement does not apply to Section 5.5.) Note that the *complexity* of a dynamic-programming algorithm is *quadratic* in the dimension of the state space and *linear* in the dimension of the action space. Note also that we have introduced the notion of *action space*. Therefore, it is the dimensionality of the state space, measurement space, action space, or all three of them that is responsible for the curse-of-dimensionality problem in dynamic programming, an issue that was discussed previously in Chapter 4.

For state $i$, for example, the available *set of actions* in the environment is denoted by $\mathcal{A}_i = \{a_{ik}\}$, where the second subscript $k$ in the action $a_{ik}$ taken by the decision-maker merely indicates the availability of more than one possible action when the environment is in state $i$. Transition of the environment from state $i$ to a new state $j$, due to action $a_{ik}$, say, is probabilistic in nature. Most importantly, however, *the transition probability from state i to state j depends entirely on the current state i and the corresponding action $a_{ik}$*. This is a restatement of the *Markov property* that was discussed in Chapter 4. This property is crucial, because it means that the current state of the environment provides the necessary information for action taken by the decision-maker.

The random variable denoting the action at time step $n$ is denoted by $A_n$. Note that the symbol $A$ representing a random variable for action at some unspecified time step is different from the symbol $\mathcal{A}$ representing a set of actions for an unspecified state. Let $p_{ij}(a)$ denote the *transition probability* from state $i$ to state $j$ due to the action taken at time step $n$, where $A_n = a$. By virtue of the Markovian assumption on state dynamics, we define the transition probability

$$p_{ij}(a) = \mathbb{P}[X_{n+1} = j \mid X_n = i, A_n = a]. \qquad (5.1)$$

The transition probability $p_{ij}(a)$ satisfies two conditions that are imposed on it by probability theory:

$$p_{ij}(a) \geq 0 \qquad \text{for all } i \text{ and } j, \qquad (5.2)$$

$$\sum_j p_{ij}(a) = 1 \qquad \text{for all } i, \qquad (5.3)$$

where both $i$ and $j$ reside in the state space.

For a given number of states and given transition probabilities, the sequence of environmental states resulting from the actions taken by the decision-maker over time forms a *Markov chain*, which was also discussed in Chapter 4.

At each transition from one state to another, the decision-maker naturally incurs a *cost*. Thus, at the $n$th transition from state $i$ to state $j$ under action $a_{ik}$, the cost incurred is denoted by $\gamma^n g(i, a_{ik}, j)$, where $g(.,.,.)$ is an *observed transition cost* and $\gamma$ is a scalar called the *discount factor* that is confined to the range $0 \leq \gamma < 1$. The discount factor reflects intertemporal preferences. By adjusting $\gamma$, we are able to control the extent to which the decision-maker is concerned with long-term versus short-term consequences of its own actions. In the limit, when $\gamma = 0$, the decision-maker is said to be *myopic*, which means that the decision-maker is concerned only with immediate consequences of its actions.

Our interest in dynamic programming is to formulate a *policy*, defined as the *mapping of states into actions*. In other words:

Policy is a rule used by the decision-maker to decide what to do, given knowledge of the current state of the environment. The policy is denoted by

$$\pi = \{\mu_0, \mu_1, \mu_2, \ldots\}, \tag{5.4}$$

where $\mu_n$ is a function that maps the state $X_n = i$ into an action $A_n = a$ at stage $n = 0, 1, 2, \ldots$. This mapping is such that we may write

$$\mu_n(i) \in \mathcal{A}_i \qquad \text{for all states } i \in \mathcal{H},$$

where $\mathcal{A}_i$ denotes the set of all possible actions available to the decision-maker in state $i$. Such policies are said to be admissible.

A policy can be nonstationary or stationary. A *nonstationary* policy is naturally time varying, as indicated in (5.4). When, however, the policy is independent of time, as shown by

$$\pi = \{\mu, \mu, \mu, \ldots\},$$

the policy is said to be *stationary*. In other words, a stationary policy specifies exactly the same action each time a particular state is visited. For a stationary policy, the underlying Markov chain may be stationary or nonstationary; it is possible to use a stationary policy on a nonstationary Markov chain, but this is *not* a wise thing to do. If a stationary policy $\mu$ is employed, then the sequence of states $\{X_n, n = 0, 1, 2, \ldots\}$ forms a Markov chain with transition probabilities $p_{ij}(\mu(i))$, where $\mu(i)$ signifies an action on state $i$ under policy $\mu$. It is indeed for this reason that the process is referred to as a *Markov decision process*.

## 5.1.1　The basic problem

A dynamic-programming problem can be of a finite-horizon or infinite-horizon kind. In a *finite-horizon problem*, the cost accumulates over a finite number of stages. In an *infinite-horizon problem*, on the other hand, the cost accumulates over an infinite number of stages. Infinite-horizon problems are of particular interest, because discounting ensures that the costs for all states are finite for any policy.

Let $g(X_n, \mu_n(X_n), X_{n+1})$ denote the *observed transition cost* incurred as a result of transition from state $X_n$ to state $X_{n+1}$ under the action of policy $\mu_n(X_n)$. The total expected cost in an infinite-horizon problem, starting from an initial state $X_0 = i$ and using a policy $\pi = \{\mu_n\}$, is defined by

$$j^\pi(i) = \mathbb{E}\left[\sum_{n=0}^{\infty} \gamma^n g(X_n, \mu_n(X_n), X_{n+1}) \mid X_0 = i\right], \tag{5.5}$$

where the expected value (signified by the operator $\mathbb{E}$) is taken over the Markov chain $\{X_1, X_2, \ldots\}$ and $\gamma$ is the discount factor. The function $J^\pi(i)$ is called the *cost-to-go function for policy $\pi$*, starting from state $i$. Its *optimal* value, denoted by $J^*(i)$, is defined by

$$J^*(i) = \min_\pi J^\pi(i). \tag{5.6}$$

The policy $\pi$ is *optimal* if, and only if, it is *greedy* with respect to $J^*(i)$. The term "greedy" is used here to describe the case when the decision-maker seeks to minimize the immediate next cost without paying any attention to the possibility that such an action may do away with access to better alternatives in the future. When the policy $\pi$ is stationary – that is, $\pi = \{\mu, \mu, \ldots\}$ – we use the notation $J^\mu(i)$ in place of $J^*(i)$ and say that $\mu$ is optimal if

$$J^\mu(i) = J^*(i) \qquad \text{for all initial states } i. \tag{5.7}$$

We may now sum up the basic problem in dynamic programming for a stationary policy as follows:

Given a stationary Markov decision process describing the interaction between a decision-maker and its environment, find a stationary policy $\pi = \{\mu, \mu, \ldots\}$ that minimizes the cost-to-go function $J^\mu(i)$ for all initial states $i$.

## 5.2 Bellman's optimality criterion

The dynamic-programming technique rests on a very simple idea known as the *principle of optimality*, due to Bellman (1957). Simply stated, this principle says the following (Bellman and Dreyfus, 1962):

An optimal policy has the property that whatever the initial state and initial decision are, the remaining decisions must constitute an optimal policy starting from the state that results from the first decision.

For example, a "decision" used in this statement may be a *choice* of control applied in the environment at a particular time, in which case a "policy" is the entire control sequence.

To formulate the principle of optimality in mathematical terms, consider a finite-horizon problem for which the cost-to-go function is defined by

$$J_0(X_0) = \mathbb{E}\left[g_L(X_L) + \sum_{n=0}^{L-1} g_n(X_n, \mu_n(X_n), X_{n+1})\right], \tag{5.8}$$

where $L$ is the *planning horizon* or *horizon depth* and $g_L(X_L)$ is the *terminal cost*. Given $X_0$, the expectation in (5.8) is over the remaining states $X_1, \ldots, X_L$. With this terminology, we may now formally state the principle of optimality as follows (Bertsekas, 2005, 2007):

Let $\pi^* = \{\mu_0^*, \mu_1^*, \ldots, \mu_{L-1}^*\}$ be an optimal policy for the basic finite-horizon problem. Assume that, when using the optimal policy $\pi^*$, a given state $X_n$ occurs with positive probability. Consider the subproblem where the environment is in state $X_n$ at stage $n$ and suppose we wish to minimize the corresponding cost-to-go function

$$J_n(X_n) = \mathbb{E}\left[ g_L(X_L) + \sum_{k=n}^{L-1} g_k(X_k, \mu_k(X_k), X_{k+1}) \right], \qquad \text{for } n = 0, \ldots, L-1. \quad (5.9)$$

Then, the truncated policy $\{\mu_0^*, \mu_1^*, \ldots, \mu_{L-1}^*\}$ is optimal for the subproblem.

We may intuitively justify the principle of optimality by the following argument: if the truncated policy $\{\mu_n^*, \mu_{n+1}^*, \ldots, \mu_{L-1}^*\}$ was not optimal as stated, then, once the state $X_n$ is reached at stage $n$, we could reduce the cost-to-function $J_n(X_n)$ simply by switching to a policy that is optimal for the subproblem.

The principle of optimality builds on the engineering notion of "divide and conquer." Basically, an optimal policy for a complex multistage planning or control problem is constructed by proceeding *backward*, stage by stage, as follows:

(1) Construct an optimal policy for the "tail subproblem" involving only the last stage of the system.
(2) Extend the optimal policy to the "tail subproblem" involving the last two stages of the system.
(3) Continue the procedure in this fashion, stage by stage, until the entire problem has been dealt with.

## 5.2.1    Dynamic-programming algorithm

On the basis of the procedure just described, we may now formulate the dynamic-programming algorithm, which runs *backward in time*, stage by stage. Let $J_{n+1}(X_{n+1})$ denote the cost-to-go function at stage $n+1$ of the interaction between the decision-maker and its environment. Then, running backward in time by one stage, there will be the observed transition cost $g_n(X_n, \mu_n(X_n), X_{n+1})$ incurred as a result of the transition from state $X_n$ to state $X_{n+1}$ under policy $\mu_n(X_n)$. The updated cost at stage $n$, therefore, is $J_{n+1}(X_{n+1})$ augmented by the observed transition cost, as shown by

$$g_n(X_n, \mu_n(X_n), X_{n+1}) + J_{n+1}(X_{n+1}). \quad (5.10)$$

Treating the state $X_{n+1}$ as a random variable, which it is, we therefore need to take the expected value of the augmented cost in (5.10) at stage $n$ with respect to $X_{n+1}$ and thus write

$$\mathbb{E}_{X_{n+1}} [g_n(X_n, \mu_n(X_n), X_{n+1}) + J_{n+1}(X_{n+1})]. \quad (5.11)$$

At this point, we go on to address the issue of finding the policy $\mu_n$ for which the expected value in (5.11) is minimized, yielding

$$J_n(X_n) = \min_{\mu_n} \underset{X_{n+1}}{\mathbb{E}} [g_n(X_n, \mu_n(X_n), X_{n+1}) + J_{n+1}(X_{n+1})]. \qquad (5.12)$$

Generalizing this formula, we may now formally state the dynamic-programming algorithm as follows (Bertsekas, 2005, 2007):

For every initial state $X_0$, the optimal cost $J^*(X_0)$ of the finite-horizon problem is equal to $J_0(X_0)$, where the function $J_0$ is obtained from the last time step of the dynamic-programming algorithm

$$J_n(X_n) = \min_{\mu_n} \underset{X_{n+1}}{\mathbb{E}} [g_n(X_n, \mu_n(X_n), X_{n+1}) + J_{n+1}(X_{n+1})], \qquad (5.13)$$

which runs backwards in time for horizon depth $L$, with

$$J_L(X_L) = g_L(X_L). \qquad (5.14)$$

Furthermore, if $\mu_n^*$ minimizes the right-hand side of (5.13) for each state $X_n$ and time step $n$, then the nonstationary policy $\pi^* = \{\mu_0^*, \mu_1^*, \ldots, \mu_{L-1}^*\}$ is optimal.

## 5.2.2 Bellman's optimality equation

In its basic form, the dynamic-programming algorithm deals with a finite-horizon problem. We are interested in extending the use of this algorithm to deal with the infinite-horizon discounted problem described by the cost-to-go function of (5.5) under a stationary policy $\pi = \{\mu, \mu, \mu, \ldots\}$. With this objective in mind, we do two things:

- First, reverse the time index of the algorithm.
- Second, define the observed transition cost $g_n(X_n, \mu(X_n), X_{n+1})$ as

$$g_n(X_n, \mu(X_n), X_{n+1}) = \gamma^n g(X_n, \mu(X_n), X_{n+1}), \qquad (5.15)$$

where, in the right-hand side, we have introduced the discount factor $\gamma$ to account for use of the transition cost $g(X_n, \mu(X_n), X_{n+1})$ and deleted the subscript $n$ in $g_n$.

We may then reformulate the dynamic-programming algorithm as

$$J_{n+1}(X_0) = \min_{\mu} \underset{X_1}{\mathbb{E}}[g(X_0, \mu(X_0), X_1) + \gamma J_n(X_1)], \qquad (5.16)$$

which starts from the initial condition $J_0(X) = 0$ for all $X$. The state $X_0$ is the initial state; $X_1$ is the new state that results from the action of policy $\mu$ on $X_0$.

Let $J^*(i)$ denote the optimal infinite-horizon cost for the initial state $X_0 = i$. We may then view $J^*(i)$ as the limit for the corresponding $L$-stage optimal cost $J_L(i)$ as the horizon depth $L$ approaches infinity; that is,

$$J^*(i) = \lim_{L \to \infty} J_L(i) \qquad \text{for all } i. \qquad (5.17)$$

This relation is the connecting link between the finite-horizon and infinite-horizon discounted problems. Substituting $n + 1 = L$ and $X_0 = i$ into (5.16) and then applying (5.17), we may go on to express the optimal infinite-horizon cost as

$$J^*(i) = \min_{\mu} \mathbb{E}_{X_1}[g(i, \mu(i), X_1) + \gamma J^*(X_1)], \qquad (5.18)$$

which is the *expectation form of Bellman's equation* for a Markov decision process. We may rewrite (5.18) by proceeding in two stages:

(1) Evaluate the expectation of the cost $g(i, \mu(i), X_1)$ with respect to $X_1$ by writing

$$\mathbb{E}_{X_1}[g(i, \mu(i), X_1)] = \sum_{j=1}^{N} p_{ij} g(i, \mu(i), j), \qquad (5.19)$$

where $N$ is the number of possible states occupied by the environment and $p_{ij}$ is the transition probability from the initial state $X_0 = i$ to the new state $X_1 = j$. The quantity defined in (5.19) is the *expected observed cost* incurred at state $i$ by following the action recommended by the policy $\mu$. Denoting this cost by $c(i, \mu(i))$, we write

$$c(i, \mu(i)) = \sum_{j=1}^{N} p_{ij} g(i, \mu(i), j). \qquad (5.20)$$

(2) Evaluate the expectation of $J^*(X_1)$ with respect to $X_1$. Here, we note that if we know the optimal cost $J^*(X_1)$ for each state $X_1$ of the finite-state system, then we may readily determine the expectation of $J^*(X_1)$ in terms of the transition probabilities of the underlying Markov chain by writing

$$\mathbb{E}_{X_1}[J^*(X_1)] = \sum_{j=1}^{N} p_{ij} J^*(j). \qquad (5.21)$$

Thus, using (5.19)–(5.21) in (5.18), we obtain the desired formula:

$$J^*(i) = \min_{\mu} \left( c(i, \mu(i)) + \gamma \sum_{j=1}^{N} p_{ij} J^*(j) \right), \qquad i = 1, 2, \ldots, N. \qquad (5.22)$$

Equation (5.22) is the *standard form of Bellman's optimality equation* for a Markov decision process. This equation should *not* be viewed as an algorithm. Rather, it represents a system of $N$ equations with one equation per state.[1] The solution of this system of equations defines the value of the optimal cost-to-go function for the $N$ states of the environment.

There are two methods for computing an optimal policy, namely policy iteration and value iteration; these two methods are described in the next two sections.

## 5.3  Policy iteration

To set the stage for a description of the policy iteration algorithm, we begin by introducing a new concept called the *Q-factor*. Consider an existing stationary policy $\mu$ for which the cost-to-go function $J^{\mu}(i)$ is known for all states $i$. For each state $i \in \mathcal{H}$ and action $a \in \mathcal{A}_i$, where $\mathcal{A}_i$ is the set of all possible actions in state $i$, the $Q$-factor is defined as the *expected observed cost plus the sum of the discounted costs of all successor states under policy $\mu$*, as shown by

$$Q^{\mu}(i,a) = c(i, a) + \gamma \sum_{j=1}^{N} p_{ij}(a) J^{\mu}(j), \tag{5.23}$$

where $a = \mu(i)$ and the term $c(i, a)$ is simply a rewrite of $c(i, \mu(i))$ in (5.20). In effect, the $Q$-factor $Q^{\mu}(i, a)$ captures the cost of being in state $i$ and taking action $a$ under policy $\mu$. Note also that the $Q$-factors denoted by $Q^{\mu}(i, a)$ for $i = 1, 2, \ldots, N$ contain more information than the cost-to-go function $J^{\mu}(i)$. For example, actions may be ranked on the basis of $Q$-factors alone, whereas ranking on the basis of cost-to-go function requires knowledge of the state-transition probabilities and costs. Note also that the optimal cost-to-go function $J^*(i)$ in (5.22) is obtained simply as $\min_{\mu} Q^{\mu}(i, a)$.

To develop insight into the meaning of the $Q$-factor, visualize a new system whose states are made up of the original states $i = 1, 2, \ldots, N$ and all the possible state–action pairs $(i, a)$, as portrayed in Figure 5.2. There are two distinct possibilities that can occur in this figure:

(1) The system is in state $(i, a)$, in which case no action is taken. Transition is made automatically to state $j$, say, with probability $p_{ij}(a)$ and cost $g(i, a, j)$ is incurred.
(2) The system is in state $i$, say, in which case action $a \in \mathcal{A}_i$ is taken; deterministically, the next state is $(i, a)$.

In light of what was said previously in Section 5.1, the policy $\mu$ is greedy with respect to the cost-to-go function $J^{\mu}(i)$ if, for all the states, $\mu(i)$ is an action that satisfies the condition

$$Q^{\mu}(i, \mu(i)) = \arg \min_{a \in \mathcal{A}_i} Q^{\mu}(i, a). \tag{5.24}$$

Two noteworthy observations follow from (5.24):

- First, it is possible for more than one action to minimize the set of $Q$-factors for some state, in which case there can be more than one greedy policy with respect to the pertinent cost-to-go function.
- Second, a policy can be greedy with respect to many different cost-to-go functions.

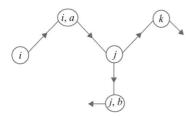

**Figure 5.2.** Illustration of two possible transitions: the transition from state $(i, a)$ to state $j$ is probabilistic, but the transition from state $i$ to $(i, a)$ is deterministic.

Moreover, the following fact is basic to all dynamic-programming methods under a greedy policy:

$$Q^{\mu^*}(i, \mu^*(i)) = \arg\min_{a \in \mathcal{A}_i} Q^{\mu}(i, a), \tag{5.25}$$

where $\mu^*$ is an optimal policy.

## 5.3.1    Formulation of the policy iteration algorithm

With the notions of $Q$-factors and greedy policy at our disposal, we are now ready to describe the policy iteration algorithm. Specifically, the algorithm operates by alternating between two steps[2]:

(1) *Policy evaluation step.* In this first step, the cost-to-go function $J^{\mu}$ for some current policy $\mu$ and corresponding $Q$-factor $Q^{\mu}(i, a)$ are computed for all the states and actions.

(2) *Policy improvement step.* In this second step, the current policy is updated in order to be greedy with respect to the cost-to-go function computed in step 1.

These two steps are illustrated in the block diagram of Figure 5.3. To be specific, we start with some initial policy $\mu_0$ and then generate a sequence of new policies $\mu_1, \mu_2, \ldots$. Given the current policy $\mu_n$, we perform the policy evaluation step by computing the cost-to-go function $J^{\mu_n}(i)$ as the solution of the linear system of equations (5.22) reproduced here in the following form:

$$J^{\mu_n}(i) = c(i, \mu_n(i)) + \gamma \sum_{j=1}^{N} p_{ij}(\mu_n(i)) J^{\mu_n}(j), \qquad i = 1, 2, \ldots, N, \tag{5.26}$$

where $J^{\mu_n}(1), J^{\mu_n}(2), \ldots, J^{\mu_n}(N)$ are the unknowns and the transition probability $p_{ij}$ is expressed as the function $p_{ij}(\mu_n(i))$ for the purpose of generality. Using these results,

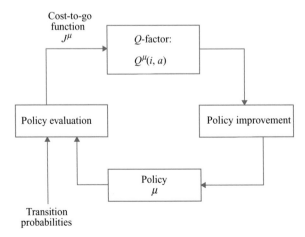

**Figure 5.3.** Block diagram of the policy iteration algorithm.

**Table 5.1.** Summary of the policy iteration algorithm

1. Start with an arbitrary initial policy $\mu_0$.
2. For $n = 0, 1, 2, \ldots$, compute the cost-to-go function $J^{\mu_{n+1}}(i)$ and $Q$-factor $Q^{\mu_{n+1}}(i, a)$ for all states $i \in \mathcal{H}$ and actions $a \in \mathcal{A}_i$.
3. For each state $i = 1, 2, \ldots, N$, compute

$$\mu_{n+1}(i) = \arg\min_{a \in \mathcal{A}_i} Q^{\mu_n}(i, a).$$

4. Repeat steps 2 and 3 until $\mu_{n+1}$ is not an improvement on $\mu_n$, at which point the algorithm terminates with $\mu_n$ as the desired policy.

we then follow (5.23) to compute the $Q$-factor for each state–action pair $(i, a)$ as follows:

$$Q^{\mu_n}(i, a) = c(i, a) + \gamma \sum_{j=1}^{N} p_{ij}(a) J^{\mu_n}(j), \qquad a = \mu_n(i) \in \mathcal{A}_i \quad \text{and} \quad i = 1, 2, \ldots, N.$$

$$(5.27)$$

Next, following (5.24), we perform the policy improvement step by computing a new policy $\mu_{n+1}$ defined by

$$\mu_{n+1}(i) = \arg\min_{a \in \mathcal{A}_i} Q^{\mu_n}(i, a), \qquad i = 1, 2, \ldots, N. \tag{5.28}$$

The two-step process just described is repeated with policy $\mu_{n+1}$ used in place of $\mu_n$ in (5.27), until we arrive at the final condition

$$J^{\mu_{n+1}}(i) = J^{\mu_n}(i) \qquad \text{for all } i,$$

at which point in time the algorithm is terminated with $\mu_n$ as the desired policy. With $J^{\mu_{n+1}} \le J^{\mu_n}$, we may then say that the policy iteration algorithm *will terminate after a finite number of iterations* because the underlying Markov decision process has a finite number of states. Table 5.1 presents a summary of the policy iteration algorithm, based on (5.26)–(5.28).

## 5.4    Value iteration

In the policy iteration algorithm, the cost-to-go function has to be recomputed entirely at each iteration of the algorithm, which can be a computationally expensive proposition, particularly when the number of states $N$ is large. Even though the cost-to-go function for the new policy may be similar to that for the old policy, there is, unfortunately, no dramatic shortcut for this computation. There is, however, another method for finding the optimal policy that avoids the burdensome task of repeatedly computing the cost-to-go function. This alternative method, based on successive approximations, is known as the value iteration algorithm.

The *value iteration algorithm* involves solving Bellman's optimality equation, given in (5.22), for each of a sequence of finite-horizon problems. In the limit, the cost-to-go function of the finite-horizon problem converges uniformly over all states to the corresponding cost-to-go function of the infinite-horizon problem as the number of iterations of the algorithm approaches infinity (Ross, 1983; Bertsekas, 2007). Simply put, the value iteration algorithm is perhaps the most widely used algorithm in dynamic programming for two reasons:

- First, it is relatively simple to implement.
- Second, it provides a natural way to solve dynamic-programming problems.

### 5.4.1    Formulation of the value iteration algorithm

Let $J_n(i)$ denote the cost-to-go function for state $i$ at iteration (time step) $n$ of the value iteration algorithm. The algorithm begins with an arbitrary guess $J_0(i)$ for $i = 1, 2, \ldots, N$. If some estimate of the optimal cost-to-go function $J^*(i)$ is available, it should be used as the initial value $J_0(i)$. Once $J_0(i)$ has been chosen, we may use (5.22) as a basis to compute the sequence of cost-to-go functions $J_1(i)$, $J_2(i)$, ..., described as follows:

$$J_{n+1}(i) = \min_{a \in \mathcal{A}_i} \left\{ c(i, a) + \gamma \sum_{j=1}^{N} p_{ij}(a) J_n(j) \right\}, \qquad i = 1, 2, \ldots, N, \qquad (5.29)$$

where, here again, the transition probability is expressed as $p_{ij}(a)$ for the purpose of generality. Application of the update to the cost-to-go function described in (5.29) for state $i$ is referred to as the *backing up of i's cost*. This backup is a direct implementation of Bellman's optimality equation. It is important to note that the values of the cost-to-go functions in (5.29) for states $i = 1, 2, \ldots, N$ are backed up *simultaneously* on each iteration of the algorithm. This method of implementation represents the traditional *synchronous* form of the value iteration algorithm. Thus, starting from arbitrary initial values $J_0(1)$, $J_0(2)$, ..., $J_0(N)$, the algorithm described in (5.29) converges to the corresponding optimal values $J^*(1)$, $J^*(2)$, ..., $J^*(N)$ as the number of iterations $n$ approaches infinity. So, ideally, the value iteration algorithm in (5.29) requires an *infinite number of iterations*.

Unlike the policy iteration algorithm, an optimal policy is not computed directly in the value iteration algorithm. Rather, the optimal values $J^*(1)$, $J^*(2)$, ..., $J^*(N)$ are first computed using (5.29). Then, a greedy policy with respect to that optimal set is obtained as an optimal policy, as shown by

$$\mu^*(i) = \arg \min_{a \in \mathcal{A}_i} Q^*(i, a), \qquad i = 1, 2, \ldots, N, \qquad (5.30)$$

where

$$Q^*(i, a) = c(i, a) + \gamma \sum_{j=1}^{N} p_{ij}(a) J_n^*(j), \qquad i = 1, 2, \ldots, N. \qquad (5.31)$$

A summary of the value iteration algorithm, based on (5.29)–(5.31), is presented in Table 5.2. This summary includes a stopping criterion for (5.29).

**Table 5.2.** Summary of the value iteration algorithm

1. Start with an arbitrary initial value $J_0(i)$ for state $i = 1, 2, ..., N$.
2. For $n = 0, 1, 2, ...$, compute the cost-to-go function

$$J_{n+1}(i) = \min_{a \in \mathcal{A}_i} \left\{ c(i, a) + \gamma \sum_{j=1}^{N} p_{ij}(a) J_n(j) \right\}, \qquad i = 1, 2, ..., N.$$

Continue this computation until we arrive at the condition

$$|J_{n+1}(i) - J_n(i)| < \varepsilon \qquad \text{for each state } i,$$

where $\varepsilon$ is a prescribed small positive number.
3. Compute the $Q$-factor

$$Q^*(i, a) = c(i, a) + \gamma \sum_{j=1}^{N} p_{ij}(a) J_n^*(j) \qquad a \in A_i, i = 1, 2, ..., N.$$

Hence, determine the optimal policy as a greedy policy for $J^*(i)$:

$$\mu^*(i) = \arg \min_{a \in \mathcal{A}_i} Q^*(i, a), \qquad i = 1, 2, ..., N.$$

## 5.5   Approximate dynamic programming for problems with imperfect state information

Thus far, up to this point in the chapter, the discussion presented on dynamic programming rests its applicability on the premise that a cognitive dynamic system satisfies two assumptions:

(1) *Assumption 1.* The state of the environment has a discrete value.
(2) *Assumption 2.* The actuator of the system has access to the exact value of the state at all times.

In certain examples of cognitive dynamic systems, exemplified by cognitive radar, both assumptions are unfortunately violated. In a radar system, regardless of its kind, the state-space model is *hybrid*, in that

- the system equation, describing evolution of the state across time, is *continuous* in time and
- the measurement equation is *discrete* in time.

Clearly, then, Assumption 1 is violated, which means that the use of dynamic programming may suffer from the *curse of dimensionality*. Moreover, Assumption 2 is also violated because the actual state of the radar environment is *hidden* from the transmitter (actuator). In fact, the only source of information about the state available to the transmitter is contained in a set of observables naturally corrupted by some kind of noise in accordance with the measurement equation of the state-space model. In this section, we address the important issue of how to get around Assumption 2. The curse-of-dimensionality problem attributed to Assumption 1 is deferred to Chapter 6 on cognitive radar.

The important point to note here is that we now have a new issue to deal with, namely *imperfect state information* about the environment. Compared with "idealized" problems with perfect state information, solutions to problems with imperfect state information are computationally more demanding and suboptimal in performance. It is in light of this latter point that dynamic-programming algorithms intended for this new class of problems are naturally "approximate"; hence the title of the section.

Notwithstanding the comments just made, problems with imperfect state information are *no* different in conceptual terms from their perfect state information counterparts (Bertsekas, 2005). Specifically, through appropriate reformulations, problems with imperfect state information can be reduced to corresponding problems with perfect state information, as discussed next.

## 5.5.1    Basics of problems with imperfect state information

Following the material presented in Chapter 4, the discrete state-space model for problems with imperfect state information is described, in its most generic form, as follows:

(1) *System equation*

$$\mathbf{x}_{n+1} = \mathbf{a}_n(\mathbf{x}_n, \mathbf{u}_n, \boldsymbol{\omega}_n),$$

where $\mathbf{u}_n$ is the control applied to the environment at time $n$ and $\boldsymbol{\omega}_n$ is the system noise. (In the case of a hybrid model, we are assuming that the continuous system equation has been discretized.)

(2) *Measurement equation*

$$\mathbf{y}_n = \mathbf{b}_n(\mathbf{x}_n, \mathbf{u}_{n-1}, \mathbf{v}_n),$$

where $\mathbf{y}_n$ is the observable at time $n$ and $\mathbf{v}_n$ is the measurement noise.

Let $\mathbf{I}_n$ denote the *information vector* available to the transmitter at time $n$, which is defined by

$$\mathbf{I}_n = [\mathbf{y}_0, \mathbf{y}_1, \ldots, \mathbf{y}_n; \mathbf{u}_0, \mathbf{u}_1, \ldots, \mathbf{u}_{n-1}], \tag{5.32}$$

where $n = 1, \ldots, L - 1$, with $L$ denoting duration of the *window* of observables being considered at time $n$. Note also that the initial value

$$\mathbf{I}_0 = \mathbf{y}_0.$$

Considering a *control* application, let $\mathbf{u}_n$ be the function that maps $\mathbf{I}_n$ into the control $\mathbf{u}_n$ that lies in the control space $\mathcal{U}_n$, as shown by

$$\mu_n(\mathbf{I}_n) = \mathbf{u}_n \in \mathcal{U}_n \qquad \text{for all } \mathbf{I}_n, n = 0, 1, \ldots, L - 1. \tag{5.33}$$

Such functions are said to constitute an *admissible policy*, for which, following (5.4), we write

$$\pi = \{\mu_0, \mu_1, \ldots, \mu_{L-1}\}. \tag{5.34}$$

Now, casting our minds back to (5.8) and adapting that equation for dealing with stochastic imperfect state-information problems as described herein, we would like to find an admissible policy $\pi$ that minimizes the cost-to-go function

$$J^\pi = \mathop{\mathbb{E}}_{\substack{\mathbf{x}_0, \boldsymbol{\omega}_n, \mathbf{v}_n \\ k=0,1,\ldots,L-1}} \left[ g_L(\mathbf{x}_L) + \sum_{k=0}^{L-1} g_k(\mathbf{x}_k, \mu_k(\mathbf{I}_k), \boldsymbol{\omega}_k) \right] \tag{5.35}$$

subject to the system equation written in an explicit form involving the information vector as well as the policy, as follows:

$$\mathbf{x}_{n+1} = \mathbf{a}_n(\mathbf{x}_n, \mu_n(\mathbf{I}_n), \boldsymbol{\omega}_n), \qquad n = 0, 1, \ldots, L-1; \tag{5.36}$$

the measurement equation is likewise written in the corresponding form:

$$\mathbf{y}_n = \mathbf{b}_n(\mathbf{x}_n, \mu_{n-1}(\mathbf{I}_{n-1}), \mathbf{v}_n), \qquad n = 1, \ldots, L-1, \tag{5.37}$$

with the initial condition

$$\mathbf{y}_0 = \mathbf{b}_0(\mathbf{x}_0, \mathbf{v}_0). \tag{5.38}$$

At this point in the discussion, it is important to note the way in which this mathematical statement differs from the *case of perfect state information* (Bertsekas, 2005):

In a perfect state-information problem, the aim is to find a policy that specifies the control $\mathbf{u}_n$ to be applied to the environment for each state $\mathbf{x}_n$ at time $n$. On the other hand, in an imperfect state-information problem, the aim is to find a policy that specifies the control to be applied to the environment for every possible information vector $\mathbf{I}_n$, which corresponds to every sequence of observables received and controls employed up to and including time $n$ for a total of $L$ iterations.

Indeed, it is this statement that distinguishes imperfect state-information problems from their perfect state-information counterparts.

## 5.5.2  Reformulation of the imperfect state-information problem as a perfect state-information problem

From the definition of the information vector $\mathbf{I}_n$ in (5.32), we readily infer the following recursion:

$$\mathbf{I}_{n+1} = [\mathbf{I}_n, \mathbf{y}_{n+1}, \mathbf{u}_n], \qquad n = 0, 1, \ldots, L-1, \tag{5.39}$$

where, as before, $\mathbf{I}_0 = \mathbf{y}_0$. According to (5.39), the information vector $\mathbf{I}_n$ is viewed as the state pertaining to a "reformulated system" with perfect information about the environment; $\mathbf{u}_n$ is the control applied to the environment at time $n$ and the observable $\mathbf{y}_{n+1}$ acts merely as a "disturbance" in evolution of the new state $\mathbf{I}_n$ across time $n$.

The conditional probability $\mathbb{P}[\mathbf{y}_{n+1} | \mathbf{I}_n, \mathbf{u}_n]$ may be expressed in the expanded form

$$\mathbb{P}[\mathbf{y}_{n+1} | \mathbf{I}_n, \mathbf{u}_n] = \mathbb{P}[\mathbf{y}_{n+1} | \mathbf{I}_n, \mathbf{u}_n, \mathbf{y}_0, \mathbf{y}_1, \ldots, \mathbf{y}_n], \tag{5.40}$$

the validity of which is justified by the following fact: the sequence $\{\mathbf{y}_k\}_{k=0}^{n}$ is, by definition, part and parcel of the information vector $\mathbf{I}_n$. In other words, the probability distribution of the observable $\mathbf{y}_{n+1}$, viewed as a "disturbance," is explicitly dependent on the new state $\mathbf{I}_n$ and the control $\mathbf{u}_n$ of the reformulated system, but not dependent on the prior sequence of disturbances $\{\mathbf{y}_k\}_{k=0}^{n}$.

Moreover, by writing the expectation

$$\mathbb{E}[g_n(\mathbf{x}_n, \mathbf{u}_n, \boldsymbol{\omega}_n)] = \mathbb{E}\left[\underset{\mathbf{x}_n, \boldsymbol{\omega}_n}{\mathbb{E}} [g_n(\mathbf{x}_n, \mathbf{u}_n, \boldsymbol{\omega}_n) | \mathbf{I}_n, \mathbf{u}_n]\right] \tag{5.41}$$

we may proceed in a manner similar to that just described and formulate the cost-to-go function for the new dynamic system described in (5.36)–(5.38). Specifically, the cost per stage of the system expressed as a function of the new state $\mathbf{I}_n$ and control $\mathbf{u}_n$ is given by the expectation

$$\tilde{g}_n(\mathbf{I}_n, \mathbf{u}_n) = \underset{\mathbf{x}_n, \boldsymbol{\omega}_n}{\mathbb{E}} [g_n(\mathbf{x}_n, \mathbf{u}_n, \boldsymbol{\omega}_n) | \mathbf{I}_n, \mathbf{u}_n]. \tag{5.42}$$

Accordingly, building on the new system equation (5.39) and the admissible policy of (5.34), we may now go on to write the following pair of equations for a problem with imperfect state information reformulated as a problem with perfect state information (Bertsekas, 2005):

$$J_{L-1}(\mathbf{I}_{L-1}) = \min_{\mathbf{u}_{L-1} \in \mathcal{U}_{L-1}} \left\{ \underset{\mathbf{x}_{L-1}, \boldsymbol{\omega}_{L-1}}{\mathbb{E}} [g_L(\mathbf{a}_{L-1}(\mathbf{x}_{L-1}, \mathbf{u}_{L-1}, \boldsymbol{\omega}_{L-1}))] \right.$$
$$\left. + g_{L-1}(\mathbf{x}_{L-1}, \mathbf{u}_{L-1}, \boldsymbol{\omega}_{L-1} | \mathbf{I}_{L-1}, \mathbf{u}_{L-1}) \right\}; \tag{5.43}$$

and for $n = 0, 1, \ldots, L-2$, we have

$$J_n(\mathbf{I}_n) = \min_{\mathbf{u}_n \in \mathcal{U}_n} \left\{ \underset{\mathbf{x}_n, \boldsymbol{\omega}_n, \mathbf{y}_{n+1}}{\mathbb{E}} [g_n(\mathbf{x}_n, \mathbf{u}_n, \boldsymbol{\omega}_n) + J_{n+1}(\mathbf{I}_n, \mathbf{y}_{n+1}, \mathbf{u}_n) | \mathbf{I}_n, \mathbf{u}_n] \right\}. \tag{5.44}$$

These two equations sum up the composition of a dynamic-programming algorithm for tackling problems with imperfect state information. To develop the optimal admissible policy, $\{\mu_0^*, \mu_1^*, \ldots, \mu_{L-1}^*\}$ for such a problem, we proceed as follows:

(1) Minimize the right-hand side of the terminal point in (5.43) for every possible value of the new state-information $\mathbf{I}_{L-1}$, obtaining the optimal policy $\mu_{L-1}^*(\mathbf{I}_{L-1})$.
(2) Simultaneously, compute the cost-to-go function $J_{L-1}(\mathbf{I}_{L-1})$ and use it to compute its next value $J_{L-2}(\mathbf{I}_{L-2})$, going backwards in time, via the minimizing equation (5.44); this computation is carried out for every possible value of $\mathbf{I}_{L-2}$.
(3) Similarly, proceed to compute $J_{L-3}(\mathbf{I}_{L-3})$ and $\mu_{L-3}^*$ and so on until the final value $J_0(\mathbf{I}_0) = J_0(\mathbf{y}_0)$ is determined.

Finally, the optimal cost-to-go function is given by the expectation

$$J^* = \underset{\mathbf{y}_0}{\mathbb{E}}[J_0(\mathbf{y}_0)] \tag{5.45}$$

In Chapter 6, on cognitive radar, we will illustrate how the dynamic-programming algorithm described in this section can be used to solve an imperfect state-information problem therein.

## 5.6 Reinforcement learning viewed as approximate dynamic programming

There is yet another practical issue with Bellman's dynamic programming that needs attention: its development assumes the availability of an explicit model that encompasses the transition probability from one state to another. Unfortunately, in many situations encountered in practice, such a model is not available. But, provided that a dynamic system is well structured and its state space has a manageable size, we may use *Monte Carlo simulation* to *explicitly estimate* the transition probabilities and corresponding observed transition costs; by its very nature, the estimation so performed is *approximate*. Accordingly, we say this kind of an approach to *approximate dynamic programming* is *direct*, because the use of simulations as described herein facilitates the "direct" application of dynamic programming methods.

Basically, the rationale behind the direct approximation of dynamic programming is to use simulations to generate *multiple system trajectories*, which lead to the construction of a *look-up table* with a separate entry for the value of each state. Naturally, the larger we make the number of system trajectories and, therefore, the size of the look-up table, the more reliable the simulation results will be at the expense of increased computational complexity. In particular, a separate variable $J(i)$ is kept in memory every time the state $i$ is visited by a trajectory of the simulated system. In so doing, we will have simulated a dynamic system with probabilistic transitions from state $i$ to state $j$ and generated the corresponding observed transition cost $g(i, j)$. Note, however, that it is quite likely in the course of simulation that some of the states are never visited; hence the reason for saying that the dynamic programming so performed is indeed approximate.

The stage is thus set for direct approximation of the two basic dynamic programming methods:[3]

- policy iteration, for which we obtain $Q$-learning, and
- value iteration, for which we obtain temporal-difference (TD) learning.

These two algorithms, discussed in the next two sections, are well known in the reinforcement learning literature for solving both finite-horizon and infinite-horizon problems. In other words, we may view reinforcement learning as *direct approximation of Bellman's dynamic programming* through the use of Monte Carlo simulations.

One final comment is in order. Naturally, the construction of a look-up table is *memory limited*. It follows, therefore, that the practical use of both algorithms, $Q$-learning and TD learning, is limited to situations where the state space is of *moderate dimensionality*.

## 5.7 Q-learning

To motivate the discussion of *Q-learning*, consider again Figure 5.1, which may also be viewed as the depiction of a reinforcement-learning system. The behavioral task of this system is how to find an optimal (i.e. minimal-cost) policy after trying out various possible sequences of actions, observing the state transitions that occur, and, finally, the corresponding observed transition costs. The policy used to generate such a behavior is called the *behavior policy*. This policy is to be distinguished from the *target policy*, the purpose of which is to *estimate* the value of a policy. With these two policies being separated from each other, $Q$-learning is said to be an *off-policy for control*. A side benefit gained from this separation may be stated as follows:

The target policy can be greedy, while the behavior policy is left to sample all the possible actions.

Thus, an off-policy method, exemplified by $Q$-learning, differs from an *on-policy method*, in that the value of a policy is being estimated while it is used for control at the same time.

To proceed with a derivation of the $Q$-learning algorithm, let

$$s_n = (i_n, a_n, j_n, g_n) \qquad (5.46)$$

denote a *four-tuple sample*, consisting of a trial action $a_n$ performed on state $i_n$ that results in a transition to the new state $j_n = i_{n+1}$ at an observed transition cost defined by

$$g_n = g(i_n, a_n, j_n), \qquad (5.47)$$

where $n$ denotes discrete time. Given such a scenario, we may now raise the following fundamental question:

Is there any on-line procedure for learning an optimal control policy through experiential interaction of a decision-maker with its environment, which is gained solely on the basis of observing samples of the form $s_n$ defined in (5.46) and (5.47)?

The answer to this question is an emphatic yes, and the solution is to be found in $Q$-learning (Watkins, 1989; Watkins and Dayan, 1992).

The *Q-learning algorithm* is an incremental dynamic-programming procedure that determines the optimal policy in a step-by-step manner. It is highly suited for solving Markov decision problems without explicit knowledge of the transition probabilities. However, successful use of $Q$-learning hinges on the assumption that the state of the environment is *fully observable*, which, in turn, means that the environment is a fully observable Markov chain.

As the name would imply, the $Q$-learning algorithm involves using the $Q$-factor defined in (5.23). Moreover, its derivation follows from (5.23) in light of Bellman's optimality equation (5.22). By so doing and using the expected observed cost $c(i, a)$ defined in (5.20) with $\mu(i) = a_i$, we may go on to write

$$Q^*(i, a) = \sum_{j=1}^{N} p_{ij}(a) \left[ g(i, a, j) + \gamma \min_{b \in \mathcal{A}_j} Q^*(j, b) \right] \qquad \text{for all } (i, a), \quad (5.48)$$

which can be viewed as a new version of Bellman's optimality equation. The solutions to the linear system of equations in (5.48) define the optimal $Q$-factors, $Q^*(i, a)$, uniquely for all state–action pairs $(i, a)$.

To simplify computational matters, we may use the idea of value iteration described in Section 5.4 in terms of the $Q$-factors to solve the linear system of equations in (5.48). Thus, for one iteration of the algorithm, we have

$$Q^+(i, a) = \sum_{j=1}^{N} p_{ij}(a) \left[ g(i, a, j) + \gamma \min_{b \in \mathcal{A}_j} Q(j, b) \right] \qquad \text{for all } (i, a),$$

where $Q^+(i, a)$ denotes the updated value of $Q(i, a)$. The *small step-size version* of this iteration is next formulated as the linear combination of two terms:

$$Q^+(i, a) = (1 - \eta)Q(i, a) + \eta \sum_{j=1}^{N} p_{ij}(a) \left[ g(i, a, j) + \gamma \min_{b \in \mathcal{A}_j} Q(j, b) \right] \quad \text{for all } (i, a),$$
(5.49)

where $\eta$ is a small *learning-rate parameter* that lies in the range $0 < \eta < 1$. In (5.49), simulations are used to generate *sample values* of the pair $(j, g(i, a, j))$ from the state–action pair $(i, a)$ in accordance with the transition probability $p_{ij}(a)$. In other words, the use of (5.49) requires the availability of prior knowledge in the form of a look-up table. Given this provision, we may then refer to (5.49) as the expectation form of $Q$-learning.

We may eliminate the need for a look-up table by formulating a *stochastic* version of (5.49). Specifically, the averaging performed in an iteration of (5.49) over all possible states is replaced by a *single sample equation*, resulting in an update formula for the $Q$-factor given as

$$Q_{n+1}(i, a) = (1 - \eta_n(i, a))Q_n(i, a) + \eta_n(i, a)[g(i, a, j) + \gamma J_n(j)] \quad \text{for } (i, a) = (i_n, a_n),$$
(5.50)

where

$$J_n(j) = \arg \min_{b \in \mathcal{A}_j} Q_n(j, b).$$
(5.51)

The index $j$ in (5.50) and (5.51) is the successor state to the state $i$ and, for the purpose of generality, $\eta_n(i, a)$ is a *learning-rate parameter* introduced at time step (i.e. iteration) $n$ for the state–action pair $(i, a)$. The update formula of (5.50) applies to the current state–action pair $(i_n, a_n)$, for which $j = j_n$ in accordance with (5.46). For all other admissible state–action pairs, the $Q$-factors remain unchanged, as shown by

$$Q_{n+1}(i, a) = Q_n(i, a), \quad \text{for all } (i, a) \neq (i_n, a_n).$$
(5.52)

Equations (5.50)–(5.52) constitute one iteration of the $Q$-*learning algorithm*.

The $Q$-learning algorithm may now be viewed in one of two equivalent ways:

- a Robbins–Monro stochastic approximation algorithm[4] or
- a combination of value iteration and Monte Carlo simulation.

In any event, the algorithm *backs up* the $Q$-factor for a *single* state–action pair at each iteration.

## 5.7.1 Summarizing remarks

(1) The $Q$-factor is defined on the basis of state–action pairs. It follows, therefore, that $Q$-learning is an *off-policy learning method* in the following twofold sense:
- First, the method learns about a *greedy* way of behaving in accordance with (5.51); this part of the method is called the *target policy*. Typically, but not always, the target policy is found to be an approximation to the optimal policy that is *deterministic*.

- Second, for its learning, the target policy uses data generated through *selective actions* in accordance with (5.50); this second part of the method is called the *behavior policy*. Typically, the behavior policy is *stochastic* because, as part of identifying the optimal policy, it involves having to *explore* all possible actions for each state.

(2) Despite the appearance of (5.50), the $Q$-learning algorithm is not a true gradient-descent method; consequently, compared with the least-mean-square (LMS) algorithm well known in adaptive filter theory (Haykin, 2002), it is *not as robust* (Baird, 1999).

## 5.8    Temporal-difference learning

The idea of TD learning was first described by Sutton (1984, 1988), preceding original formulation of the $Q$-learning algorithm by Watkins (1989). Just as the $Q$-factor plays a dominant role in the $Q$-learning algorithm, so it is with the notion of TD – to be defined – that plays a dominant role in formulation of the TD-learning algorithm.

Let $J(i_n)$ and $J(i_{n+1})$ denote the cost-to-go function at time steps $n$ and $n + 1$ respectively. Correspondingly, let the observed cost incurred through transition from time step $n$ to $n + 1$ be denoted by $g(i_n, i_{n+1})$. On this basis, the *TD error*, denoted by $d_n$, is defined by

$$d_n = g(i_n, i_{n+1}) + J(i_{n+1}) - J(i_n), \qquad n = 0, 1, \ldots, N - 1. \qquad (5.53)$$

For the *one-step* transition from state $i_n$ to $i_{n+1}$, (5.53) represents the difference between estimates of the cost-to-go function at two different times:

- The sum term, $g(i_n, i_{n+1}) + J(i_{n+1})$, represents the *sample observation* of the cost-to-go function resulting from the move from state $i_n$ to $i_{n+1}$.
- The remaining term, $J(i_n)$, represents the *current estimate* of the cost-to-go function for being in state $i$.

In effect, the TD error $d_n$ provides the "signal" to determine whether the current estimate $J(i_n)$ should be *increased* or *decreased*. Using the definition of (5.53), the one-step TD algorithm is expressed as follows:

$$J^+(i_n) = J(i_n) + \eta d_n, \qquad (5.54)$$

where $\eta$ is a learning-rate parameter, $J(i_n)$ is the *current estimate*, $J^+(i_n)$ is the *updated estimate*, and the product term $\eta d_n$ is the *correction* applied to the current estimate in order to produce the updated one. The one-step update rule of (5.54) is commonly referred to as the TD(0) *algorithm*, in which, as before, the abbreviation TD stands for "temporal difference."

To develop a physical insight into what goes on in TD(0), let

$$J(i_n') = g(i_n, i_{n+1}) + J(i_{n+1}),$$

in light of which we may depict a graphical representation of (5.53) and (5.54) as shown in Figure 5.4. According to this figure, we are attempting to "back up" the cost-to-go

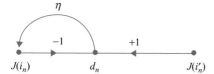

**Figure 5.4.** Signal-flow graph representation of the TD(0) algorithm; the figure illustrates the notion of sample backup, shown by the feedback labeled $\eta$ from $d_n$ to $J(i_n)$.

function of being in the state $i_n$ by an incremental amount equal to the TD error $d_n$ multiplied by the learning-rate parameter $\eta$. To be more precise, the current cost-to-go function for being in the previous state $i_n$ is adjusted so as to assume a new value closer to that of the subsequent state $i'_n$, namely $J(i'_n)$. This adjustment constitutes a *sample backup* from the state $i'_n$ to $i_n$.

### 5.8.1    Multistep TD learning

A more elaborate version of the iterative TD algorithm, also originated by Sutton (1988), incorporates a new parameter, denoted by $\lambda$, which lies in the range $0 < \lambda \leq 1$. To be specific, the algorithm is formulated as follows (Bertsekas and Tsitsiklis, 1996):

$$J^+(i_n) = J(i_n) + \eta \left( \sum_{k=n}^{\infty} \lambda^{k-n} d_k \right). \tag{5.55}$$

The expanded iteration algorithm of (5.55) is commonly referred to as TD($\lambda$). With $\lambda$ lying in the range $0 < \lambda \leq 1$, it plays the role of an exponential weighting factor that decreases in time. The summation term on the right-hand side of (5.55) is called the *multistep TD error*. In particular, we see that, at time step $n$, the correction in updating the current estimate $J(i_n)$ to the new value $J^+(i_n)$ involves an infinite summation of exponentially weighted values of the TD error $d_n$; hence the reference to (5.55) as the *infinite-horizon version* of TD learning.

To obtain the corresponding *finite-horizon version*, we may simply truncate the summation on the right-hand side of (5.55), obtaining

$$J^+(i_n) = J(i_n) + \eta \left( \sum_{k=n}^{T} \lambda^{k-n} d_k \right), \tag{5.56}$$

where $T$ is some prescribed time limit that defines the duration of an *episode*. By "episodes" we mean the subsequences into which the interaction between a decision-maker and its surrounding environment split naturally.

In both (5.55) and (5.56), we have ignored the use of discount factor $\gamma$. The inclusion of $\gamma$ in the learning process has the effect of replacing $\lambda$ in (5.55) and (5.56) with the product term $\lambda\gamma$. Therefore, we may look upon $\lambda$ in both equations (5.55) and (5.56) as an "algorithmic device" that could have exactly the same effect as $\gamma$. To this end, $\lambda$ is commonly set to be less than the desired $\gamma$ to produce a form of heuristic discounting (Powell, 2007).

## 5.8.2    Eligible traces

Sutton and Barto (1998) introduced the idea of *eligible traces* in the context of a *backward view* of TD learning, in which time is oriented backward. As illustrated in Figure 5.5a, the computation starts with state $i_n$, for example, then it proceeds to $i_{n-1}$, $i_{n-2}$, and so on. Accordingly, each state is associated with an additive memory variable, called the *eligibility trace*, that has a positive value. In the absence of a discount factor, for which a case was made in the preceding subsection, the eligibility traces for all states decay by a factor equal to $\lambda$. Specifically, the eligibility trace for state $i$ at time $n$ is defined by

$$e_n(i) = \begin{cases} \lambda e_{n-1} & \text{if } i \neq i_n \\ \lambda e_{n-1} + 1 & \text{if } i = i_n \end{cases}. \tag{5.57}$$

It is for this reason that $\lambda$ is appropriately called the *trace-decay parameter*. Moreover, the *accumulative* nature of the eligible trace may be explained as follows: each time a particular state is visited, the eligible trace accumulates in accordance with the number of visits made and then decays gradually when the visits end, as illustrated in Figure 5.5b.

The idea of eligible traces has practical significance: they indicate the degree to which each state is *eligible* to perform its learning process whenever a reinforcement event occurs. As pointed out previously, the occurrence of a TD error provides the signal for the estimate of the cost-to-go function to increase or decrease. To be specific, the incremental change in the cost-to-go function $J(i_n)$, pertaining to state $i$ visited at time $n$, is defined by

$$\Delta J(i_n) = \eta d_n e_n(i_n), \tag{5.58}$$

where $\eta$ is the learning-rate parameter and $d_n$ is the TD error.

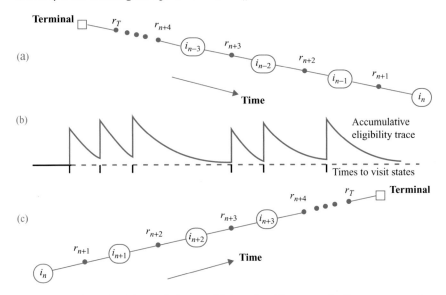

**Figure 5.5.** Backward and forward views ot TD learning. Note: an episodic sequence ends at the "Terminal." (a) Backward view of TD learning, with each update depending on the current TD error combined with eligibility of past traces. (b) Illustrating the idea of accumulative eligibility trace in backward view of TD learning. (c) Forward view of TD learning, with each state updated by looking forward to future rewards and states. Note that the time scales of parts (a), (b), and (c) are not intended to be coincident with each other.

Thus far, the discussion has focused on the backward view of TD learning. As an alternative, we may adopt a *forward view* of TD learning by orienting time in the forward direction, as illustrated in Figure 5.5c. In this third part of the figure, the computation starts with state $i_n$, then it proceeds to $i_{n+1}$, $i_{n+2}$, and so on. Unlike the backward view, there is naturally *no* memory associated with states in the forward view. Nevertheless, Sutton and Barto (1998) showed that the forward and backward views of TD learning are in actual fact equivalent.

However, these two views of TD learning are radically different from each other:

- The forward view is noncausal, which means that TD learning cannot be implemented in real time.
- On the other hand, the backward view is causal and TD learning is, therefore, implementable in real time. Just importantly, this viewpoint is simple to understand conceptually and also relatively simple to implement computationally. But it requires the use of memory.

### 5.8.3    Two limiting cases of TD learning

From a theoretical perspective, the eligibility trace for varying $\lambda$ in the range $[0, 1]$ provides a bridge from TD to Monte Carlo methods. To elaborate on this statement, consider the following two limiting cases:

(1) If we let $\lambda = 0$ and use the convention that $0^0 = 1$, then (5.55) reduces to

$$J^+(i_n) = J(i_n) + \eta d_n,$$

which is a repeat of (5.54). Indeed, it is for this reason that the algorithm of (5.54) is commonly referred to as TD(0), as pointed out previously. With $\lambda = 0$, the eligibility trace defined in (5.57) reduces to zero.

(2) For the other limiting case, if we let $\lambda = 1$, then (5.55) reduces to

$$J^+(i_n) = J(i_n) + \eta \sum_{k=0}^{N-n-1} d_{k+n},$$

which, except for the scaling factor $\eta$, is recognized as a *Monte Carlo simulation procedure* for updating the current estimate $J(i_n)$ to its new value $J^+(i_n)$. One practical disadvantage of the Monte Carlo method is the inability to learn from an *episode* until it is over.

There is one other perspective for viewing the eligibility trace. Specifically, we may think of it as a mechanism for dealing with the *temporal credit assignment problem* in the following sense. When a TD error occurs, credit or blame is assigned to states or actions that are eligible.

### 5.8.4    Summarizing remarks

(1) In contrast to $Q$-learning, which is an *off-policy* method well suited for control, TD learning is an *on-policy* method well suited for prediction; by "on-policy" we mean that it involves a single policy, namely the target policy.

(2) Through augmentation of TD learning with different values of the trace-decay parameter $\lambda$ inside the range [0, 1], we will have generated a family of methods that span a wide spectrum occupied by the one-step method (i.e. TD(0)) at one end of the spectrum and the Monte Carlo method (i.e. TD(1)) at the other end.

(3) As with $Q$-learning, the TD learning algorithm of (5.56) is *not* a true gradient-descent method, with the result that it is not as robust as the LMS algorithm whose formulation involves the use of stochastic gradients.

(4) In a neurobiological context, there is evidence that reward signals resulting from interactions of an animal with the environment are proceeded by midbrain neurons known as *dopamine neurons* (Schultz, 1998). Viewing these neurons as "a retina of the rewards system," the responses produced by dopamine neurons may be considered as teaching signals for TD($\lambda$) learning.

## 5.9    On the relationships between temporal-difference learning and dynamic programming

Previously, in Section 5.6, we stressed the fact that reinforcement learning, exemplified by TD learning, is an *approximate* form of dynamic programming. The essence of this approximation may be summed up as follows:

Dynamic programming makes full use of backups, whereas TD learning uses a sample backup.

Actually, the relationship between TD learning and dynamic programming is much deeper than just the issue of approximation, in that it manifests itself in other important ways. For example, in Section 5.11 dealing with a new generation of reinforcement learning algorithms and beyond, we will describe a ground-breaking algorithm, called linear GQ($\lambda$), which embodies ideas rooted in traditional reinforcement learning methods, but, most importantly, it looks to Bellman's equation of optimality (5.22) for the mathematical framework to formulate the objective function to be optimized.

However, there is an issue that needs attention, namely that of terminology. To elaborate, the terminology used in (5.54) for TD(0), the simplest form of TD learning, follows that in the classical dynamic-programming literature. Yet, TD learning was originally formulated as a reinforcement-learning algorithm. It would, therefore, be instructive to use TD learning as the basis for establishing the relationships between the dynamic-programming and reinforcement-learning terminologies.

To begin, the reinforcement-learning literature emphasizes the *reward function*, defined as

The reward function, denoted by $r_{n+1}$, is the reward (contribution) given to a decision-maker by the environment at time $n + 1$ when the decision-maker is in state $i_n$ for having taken an action on the environment.

In other words, the reward function indicates what is "good" in an immediate sense. The dynamic-programming function that corresponds to the reward function is the observed transition cost $g(i_n, i_{n+1})$.

The next important function to be considered in reinforcement learning is the *value function*. With the reward being typically random, the value function operating under policy $\pi$ is defined in terms of the reward function as follows:

$$V^{\pi}(i) = \mathbb{E}_{\pi}\left[\sum_{k=0}^{\infty} \gamma^k r_{n+k+1} \,|\, i_n = i\right], \tag{5.59}$$

where, as usual, $\gamma$ is the discount factor. In words, the value function $V^{\pi}(i)$ specifies what is "good" as a result of interactions with the environment for a long period of time.

Comparing (5.59) with (5.5), we see that the value function $V^{\pi}(i)$ in reinforcement learning corresponds to the cost-to-go function $J^{\pi}(i)$ in dynamic programming. More to the point, whereas $J^{\pi}(i)$ is *minimized* in dynamic programming, $V^{\pi}(i)$ is *maximized* in reinforcement learning.

Table 5.3 summarizes the correspondences between dynamic programming on the left and TD learning on the right. For the sake of interest, this table also includes the correspondences for the error in TD(0).

### 5.9.1 $\lambda$-return

However, there is another concept, called the *return*, widely used in the reinforcement learning literature, and for which there is no apparent counterpart in dynamic programming. The return, defined as some function of the reward sequence, signifies whether a reinforcement-learning task is episodic or continued, discounted or nondiscounted. To be specific, consider TD($\lambda$) for which the $\lambda$-return at time $n$ is defined in terms of a weighted sequence as follows (Sutton and Barto, 1998):

$$\begin{aligned} R_n^{(\lambda)} &= (1-\lambda)\sum_{k=1}^{\infty} \lambda^{k-1} R_n^{(k)} \\ &= (1-\lambda)(R_n^{(1)} + \lambda R_n^{(2)} + \lambda^2 R_n^{(3)} + \cdots + \lambda^{k-1} R_n^{(k)} + \cdots). \end{aligned} \tag{5.60}$$

The normalization factor $(1 - \lambda)$ is introduced in (5.60) to ensure that the sum of the weights, all the way from $k = 1$ to $k = \infty$, add up to unity. Moreover, from (5.60) we see that the weight *fades* by the amount $\lambda$ (i.e. trace-decay parameter) after each additional

**Table 5.3.** Correspondences between dynamic-programming and reinforcement-learning terminologies

| Dynamic programming | Reinforcement learning |
|---|---|
| Observed transition cost: $g(i_n, i_{n+1})$ | Immediate reward: $r_{n+1}$ |
| Cost-to-go function: $J(i_n)$ | Value function: $V(i_n)$ |
| TD(0) error: $g(i_n, i_{n+1}) + J(i_{n+1}) - J(i_n)$ | TD(0) error: $r_{n+1} + V(i_{n+1}) - V(i_n)$ |

time step. Note also that after a terminal step is reached at time $T$ in an episodic scenario, all the subsequent returns are equal to $R_n$, as shown by

$$R_n^{(\lambda)} = (1-\lambda) \sum_{k=1}^{T-n-1} \lambda^{k-1} R_n^{(k)} + \lambda^{T-n-1} R_n. \qquad (5.61)$$

Based on (5.61), we may distinguish two limiting cases:

(1) For $\lambda = 0$, for which $R_n^{(0)}$ reduces to one. That is, the $\lambda$-return for zero $\lambda$ is the same as the one-step TD method (i.e. TD(0)).
(2) For $\lambda = 1$, for which $R_n^{(1)}$ reduces to the conventional return $R_n$. That is, the $\lambda$-return for $\lambda = 1$ is the same as the Monte Carlo method.

These two limiting cases are in perfect accord with those discussed in Section 5.8.

## 5.10    Linear function approximations of dynamic programming

Typically, large-scale cognitive dynamic systems have a state space of high dimensionality. As a consequence, when we deal with such systems we may experience the *curse-of-dimensionality problem*, which, as discussed previously in Chapter 4, refers to the exponential growth of computational complexity with increasing dimensionality of the state space. In situations of this kind, we are compelled to abandon the idealized notion of optimality and be content with a suboptimal solution. In effect, *performance optimality is traded off for computational tractability*. This kind of strategy is precisely what the human brain does on a daily basis: faced with a difficult decision-making problem, the brain provides a suboptimal solution that is the "best" in terms of reliability and available resource allocation. Inspired by the human brain, we may go on to make the following statement in solving difficult dynamic programming problems (Werbos, 2004):

Do as well as possible, and not more.

With suboptimality in mind, several *approximate dynamic-programming algorithms* have been described in the literature. In most cases of practical interest, they all follow a linear approach to function approximation. The basic idea behind the *linear approach* is summed up as follows:

Algorithmic iterations are performed within a chosen subspace whose dimensionality is lower than that of the original space of interest and, in mathematical terms, the subspace is spanned by a prescribed set of basis functions.

The basis functions are commonly referred to as *feature vectors*. The adoption of a linear approach for dynamic-programming approximation is motivated by the fact that it is insightful, easy to formulate and analyze in mathematical terms, and equally easy to implement in computational terms. The iterations performed in formulating the approximate dynamic-programming algorithm may be value iterations, in which case the original space of interest is the state space. Alternatively, the iterations may be policy iterations, in which case the original space to be approximated is a joint state–action space.

During the past two decades, much has been written on approximate dynamic pro-gramming using the linear approach. In this context, we may identify two schools of thought in addressing the issue of adaptation:

- one using the method of least squares and
- the other using the method of gradient descent.

For reasons that will become apparent in the next section, we will confine the discussion to two recent contributions to the literature based on the method of steepest descent, as summarized here:

(1) *GQ($\lambda$) for prediction* (Maei and Sutton, 2010). This first algorithm is a predictive-learning algorithm of the off-policy kind, embracing target policy as well as behavior policy. It is applicable to problems required to connect low-level experience to high-level representations:
  - *Low-level experience* refers to rich signals received back and forth between a cognitive dynamic system (e.g. cognitive radar) and the surrounding environment.
  - *High-level representations* refer to experiential knowledge gained from succes-sive interactions of the system with the environment.
(2) *Greedy GQ for control* (Maei *et al.*, 2010). This second algorithm is an extension of GQ($\lambda$), in which the target policy is *greedy* with respect to a linear function approxi-mation of the $Q$-factor: $Q(i, a)$ for state–action pair $(i, a)$.

The GQ($\lambda$) and Greedy GQ algorithms constitute two ground-breaking methods in a new family of *modern reinforcement-learning* algorithms that are collectively called *linear GQ methods*. Details of the GQ($\lambda$) for predictive learning are presented in the next section. Highlights of the greedy GQ for control are presented in Section 5.12 "Summary and discussion," at the end of the chapter.

## 5.11 Linear GQ($\lambda$) for predictive learning

The GQ($\lambda$) algorithm, first described by Maei and Sutton (2010), distinguishes itself from the traditional $Q$-learning and TD-learning algorithms in three important ways:

(1) Unlike $Q$-learning and TD learning, when it comes to adaptation, the GQ($\lambda$) algo-rithm exploits the *method of gradient descent* in exactly the same way as it is done in the LMS algorithm, known for its computational simplicity as well as robustness (Haykin, 2002). So, the letter G in GQ($\lambda$) is attributed to gradient descent.
(2) The letter Q in GQ($\lambda$) indicates that the GQ($\lambda$) algorithm is an extension of $Q$-learning. Accordingly, the GQ($\lambda$) algorithm inherits the unique characteristic of *off-policy learning*, whereby the target policy, denoted by $\pi$, takes care of policy estimation using data generated by the behavior policy, denoted by $b$, through selec-tive actions on the environment.
(3) The use of trace-decay parameter $\lambda$ in the argument of GQ($\lambda$) is intended to remind us of the fact that, when function approximation is the issue of interest, one of the key features of TD learning is the ability to *generalize* predictions to states that

may not have been visited during the learning process. It follows, therefore, that TD learning is also a core ingredient in formulation of the GQ($\lambda$) algorithm.

From what has just been described, the GQ($\lambda$) algorithm has the making of a "universal" algorithm for *predictive learning*.

To be more to the point, we may go on to say that this new algorithm based on the linear approach has the built-in methodologies to address the challenge that we have been faced with for a long time:

In the context of predictive learning, how do we approximate Bellman's dynamic programming for large-scale applications in a computationally feasible and robust manner?

The GQ($\lambda$) algorithm appears to have the potential to rise to this challenge by offering four key attributes:

(1) The algorithm has the ability to *learn about temporally abstract predictions*, which, in turn, makes it possible to learn experientially grounded knowledge.
(2) The learning process is performed in an *on-line* manner.
(3) Just as importantly, the computational complexity of the algorithm *scales linearly* with the dimensionality of the approximating feature space, which is made possible through the adoption of gradient descent for adaptation.
(4) Guaranteed convergence of the GQ($\lambda$) algorithm to a fixed point in TD($\lambda$) with *probability one* is realized by the adoption of off-policy learning.

These four attributes, viewed together, are of profound practical importance. In particular, attributes (3) and (4) make all the difference when GQ($\lambda$) is compared with other linear approaches to approximate dynamic programming that resort to *second-order methods* for guaranteed convergence to a fixed point with probability one. Specifically, we may mention two algorithms:

- The least-squares temporal difference (LSTD($\lambda$)) algorithm described by Bradtke and Barto (1996).
- The least-squares policy evaluation (LSPE($\lambda$)) algorithm described by Bertsekas (2007).

Although these two algorithms are different in their approaches, they both rely on the method of least squares for adaptation. With dimensionality of the features space viewed as the frame of reference, the method of least squares scales according to a *square law*, whereas the method of gradient descent scales according to a *linear law* (Haykin, 2002).

With computational complexity being of overriding importance in large-scale applications, the conclusion to be drawn from the four attributes of linear GQ($\lambda$) and the follow-up remarks on attributes (3) and (4) may be summarized as follows:

With large-scale applications of approximate dynamic programming for predictive learning as the challenge and guaranteed algorithmic convergence as the requirement, in light of what we know currently, the linear GQ($\lambda$) algorithm is the preferred method of choice.

Now that we have an appreciation for practical benefits of the linear GQ($\lambda$), we may proceed with the task of deriving this novel algorithm.

## 5.11.1   Objective function setting the stage for approximation

First and foremost, the key element in deriving the linear GQ($\lambda$) is how to formulate an *off-line, $\lambda$-weighted objective function* that accounts for two of its three algorithmic ingredients: Q-learning and TD($\lambda$) learning.

To this end, we begin the formulation with Bellman's optimal equation for a Markov decision process, namely (5.22). Referring to the *operator notation* described in Note 1 at the end of the chapter, this equation is rewritten in the form

$$\mathcal{M}^{\pi}(\mathbf{j}) = \mathbf{c} + \mathbf{pj}, \tag{5.62}$$

where we have simplified matters by putting aside the discount factor $\gamma$. The vectors $\mathbf{c}$ and $\mathbf{j}$ respectively represent the expected observed cost $\{c(i, \mu(i))\}_{i=1}^{N}$ and the cost-to-go function $\{J^{\pi}(i)\}_{i=1}^{N}$, and the matrix $\mathbf{P}$ represents the transition probabilities $\{p(i, j)\}_{i,\ j=1}^{N}$; as usual, $N$ is the number of states. The "min" operator $\mathcal{M}^{\pi}(j)$ is *linear*, since the operations it performs are all additions and multiplications. For the task at hand, however, we find it more convenient to work with the much simplified definition

$$\mathbf{j} = \mathcal{T}^{\pi}\mathbf{j}, \tag{5.63}$$

where $\mathcal{T}^{\pi}$ is called the *Bellman operator for target policy $\pi$*.

We are still not quite where we would like to be in formulating the objective function for the linear GQ($\lambda$). To get there, let $\mathbf{q}$ denote the vector of Q-factors under action $a$, as shown by

$$\mathbf{q} = \begin{bmatrix} Q(1,a) \\ Q(2,a) \\ \vdots \\ Q(N,a) \end{bmatrix}. \tag{5.64}$$

In a corresponding way to (5.63), we may define the *off-policy, $\lambda$-weighted version of the Bellman operator for target policy $\pi$* as follows:

$$\mathbf{q} = \mathcal{T}_{\lambda\beta}^{\pi}\mathbf{q}, \tag{5.65}$$

where the two subscripts ($\lambda$ for *trace-decay parameter* and $\beta$ for *termination probability of the target policy $\pi$*) are intended to remind us that (5.65) pertains to an off-policy learning algorithm with $\lambda$-weighting. Two noteworthy points:

(1) By definition, the whole term $\mathcal{T}_{\lambda\beta}^{\pi}\mathbf{q}$ involves expectations in the form of transition probabilities; therefore, we may perform *direct sampling* on this term and obtain $\lambda$-returns.

(2) If we were to put the two parameters $\lambda$ and $\beta$ aside, then (5.65) may be viewed as the Bellman operator version of (5.48).

Thus, equipped with the definition in (5.65), the stage is now set for introducing feature vectors into the discussion, as shown by the $N$-by-$s$ matrix

$$\mathbf{\Phi} = \begin{bmatrix} \boldsymbol{\phi}^{\mathrm{T}}(1,\ a) \\ \boldsymbol{\phi}^{\mathrm{T}}(2,\ a) \\ \vdots \\ \boldsymbol{\phi}^{\mathrm{T}}(N,\ a) \end{bmatrix}, \tag{5.66}$$

where $s$ denotes the dimension of feature vector $\phi(i)$ for state $i$ that extends from 1 to $N$. The *approximate Q-factor* for state–action pair $(i, a)$ is defined by the inner product

$$Q_{\mathbf{w}}(i) = \phi^{\mathrm{T}}(i)\mathbf{w} \qquad i = 1, 2, \ldots, N, \qquad (5.67)$$

which includes the use of a *weight vector* $\mathbf{w}$ introduced herein for reasons that will become apparent momentarily. Equivalently, we may express the vector of approximate $Q$-factors in the compact matrix form

$$\mathbf{q}_{\mathbf{w}} = \begin{bmatrix} Q_{\mathbf{w}}(1, a) \\ Q_{\mathbf{w}}(2, a) \\ \vdots \\ Q_{\mathbf{w}}(N, a) \end{bmatrix} \qquad (5.68)$$
$$= \Phi\mathbf{w}.$$

The weight vector $\mathbf{w}$ is to be *adapted* in such a way that the $\lambda$ returns provide the "best" fit to the actual vector $\mathbf{q}$ in some statistical sense.

Following the format of (5.65), we may equally well define the *off-policy, $\lambda$-weighted version of the Bellman operator* for the approximate vector $\mathbf{q}_{\mathbf{w}}$, as shown by

$$\mathbf{q}_{\mathbf{w}} = \mathcal{T}_{\lambda\beta}^{\pi}\mathbf{q}_{\mathbf{w}}, \qquad (5.69)$$

which provides the basis to formulate the objective function that we have been seeking for the linear GQ($\lambda$). To elaborate on (5.69), the operator $\mathcal{T}_{\lambda\beta}^{\pi}$ takes the approximate vector $\mathbf{q}_{\mathbf{w}}$ as input and for each state–action pair it returns the expected corrected $\mathbf{q}_{\mathbf{w}}$ as output under the following premise: the Markov decision process starts on the state–action pair under study, actions are taken according to policy $\pi$, and the vector $\mathbf{q}_{\mathbf{w}}$ is used to correct the return truncations.

To provide a Euclidean basis for how close the corrected vector $\mathcal{T}_{\lambda\beta}^{\pi}\mathbf{q}_{\mathbf{w}}$ is to the approximate vector $\mathbf{q}_{\mathbf{w}}$, we introduce a matrix $\Pi$ that *projects* points in the $Q$-factor space into the linear space occupied by $\mathbf{q}_{\mathbf{w}}$. The stage is now set for formulating the objective function as an off-policy, $\lambda$-weighted version of the Bellman error function (Maei and Sutton, 2010):

$$\mathcal{E}(\mathbf{w}_n) = \left\| \mathbf{q}_{\mathbf{w},n} - \Pi\mathcal{T}_{\lambda\beta}^{\pi}\mathbf{q}_{\mathbf{w},n} \right\|_{\mathbf{D}}^2, \qquad (5.70)$$

where $n$ denotes discrete time and the matrix $\mathbf{D}$ is an $N$-by-$N$ diagonal matrix whose only nonzero diagonal entries correspond to *relative frequency* (in probabilistic terms) with which each state–action pair is visited under the behavior policy $b$. The squared Euclidean norm in (5.70) is defined by the *quadratic form*

$$\left\| \mathbf{v}_n \right\|^2 = \mathbf{v}_n^{\mathrm{T}}\mathbf{D}\mathbf{v}_n,$$

where

$$\mathbf{v}_n = \mathbf{q}_{\mathbf{w},n} - \Pi\mathcal{T}_{\lambda\beta}^{\pi}\mathbf{q}_{\mathbf{w},n}.$$

Figure 5.6 illustrates the comparison of two different criteria:

(1) One criterion is based on the objective function of (5.70), involving the composite operator $\Pi \mathcal{T}_{\lambda\beta}^{\pi}$ that consists of the off-policy, $\lambda$-weighted version of the Bellman operator $\mathcal{T}_{\lambda\beta}^{\pi}$ multiplied by the projection matrix $\Pi$.
(2) The other criterion is based on a simpler objective function that involves the operator $\mathcal{T}_{\lambda\beta}^{\pi}$ by itself.

This figure clearly illustrates superiority of the criterion based on $\Pi \mathcal{T}_{\lambda\beta}^{\pi}$ as in (5.70) over the simpler criterion; hence the justification for concentrating on the first criterion.

The symbol $\mathcal{E}$ used to denote the objective function in (5.70) is intended to distinguish it from the symbol $J$ used previously to denote the cost-to-go function.

Resuming the discussion on the objective function in (5.70), we find it desirable to reformulate this objective function in a way that would involve the use of statistical expectations. This reformulation is doable by proceeding in two stages.

> *Stage 1.* In this first stage, the objective function is expressed as a quadratic form defined in terms of two expectations. The outer expectations in the quadratic form are conditional on the target policy $\pi$ and the expectation in the middle is conditional on the behavior policy $b$. A drawback of this quadratic form, however, is that it is difficult to work with expectations conditional on the target policy.
> *Stage 2.* This second stage overcomes the difficulty by conditioning the statistical expectations solely on the behavior policy.

Thus, the objective function in (5.70) takes the desired quadratic form (Maei and Sutton, 2010):

$$\mathcal{E}(\mathbf{w}_n) = \mathbb{E}_b[\delta_{\text{off},n}\boldsymbol{\phi}_n^{\mathrm{T}}]\mathbb{E}_b[\boldsymbol{\phi}_n\boldsymbol{\phi}_n^{\mathrm{T}}]^{-1}, \mathbb{E}_b[\delta_{\text{off},n}\boldsymbol{\phi}_n], \qquad (5.71)$$

where we have introduced the new term $\delta_{\text{off}}$, denoting the *off-policy version of the multistep TD error*, and $\mathbb{E}_b$ denotes the expectation with respect to the behavior policy. The error $\delta_{\text{off}}$ is dependent on the trace-decay parameter $\lambda$, termination probability $\beta$, and a new parameter $\rho$ resulting from stage 2 of reformulating the objective function.

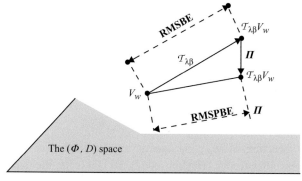

**Figure 5.6.** Illustrating the comparison between two criteria for selecting the objective function of GQ($\lambda$). RMSBE: root mean-square of Bellman error; RMSPBE: root mean-square *projected* Bellman error. The term $V_w$ denotes the value function for any $\mathbf{w}$ and the $\mathbf{q}_w$ in (5.68) lies in the ($\Phi$, $\mathbf{D}$) space.

## 5.11.2    The GQ($\lambda$) algorithm

To proceed with derivation of the GQ($\lambda$) algorithm, we now look to the third ingredient of the linear GQ($\lambda$): the *method of gradient descent*, the application of which requires evaluation of the *gradient vector* defined as the partial derivative of the objective function $\mathcal{E}(\mathbf{w}_n)$ with respect to the weight vector $\mathbf{w}_n$. Specifically, using (5.71), we obtain

$$\frac{1}{2}\nabla\mathcal{E}(\mathbf{w}_n) = \frac{1}{2}\nabla(\mathbb{E}_b[\delta_{\mathrm{off},n}\boldsymbol{\phi}_n^{\mathrm{T}}]\mathbb{E}_b[\boldsymbol{\phi}_n\boldsymbol{\phi}_n^{\mathrm{T}}]^{-1}\mathbb{E}_b[\delta_{\mathrm{off},n}\boldsymbol{\phi}_n]),$$

where the scaling factor 1/2 has been introduced merely for mathematical convenience and $\nabla$ denotes the *gradient operator*. Bearing in mind that

- $\delta_{\mathrm{off},n}$ is the only term dependent on the weight vector $\mathbf{w}_n$,
- the scaling factor 1/2 is canceled as there is a pair of $\delta_{\mathrm{off}}$ to be differentiated, and
- the order of gradient operator $\nabla$ and expectation operator $\mathbb{E}_b$ is interchangeable, as they are both linear,

we may then go on to write

$$\frac{1}{2}\nabla\mathcal{E}(\mathbf{w}_n) = \mathbb{E}_b[\nabla\delta_{\mathrm{off},n}\boldsymbol{\phi}_n^{\mathrm{T}}]\mathbb{E}_b[\boldsymbol{\phi}_n\boldsymbol{\phi}_n^{\mathrm{T}}]^{-1}\mathbb{E}_b[\delta_{\mathrm{off},n}\boldsymbol{\phi}_n]. \tag{5.72}$$

By definition, the off-policy version of the multistep TD error is related to the corresponding off-policy $\lambda$-return, denoted by $R_{\mathrm{off},n}$, as follows:

$$\delta_{\mathrm{off},n} = R_{\mathrm{off},n} - \mathbf{w}^{\mathrm{T}}\boldsymbol{\phi}_n. \tag{5.73}$$

The inner product $\mathbf{w}_n^{\mathrm{T}}\boldsymbol{\phi}_n$ is an estimate of the value function, which is meant to equal the off-policy $\lambda$-return in expected value. Hence, applying the gradient operator $\nabla$ to the error $\delta_{\mathrm{off},n}$ in (5.73) with respect to $\mathbf{w}$, yields the following relationship between $\nabla\delta_{\mathrm{off},n}$ and $\nabla R_{\mathrm{off},n}$:

$$\nabla\delta_{\mathrm{off},n} = \nabla R_{\mathrm{off},n} - \boldsymbol{\phi}_n. \tag{5.74}$$

Thus, substituting (5.74) into (5.72) and simplifying terms, we obtain

$$\frac{1}{2}\nabla\mathcal{E}(\mathbf{w}_n) = -(\mathbb{E}_b[\boldsymbol{\phi}_n\boldsymbol{\phi}_n^{\mathrm{T}}] + \mathbb{E}_b[\nabla R_{\mathrm{off},n}\boldsymbol{\phi}_n^{\mathrm{T}}])\mathbb{E}_b[\boldsymbol{\phi}_n\boldsymbol{\phi}_n^{\mathrm{T}}]^{-1}\mathbb{E}_b[\delta_{\mathrm{off},n}\boldsymbol{\phi}_n] \tag{5.75}$$

$$= -\mathbb{E}_b[\delta_{\mathrm{off},n}\boldsymbol{\phi}_n] + \mathbb{E}_b[\nabla R_{\mathrm{off},n}\boldsymbol{\phi}_n^{\mathrm{T}}]\mathbb{E}_b[\boldsymbol{\phi}_n\boldsymbol{\phi}_n^{\mathrm{T}}]^{-1}\mathbb{E}_b[\delta_{\mathrm{off},n}\boldsymbol{\phi}_n].$$

To simplify the formula of (5.75), we now introduce a *secondary weight vector* defined as

$$\mathbf{w}_{\mathrm{sec},n}^{*} = \mathbb{E}_b[\boldsymbol{\phi}_n\boldsymbol{\phi}_n^{\mathrm{T}}]^{-1}\mathbb{E}_b[\delta_{\mathrm{off},n}\boldsymbol{\phi}_n]. \tag{5.76}$$

The mathematical structure of $\mathbf{w}_{\mathrm{sec},n}$ is exactly similar to that of the well-known *Wiener filter* in linear filter theory (Haykin, 2002). To elaborate, the expectation term $\mathbb{E}_b[\boldsymbol{\phi}_{\mathrm{off},n}\boldsymbol{\phi}_n^{\mathrm{T}}]$ represents the *correlation matrix* of the state–action feature vector $\boldsymbol{\phi}_n$, being viewed as an "input vector." The second expectation term $\mathbb{E}_b[\delta_{\mathrm{off},n}\boldsymbol{\phi}_n]$ represents the *cross-correlation vector* of $\boldsymbol{\phi}_n$ with the TD error $\delta_{\mathrm{off},n}$, being viewed as the "desired response." Thus, the use of the secondary weight vector $\mathbf{w}^*_{\text{sec},n}$ in (5.75) results in the much simplified expression

$$\frac{1}{2}\nabla\mathcal{E}(\mathbf{w}_n) = -\mathbb{E}_b[\delta_{\text{off},n}\boldsymbol{\phi}_n] + \mathbb{E}_b[\nabla R_{\text{off},n}\boldsymbol{\phi}_n^{\mathrm{T}}]\mathbf{w}^*_{\text{sec},n}.\qquad(5.77)$$

Derivation of the gradient vector in (5.77) is based on the forward view of TD learning. From a practical perspective, it is instructive to convert the forward view to its backward-view equivalent that is more convenient for low-memory mechanistic implementation. (The forward and backward views of TD learning were discussed in Section 5.8.) With this conversion in mind, we make use of the following expectation pair of TD forward-view and backward-view equivalences (Maei and Sutton, 2010):

(1)

$$\mathbb{E}_b[\delta_{\text{off},n}\boldsymbol{\phi}_n] = \mathbb{E}_b[\delta_n\mathbf{e}_n],\qquad(5.78)$$

where $\delta_n$ is the one-step TD error and the eligibility trace vector $\mathbf{e}_n$ is defined by

$$\mathbf{e}_n = \boldsymbol{\phi}_n + (1 - \beta_n)\lambda_n\rho_n\mathbf{e}_{n-1}.\qquad(5.79)$$

(2)

$$\mathbb{E}_b[\nabla R_{\text{off},n}\boldsymbol{\phi}_n^{\mathrm{T}}] = \kappa_{n+1}\mathbb{E}_b[\bar{\boldsymbol{\phi}}_{n+1}\mathbf{e}_n^{\mathrm{T}}],\qquad(5.80)$$

where

$$\kappa_n = (1 - \beta_n)(1 - \lambda_n).\qquad(5.81)$$

The vector

$$\bar{\boldsymbol{\phi}}_n = \sum_a \pi(i_n, a)\boldsymbol{\phi}(i_n, a)\qquad(5.82)$$

describes the composite feature vector $\boldsymbol{\phi}(i_n, a)$ weighted by the corresponding target policy $(i_n, a)$ over the entire action space $\mathcal{A}$ at time $n$. Thus, substituting the pair of equivalences in (5.78) and (5.80) into the formula for the gradient vector of (5.77), we get

$$\frac{1}{2}\nabla\mathcal{E}(\mathbf{w}_n) = -\mathbb{E}_b[\delta_n\mathbf{e}_n] + \kappa_{n+1}\mathbb{E}_b[\bar{\boldsymbol{\phi}}_{n+1}\mathbf{e}_n^{\mathrm{T}}]\mathbf{w}^*_{\text{sec},n}.\qquad(5.83)$$

However, a practical difficulty in using the gradient vector of (5.83) is the fact the two expectations embodied in this formula are typically not calculable in practice, and so it is with the secondary weight vector $\mathbf{w}^*_{\text{sec},n}$. In situations of this kind, we may follow the normal procedure adopted in adaptive filter theory (Widrow and Stearns, 1985; Haykin, 2002):

Abandon the method of gradient descent in favor of its stochastic version.

That is, in the context of (5.83), we do the following: remove the two expectations $\mathbb{E}_b$ with respect to the behavior policy. Thus, using the symbol $\mathbf{g}(\mathbf{w}_n)$ for the stochastic gradient vector, we may write

$$\mathbf{g}(\mathbf{w}_n) = -\delta_n\mathbf{e}_n + \kappa_{n+1}\bar{\boldsymbol{\phi}}_{n+1}\mathbf{e}_n^{\mathrm{T}}\mathbf{w}_{\text{sec},n},\qquad(5.84)$$

which involves both $\mathbf{w}_n$ and $\mathbf{w}_{\text{sec},n}$.

To apply gradient descent, we obviously use the *negative* of the stochastic gradient vector $g(w_n)$ to compose the correction term in the adaptation rule for the primary weight vector. Accordingly, we may express this rule as follows:

$$\begin{aligned}
\mathbf{w}_{n+1} &= \mathbf{w}_n - \eta_{\mathbf{w},n}\mathbf{g}(\mathbf{w}_n)\\
&= \mathbf{w}_n + \eta_{\mathbf{w},n}(\delta_{\mathbf{w},n}\mathbf{e}_{off,n} - \kappa_{n+1}\mathbf{e}_n^{\mathrm{T}}\mathbf{w}_{sec,n}\bar{\boldsymbol{\phi}}_{n+1}),
\end{aligned}$$

(5.85)

where $\eta_{\mathbf{w},n}$ is the first of two learning-rate parameters in GQ($\lambda$).

To formulate the corresponding adaptation rule for the secondary weight vector, we may proceed as follows:

(1) As pointed out previously, the secondary weight vector in (5.76) may be viewed as a Wiener filter with the state–action feature vector $\boldsymbol{\phi}_n$ as the input vector and $\delta_{off,n}$ as the desired response. Such a filter is realized by *minimizing the mean-square error*

$$\frac{1}{2}\mathbb{E}_b(\delta_{off,n} - \mathbf{w}_{sec,n}^{\mathrm{T}}\boldsymbol{\phi}_n)^2.$$

The stochastic gradient vector for the secondary weight vector is therefore defined by

$$-\delta_{off,n}\boldsymbol{\phi}_n + (\mathbf{w}_{sec,n}^{\mathrm{T}}\boldsymbol{\phi}_n)\boldsymbol{\phi}_n.$$

Following LMS filter theory (Haykin, 2002), we may formulate the adaptation rule:

$$\mathbf{w}_{sec,n+1} = \mathbf{w}_{sec,n} + \eta_{\mathbf{w}_{sec,n}}(\delta_{off,n}\boldsymbol{\phi}_n - (\mathbf{w}_{sec,n}^{\mathrm{T}}\boldsymbol{\phi}_n)\boldsymbol{\phi}_n).$$

(5.86)

(2) Referring to the stochastic version of the equivalence in (5.78), we may make two observations for the secondary weight vector:
- $\delta_{off,n}\boldsymbol{\phi}_n$ is the forward view of TD update and
- $\delta_n\mathbf{e}_n$ is the corresponding backward view of TD update.

Thus, substituting $\delta_n\mathbf{e}_n$ for $\delta_{off,n}\boldsymbol{\phi}_n$ in (5.86), we obtain the desired update rule for the secondary weight vector:

$$\mathbf{w}_{sec,n+1} = \mathbf{w}_{sec,n} + \eta_{\mathbf{w}_{sec,n}}(\delta_n\mathbf{e}_n - (\mathbf{w}_{sec,n}^{\mathrm{T}}\boldsymbol{\phi}_n)\boldsymbol{\phi}_n),$$

(5.87)

where $\eta_{\mathbf{w}_{sec,n}}$ is the second one of the two learning-rate parameters in GQ($\lambda$).

## 5.11.3    Weight-doubling trick

The pair of adaptation rules of (5.85) and (5.87) is unique to approximate dynamic programming: the combination of these two rules working together is referred to as the *weight-doubling trick* or *Sutton trick*, acknowledging its originator.

To elaborate, the main trick needed to make gradient-TD methods work in linear complexity is the introduction of a secondary weight vector that is updated in parallel with the primary weight vector for the approximate $Q$-function. The secondary weight vector is updated according to the classical LMS algorithm using the state–action feature vector as input and the one-step TD error as the desired output. The actual output, linear in the features, becomes an estimate of the expected TD error in the presence of the given feature vector. This estimate is then used in the update of the primary weight vector, such that this update is in the direction of the gradient of mean-square projected Bellman error.

### 5.11.4 Eligibility traces vector

The idea of eligibility traces in TD learning was discussed in Section 5.8. Equation (5.79) expands on that idea by defining the GQ($\lambda$)'s update rule for the eligibility traces vector $\mathbf{e}_n$ for state–action pair $(i_n, a_n)$ at time $n$. The vector $\mathbf{e}_n$ represents a sort of memory variable associated with each state–action pair. On each step at time $n$, this memory variable is incremented by the amount $\boldsymbol{\phi}_n$ for state–action pair $(i_n, a_n)$ that is just visited and is decayed by the amount of $\gamma_n \lambda_n \rho_n$, where the discount factor $\gamma_n = 1 - \beta_n$. The reason we need these traces is because in the forward view we could not store the computed error updates – simply put, it is not practical. By the forward view to backward view transition trick, eligibility traces help us to estimate the error updates based on the whole history of input data. The eligibility traces are particularly helpful to bridge temporal gaps when the experience is processed at a temporally fine resolution.

### 5.11.5 New action–state feature vector

The $\overline{\boldsymbol{\phi}}_n$ is an action–state feature vector for state $i_n$ weighted by the corresponding target policy over the entire action space; thereby, the resulting vector is independent of any action. The reason we have it in the GQ($\lambda$) algorithm is because – before we do any direct sampling from the gradient term ($-1/2$ gradient of the projected Bellman error) and come up with stochastic updating rules – expectation of an update term is over the target policy for actions taken from the next state.

### 5.11.6 Summarizing remarks

Table 5.4 provides a summary of the GQ($\lambda$) algorithm. In a related context, it is instructive that we describe the meaning of each of the related parameters, namely $\beta$, $\kappa$, $r$, $z$, $\rho$, and $\lambda$ as follows.

*Parameter $\beta$* is called the *probability of termination* from a given state; it lies in the range $[0, 1]$. Moreover, the difference term, $1 - \beta$, plays the role of discount factor $\gamma$. This relationship between $\beta$ and $\gamma$ tells us that the time-varying discount factor $\gamma_n$ may be considered as the "continuation probability."

*Parameter $\kappa$* is defined in terms of the termination probability $\beta$ and trace-decay parameter $\lambda$ as follows:

$$\kappa_n = (1 - \beta_n)(1 - \lambda_n).$$

With $0 \le \beta \le 1$ and $0 \le \lambda \le 1$, it follows that $0 \le \kappa \le 1$ for all $n$.

*Off-policy reward $r$.* Suppose the decision-maker takes action $a$ at state $i$ according to behavior policy $b$; and it receives the reward $r$ and moves to the next state $i'$. In so doing, nothing on target policy $\pi$ has been used. It is from the state $i'$ that the target policy $\pi$ is considered. In other words, in linear GQ($\lambda$), the reward $r$ is independent of the target policy $\pi$.

*Relation of $z$ to $r$.* In contrast to $r$, which plays the role of *transient signals* while the decision-maker is behaving, the parameter $z$ is the final outcome of target policy $\pi$.

**Table 5.4** Summary of the GQ($\lambda$) algorithm

*Initialization of the algorithm*
- The primary weight vector $\mathbf{w}$ is initialized in an arbitrary manner.
- The secondary weight vector $\mathbf{w}_{\text{sec}}$ is initialized to zero.
- The auxiliary memory vector $\mathbf{e}$ is also initialized to zero.

*Defining parameters of the algorithm*

$$\rho_n = \frac{\pi(i_n, a_n)}{b(i_n, a_n)},$$

where $(i_n, a_n)$ denotes the state–action pair at time $n$.

$$\kappa_n = (1 - \beta_n)(1 - \lambda_n),$$

where $0 \leq \beta_n \leq 1$ and $0 \leq \lambda_n \leq 1$ for all $n$.

$\eta_{\mathbf{w},n}$ = fixed or decreasing learning-rate parameter for updating the primary weight vector $\mathbf{w}_n$.
$\eta_{\mathbf{w}_{\text{sec}},n}$ = fixed or decreasing learning-rate parameter for updating the secondary weight vector $\mathbf{w}_{\text{sec},n}$.

*Feature vectors summed over the action space*

$$\overline{\boldsymbol{\phi}}_n = \sum_a \pi(i_n, a) \boldsymbol{\phi}(i_n, a)$$

*One-step TD error*

$$\delta_n = r_{n+1} + \beta_{n+1} z_{n+1} + (1 - \beta_{n+1}) \mathbf{w}_n^\mathrm{T} \overline{\boldsymbol{\phi}}_{n+1} - \mathbf{w}_n^\mathrm{T} \boldsymbol{\phi}_n$$

*Update rules of the algorithm*
1. Primary weight vector

$$\mathbf{w}_{n+1} = \mathbf{w}_n + \eta_{\mathbf{w},n}(\delta_n \mathbf{e}_n - \kappa_{n+1}(\mathbf{e}_n^\mathrm{T} \mathbf{w}_{\text{sec},n}) \overline{\boldsymbol{\phi}}_{n+1})$$

2. Secondary weight vector

$$\mathbf{w}_{\text{sec},n+1} = \mathbf{w}_{\text{sec},n} + \eta_{\mathbf{w}_{\text{sec}},n}(\delta_n \mathbf{e}_n - (\boldsymbol{\phi}_n^\mathrm{T} \mathbf{w}_{\text{sec},n}) \boldsymbol{\phi}_n)$$

3. Eligibility trace vector
$$\mathbf{e}_n = \boldsymbol{\phi}_n + (1 - \beta_n) \lambda_n \rho_n \mathbf{e}_{n-1}$$

*Parameter* $\rho$ represents an *importance sampling ratio*. More importantly, for any given state and action pair $(i, a)$, it is defined as the ratio of
- the probability of taking action $a$ according to the target policy $\pi$ to
- the behavior policy $b$, given the state $i$.

That is,

$$\rho = \frac{\pi(i, a)}{b(i, a)}.$$

It follows, therefore, that $\rho$ will vary from time to time.

*Parameter* $\lambda$ is the trace-decay parameter in TD learning; its value lies in the range $[0, 1]$.

### 5.11.7    Practical considerations

The GQ($\lambda$) algorithm summarized in Table 5.4 is limited to policy evaluations. As such, in its current form, it is only applicable to predictive learning.

Another point of practical interest: Maei and Sutton (2010) show that the GQ($\lambda$) algorithm converges with *probability one* to the TD($\lambda$) fixed point under the standard assumptions:

$$\eta_{\mathbf{w},n} > 0 \text{ and } \eta_{\mathbf{w}_{\mathrm{sec},n}} > 0 \qquad \text{for all } n,$$

$$\sum_{n=\infty}^{\infty} \eta_{\mathbf{w},n} = \sum_{n=\infty}^{\infty} \eta_{\mathbf{w}_{\mathrm{sec}},n} = \infty,$$

$$\sum_{n=\infty}^{\infty} \eta_{\mathbf{w},n}^2 < \infty,$$

$$\sum_{n=\infty}^{\infty} \eta_{\mathbf{w}_{\mathrm{sec},n}} < \infty,$$

$$\frac{\eta_{\mathbf{w},n}}{\eta_{\mathbf{w}_{\mathrm{sec},n}}} \to 0 \quad \text{as} \quad n \to \infty.$$

## 5.12    Summary and discussion

### 5.12.1    Bellman's dynamic programming

In this chapter we studied Bellman's dynamic programming for decision-making and control of the environment. Simply put, the dynamic-programming technique provides a decision-maker (e.g. transmitter in a cognitive dynamic system) with the means to improve long-term performance at the expense of short-term performance. With this goal in mind, interest in dynamic programming is to formulate a *policy that maps states into actions*, for which *Markov decision processes* provide a well-suited *model*.

The dynamic-programming technique rests on the *principle of optimality*, the essence of which is that an optimal policy for decision-making can be constructed in a piecemeal fashion. Given a complex multistage planning or control problem, we first construct an optimal policy for the tail subproblem involving the last stage, then extend the optimal policy to the tail subproblem involving the last two stages, and continue in this manner until the entire problem has been tackled. Given a dynamic system with $N$ states, the principle of optimality provides the framework for developing *Bellman's optimality equation*, which represents a system of $N$ equations with one equation per state.

To compute the optimal policy, we may use policy iteration or value iteration. In the *policy iteration algorithm*, a sequence of stationary policies is generated, each of which improves the cost-to-go function over the preceding one. On the other hand, *value iteration* is a successive approximation algorithm, which, in theory, requires an infinite number of iterations; a stopping criterion is usually built into the algorithm to terminate it when the change in the cost-to-go function from one iteration to the next is small enough to be ignored. Because of its simplicity, the value iteration algorithm is the most widely used algorithm in dynamic programming.

### 5.12.2    Imperfect state information

Dynamic programming requires access to the exact value of the state of the environment. Unfortunately, this requirement is violated in cognitive radar, to name just one important example. In Section 5.5 we described a procedure whereby, through certain reformulations, problems with imperfect state information can be reduced to corresponding problems with perfect state information, thereby paving the way for the application of dynamic programming. We will have more to say on this issue in Chapter 6 on cognitive radar.

### 5.12.3    Reinforcement learning

The policy and value iteration algorithms are applicable when we have a mathematical model of transition probabilities from one state to the next, including the cost incurred as a result of each transition. In practice, however, it is difficult to construct such a model, in which case we have to resort to approximation in one form or another. What is truly remarkable is that *reinforcement learning*, developed in neural computation as an intermediate learning procedure that lies between supervised learning and unsupervised learning, is, in reality, an approximate form of dynamic programming. In *supervised learning*, we have a "teacher" that supplies a *desired response* for every input applied to a neural network, for example, with the objective of adjusting the free parameters of the network such that the resulting "output" is as close as possible to a desired response in some statistical sense. In *unsupervised learning*, there is no teacher, but provision is made for a *task-dependent measure* of the "quality" of representation that the network is required to learn, and free parameters of the network are optimized with respect to that particular measure. In reinforcement learning, on the other hand, the network learns in on-line continuous *interactions* with the environment in a manner similar to that in approximate dynamic programming.

Reinforcement learning embodies two well-known algorithms: *Q-learning* and *TD learning*, both of which bypass the need for transition probabilities. *Q*-learning is an off-policy algorithm well suited for control. On the other hand, TD learning is an on-line prediction method that learns how to compute an estimate, partly, on the basis of other estimates through boot-strapping. Both *Q*-learning and TD learning are limited in their application to situations where the state space is of moderate dimensionality.

### 5.12.4    Linear GQ($\kappa$) algorithm

To deal with more difficult situations involving large-scale applications, we may look to the GQ($\lambda$) algorithm. This new approximate dynamic-programming algorithm is rooted in *Q*-learning, TD learning, and stochastic gradients for algorithmic adaptation; the approximation follows a linear approach. Insofar as the general setting of the algorithm is concerned, there are two mechanisms to be considered, namely varying eligibility traces and off-policy learning, the combination of which is aimed at temporally abstract predictions. Most importantly, its objective function is based on the off-policy, $\lambda$-weighted version of the Bellman error function.

The algorithm has the potential to make a significant difference in prediction modeling, for which it is well suited. Its algorithmic uniqueness lies in four distinctive attributes of practical importance:

- first, ability to learn experientially grounded knowledge;
- second, on-line learning;
- third, linear computational scalability with respect to dimensionality of the feature space; and
- fourth, convergence of the algorithm to the TD fixed point with *probability one* under standard assumptions.

## 5.12.5    Greedy-GQ

Whereas $GQ(\lambda)$ is well suited for prediction, Greedy-GQ is well suited for control. Specifically, the Greedy-GQ algorithm (Maei *et al.*, 2010) is the first TD method for *off-policy control* with unrestricted linear function approximation. Greedy-GQ uses linear function approximation and possesses a number of important and desirable algorithmic features:

- on-line, incremental adaptations of primary and secondary weight vectors;
- linear complexity in terms of per-time-step computational costs and memory;
- no restriction imposed on the features used in the linear approximation; and
- guaranteed convergent behavior.

In an algorithmic context, Greedy-GQ can be seen as a generalization of $GQ(\lambda)$ applied to a *control* setting by allowing certain changes in the target policy. In particular, similar to $Q$-learning, the target policy is greedy with respect to a linear approximation to the optimal $Q$-function; hence its applicability to control. The update rules for adaptations in Greedy-GQ are similar to $GQ(\lambda)$ for $\lambda = 0$, and they are, therefore, analogous to those of $Q$-learning with linear function approximation, except that we have an additional correction term.

The main feature that distinguishes Greedy-GQ from popular and conventional off-policy algorithms such as $Q$-learning is that it is based on minimizing a Bellman error objective function and *it has a stability guarantee*.

Although the objective function in Greedy-GQ resembles that of $GQ(\lambda)$, the techniques used for derivations and stability analysis of the algorithm are substantially different. This is due to the greedy policy, which highly affects the properties of the objective function. It turns out that the objective function is neither convex nor differentiable. However, it has some unique properties: it is in the form of a piecewise quadratic function with respect to the primary weight vector and its global minimum matches the $Q$-learning fixed-point (assuming that such a fixed-point exists).

An important future open problem is to study the properties of local minima of the Greedy-GQ that lie in the border of quadratic functions. Because the Greedy-GQ objective function is nondifferentiable, the algorithm is derived on the basis of a subgradient method.

### 5.12.6 New generation of approximate dynamic programming algorithms: linear GQ methods

The GQ($\lambda$) and Greedy-GQ represent a new generation of approximate dynamic programming/modern reinforcement-learning algorithms. Though they are intended for different applications, one for predictive modeling and the other for control, they do share several distinctive characteristics:

- linear function approximation;
- incorporation of both $Q$-learning and TD learning;
- common use of the Bellman error objective function;
- linear scaling of computational dimensionality with respect to dimensionality of the feature space achieved by exploiting the use of stochastic gradients; and
- guaranteed convergence.

Between them, therefore, we may say that the GQ($\lambda$) and Greedy-GQ hold the potential for addressing the nagging issue of how to deal with the curse-of-dimensionality problem when we are challenged with large-scale reinforcement-learning applications. To substantiate this potential processing power, it is important that they are both tested for their respective application areas under large-scale operating conditions.

### Notes and practical references

1. *Bellman's optimality equation using operator notation*

   In (5.22) we have the description of Bellman's optimality equation for a Markov decision process as a system of $N$ equations with one operation per state. This optimality equation may be put into a compact mathematical form using operator notation (Powell, 2007).

   To this end, we first introduce the following vector and matrix notations:

$$\mathbf{j} = \{J^*(i)\}_{i=1}^{N},$$

$$\mathbf{c} = \{c(i, \mu(i))\}_{i=1}^{N},$$

$$\mathbf{P} = \{p_{ij}\}_{i,j=1}^{N},$$

   where $N$ is the number of states. We may then rewrite (5.22) in the simplified matrix form:

$$\mathbf{j} = \min_{\mu}(\mathbf{c} + \gamma \mathbf{P} \mathbf{j}),$$

where $\gamma$ is the discount factor. To proceed further, we introduce two more notations:

(1) The $\mathcal{M}$ to denote the "min" operator.
(2) The $\mathcal{J}$ to denote the space of all possible cost-to-go functions.

For some nonstationary policy $\pi = \{\mu_0, \mu_1, \mu_2, \ldots\}$ that includes the stationary policy as a special case, we may finally go on to write

$$\mathcal{M}^{\pi}(\mathbf{j}) = \mathbf{c} + \gamma \mathbf{P} \mathbf{j} \tag{5.88}$$

for some $\mathbf{j} \in \mathcal{J}$. In mathematical terms, $\mathcal{M}^{\pi}$ is referred to as a *linear operator*, since the operations it performs are all additive and multiplicative. Moreover, the function composed of the sum term, $\mathbf{c} + \gamma \mathbf{P} \mathbf{j}$, is said to be an *affine function*. Thus, we may view (5.88) as a descriptor of the affine $N$-by-$N$ *Bellman operator* for target policy $\pi$.

2. *Actor–critic interpretation of policy iteration*

In the reinforcement-learning literature, the policy iteration algorithm is referred to as an *actor–critic architecture* (Barto *et al.*, 1983). In this context, the policy improvement in Figure 5.3 assumes the role of *actor*, because it is responsible for the way in which a decision-maker (i.e. agent) acts. By the same token, the policy evaluation assumes the role of *critic*, because it is responsible for criticizing the action taken by the decision-maker.

3. *Sarsa*

The study of direct approximate dynamic-programming algorithms, namely reinforcement learning, would be incomplete without the inclusion of *Sarsa* (Sutton and Barto, 1998).

The name "Sarsa" is used for both a prediction algorithm and a control algorithm. In both cases, the update of the weights of the $Q$-function is based on a short segment of experience consisting of a State, the corresponding Action, the resultant Reward and next State, and finally the next Action, hence the name S-A-R-S-A. Sarsa is the only algorithm whose TD error and update is based on exactly these five real, experience events. The one-step tabular form of the Sarsa update is described by

$$Q^{+}(i_n, a_n) = Q(i_n, a_n) + \eta r_{n+1} + \eta (Q(i_{n+1}, a_{n+1}) - Q(i_n, a_n)).$$

The difference term inside the parentheses on the extreme right-hand side is the TD error. In the prediction form of the algorithm, the actions are selected according to a fixed policy $\pi_i$ and the approximate $Q$-function converges to $(\hat{Q}|i_n)$ if the step-size $\eta$ is reduced over time in a consistent manner with the standard stochastic approximation conditions. Note that, in this case, the algorithm is formally equivalent to the one-step tabular TD algorithm because the process transitioning from state–action pair to state–action pair is simply an uncontrolled Markov process. Because of this, all the results for TD, including the use of eligibility traces and function approximation, carry over to Sarsa.

The version of Sarsa with eligibility traces and linear function approximation, known as *linear Sarsa* $\lambda$, is widely used in reinforcement learning. For the control case, the same update is used only with the policy not held fixed, but allowed to change; for example, to be the $\varepsilon$-greedy policy for the current approximate

$Q$-function. Thereby, Sarsa becomes a simple and effective algorithm for on-policy control.

## 4. Robbins–Monro stochastic approximation

In the pioneering classic paper by Robbins and Monro (1951), the following stochastic approximation problem is addressed.

Let $M(x)$ be a function of the real-valued parameter $x$. Suppose that, for each value of $x$, there is a random variable $Y = y(x)$ with the conditional distribution function $p(y|x)$, such that

$$M(x) = \int_{-\infty}^{\infty} y p(y|x) \, dx$$

is the expected value of $Y$ given $x$. Neither the conditional distribution function $p(y|x)$ nor the function $M(x)$ is known. But it is assumed that, for some prescribed $\alpha$, the equation

$$M(x) = \alpha$$

has a unique root denoted by $\theta$. The problem of interest is to estimate $\theta$ by making successive observations on the random variable $Y$ at levels denoted by $x_1, x_2, \ldots$, which are determined in accordance with some definite algorithm.

To this end, it is postulated that the following two standard conditions hold for some parameter $\eta_n$:

(1) $\displaystyle\sum_{n=1}^{\infty} \eta_n = \infty$,

(2) $\displaystyle\sum_{n=1}^{\infty} \eta_n^2 < \infty$.

The *Robbins–Monro algorithm* is thus defined recursively for all $n$ as follows:

$$x_{n+1} = x_n + \eta_n(\alpha - y_n),$$

where $\eta_n$ plays the role of a step-size parameter that decreases with $n$.

Convergence behavior of the Robbins–Monro algorithm has been well studied in the literature. In particular, it has been shown (Robbins and Monro, 1951; Wolfowitz, 1952) that $x_n$ converges stochastically to the desired $\theta$ provided that, in addition to conditions (1) and (2), two other conditions also hold:

(3) The absolute value of $M(x)$ is finite; that is, $|M(x)| \leq C < \infty$ for some constant $C$.
(4) The variance of the random value $Y$, denoted by $\sigma^2$, is also finite; that is,

$$\int_{-\infty}^{\infty} (y - M(x))^2 \, p(y|x) \, dx \leq \sigma^2 < \infty.$$

# 6 Cognitive radar

With the material presented in the previous four chapters dealing with fundamental aspects of cognitive dynamic systems at our disposal, the stage is now set for our first application: cognitive radar, which was described for the first time by Haykin (2006a).

*Radar* is a remote-sensing system with numerous well-established applications in surveillance, tracking, and imaging of targets, just to name a few. In this chapter, we have chosen to focus on *target tracking* as the area of application. The message to take from the chapter is summed up as follows:

Cognition provides the basis for a "transformative software technology" that enables us to build a new generation of radar systems with reliable and accurate tracking capability that is beyond the reach of traditional radar systems.

This profound statement has many important practical implications, which may be justified in light of what we do know about the echolocation system of a bat. The echolocation system (i.e. sonar) of a bat provides not only information about how far away a target (e.g. flying insect) is from the bat, but also information about the relative velocity of the target, the size of the target, the size of various features of the target, and azimuth and elevation coordinates of the target. The highly complex neural computations needed to extract all this information from target echoes are performed within the "size of a plum." Indeed, an echolocating bat can capture its target with a facility and success rate that would be the envy of a radar or sonar engineer (Suga, 1990; Simmons *et al.*, 1992). How, then, does the brain of a bat do it? The bat is endowed with the ability to build up its own *rules of behavior* over the course of time, through what we usually call "experience." To be more precise, experience, gained through bat's interations with the environment, is built up over time with the development continuing well beyond birth. Simply put, the "developing" nervous system of a bat, or for that matter a human, is synonymous with a plastic brain: *plasticity* permits the developing nervous system to *adapt* to its surrounding environment (Haykin, 2009). We may, therefore, go on to say:

The echolocating bat is a living biological proof of existence for cognitive radar.

This statement is important because it recognizes that the echolocating bat is an *active sensor* just like the radar is; we say "active" in the sense that they both transmit a signal to *illuminate* the environment and then "listen" to the echo from an unknown target to extract from it valuable information about the target.

Before we proceed with the study of cognitive radar, it is instructive that we distinguish between three classes of radars, discussed next.

## 6.1     Three classes of radars defined

Given what we already know about radar systems already in existence or those described in the literature, and how cognitive radar fits into the scheme of things, we may identify three classes of radars, defined as follows:

(1) *Traditional active radar*

This class of radars refers to traditional radar systems, in which the transmitter illuminates the environment and the receiver processes the radar returns (echoes) produced as a result of reflections from an unknown target embedded in the environment. In other words, a traditional active radar is a *feedforward* information-processing system. The classic handbook on radar by Skolnik (2008) covers the many types of radar that belong to this class.

(2) *Fore-active radar*

Radars belonging to this second class are intended to *manage the allocation of available resources for the purpose of control in an on-line adaptive manner* (Krishnamurthy and Djonin, 2009). The resources residing in the transmitter may include, for example,
- library of transmit waveforms for target tracking or
- set of scan times for environmental surveillance.

To facilitate control of the receiver by the transmitter via the radar environment, there has to be *feedback information* from the receiver to the transmitter. In other words, a fore-active radar is essentially a *closed-loop feedback control system*.[1] What we have just described here is simply another way of referring to the perception–action cycle. In other words, a fore-active radar may be viewed as the first stage towards the cognition of radar. Indeed, it is for this reason that basically a fore-active radar acts as a perception–action cycle, which is a viewpoint that we will follow in later sections of the chapter dealing with simulations.

(3) *Cognitive radar*

For a radar to be cognitive, it has to satisfy four processes (putting language aside):
- perception–action cycle for maximizing information gain about the radar environment computed from the observable data:
- memory for predicting the consequences of actions involved in illuminating the environment and parameter selection for environmental modeling;
- attention for prioritizing the allocation of available resources in accordance with their importance; and
- intelligence for decision-making, whereby intelligent choices are made in the face of environmental uncertainties.

As pointed out in Chapter 2 commenting on the visual brain, perception is performed in one part of the brain and action is performed in another part of the brain. And so it is with cognitive radar: perception of the environment is performed in the receiver, and action aimed at illuminating the environment is taken in the transmitter. Memory is physically distributed throughout the radar system. Attention does not have a physical location of its own; rather, its algorithmic mechanisms are based on perception and memory. As for intelligence, its algorithmic decision-making and a control algorithms are based on perception, memory, and attention; as such, among all four cognitive processes, intelligence, therefore, is the most computationally complex and the most profound in terms of information-processing power. Simply put, for its existence, intelligence relies on local and global feedback loops distributed throughout the radar system.

   Now that we have identified the four processes that collectively define a cognitive radar and recognize the fact that the perception–action cycle is the very *baseline* upon which memory is built, it is logical that we begin the study of cognitive radar with the perception–action cycle in the next section. Indeed, this process will occupy our attention up to and including Section 6.12, simply because the perception–action cycle of a cognitive radar has many facets, all of which need detailed attention. In so doing, we will have then paved the way for building memory on top of the perception–action cycle later on in the chapter.

## 6.2   The perception–action cycle

The perception–action cycle of a cognitive dynamic system was first described in Chapter 1. With emphasis on cognitive radar in this chapter, we may reproduce Figure 1.1 in that introductory chapter in the form shown in Figure 6.1 representing the first basic process in cognitive radar.

   Moreover, we have chosen *target tracking* as the application on which the study of cognitive radar will be focused in this chapter. In this context, the primary function of the *environmental-scene analyzer*, the only fundamental block in the receiver shown in Figure 6.1, is, therefore, to provide an *estimate of the state* of the radar environment by processing the observables. The term *radar environment* is used here to refer to the electromagnetic medium in which a target of interest is embedded. The observables refer to the *radar returns* produced by reflections from the target due to illumination of the radar

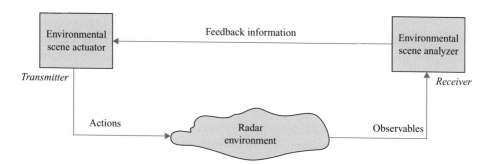

**Figure 6.1.** The perception–action cycle of radar with global feedback as the first step towards cognition.

environment by a signal radiated from the transmitter. In effect, state estimation serves as "perception" of the environment in the perception–action cycle of Figure 6.1.

Insofar as this cycle is concerned, another function of the receiver is to compute *feedback information* that provides a compressed measure of information contained in the radar returns about the unknown target. Typically, the transmitter and receiver of the radar are collocated,[2] in which case delivery of the feedback information to the transmitter by the receiver is accomplished simply through a direct linkage, thereby simplifying the radar system design.

Turning next to the *environmental scene actuator*, which is the only functional block in the transmitter shown in Figure 6.1, its primary function is to minimize a *cost-to-go function* based on feedback information from the receiver, so as to act in the radar environment optimally in some statistical sense. This optimization manifests itself in the selection of a transmitted signal whose waveform controls the receiver via the environment on a cycle-by-cycle basis. In this sense, therefore, we may look to the environmental scene actuator as a *controller*.

With emphasis on the term "information" in what we have just discussed here, the perception–action cycle in Figure 6.1 provides the basis for *cyclic directed information–flow* across the entire radar system, inclusive of the environment.

## 6.3    Baseband model of radar signal transmission

With *control* of the receiver by the transmitter via the environment as a basic design objective of cognitive radar, we need an *agile mechanism for waveform selection*, which, desirably, is digitally implementable. To this end, the transmitter may be equipped with a *library* composed of a prescribed set of waveforms, hereafter referred to simply as the *waveform generator*.[3]

The key question is: What kind of a waveform do we choose for the library? With target tracking as the task of interest in this chapter, there is, unfortunately, no single waveform that can satisfy the joint estimation of both *range* (i.e. distance of the target from the radar) and *range-rate* (i.e. radial velocity of the target) (Wicks *et al.*, 2010). To elaborate, it is well known that a *constant-frequency* (CF) pulse exhibits good range-rate resolution but relatively poor range resolution. On the other hand, a *linear frequency-modulated* (LFM) pulse exhibits the converse resolution characteristics: good range resolution but poor range-rate resolution. As a compromise between these two conflicting needs, we propose to opt for a transmit-waveform that combines two forms of modulation: Gaussian envelope for amplitude modulation and LFM for frequency modulation.

This composite modulation is imposed on a sinusoidal *carrier* whose frequency is compatible with the propagation properties of the electromagnetic medium, thereby facilitating a wireless link from the transmitter to the receiver. Throughout the chapter, it is assumed that the transmitted radar signal is *narrowband*, which means that the bandwidth occupied by the composite modulated pulse is a small fraction of the carrier frequency.

### 6.3.1      Baseband models of the transmitted and received signals

The term "baseband" refers to the original frequency content of the composite modulated pulse produced by the waveform generator. In baseband modeling, we purposely
dispense with the carrier frequency analytically in such a way that there is no loss of
information contained in the composite modulated pulse. This objective is realized
through a *band-pass-to-low-pass transformation*, whereby a real-valued band-pass
signal is transformed into a complex-valued low-pass signal (Haykin, 2000).

To proceed then, let $s_T(t)$ denote the narrowband transmitted radar signal, the spectrum of which is centered on the carrier frequency $f_c$; and let $\tilde{s}(t)$ denote the low-pass
*complex envelope* of $s_T(t)$. Then, by definition, we write

$$s_T(t) = \sqrt{2E_T}\, \mathrm{Re}[\tilde{s}(t)\exp(j2\pi f_c t)] \qquad (6.1)$$

where $E_T$ is *energy of the transmitted signal* $s_T(t)$ and the operator $\mathrm{Re}[\cdot]$ extracts the
real part of the complex quantity enclosed inside the square brackets. According to this
definition, the complex envelope $\tilde{s}(t)$ has unit energy, as shown by

$$\int_{-\infty}^{\infty} |\tilde{s}(t)|^2\, dt = 1 \qquad (6.2)$$

Consider next the radar returns (i.e. received signal) at the receiver input. Let $\tau$
denote the *round-trip delay time from the target* and $f_D$ denote the *Doppler frequency*
resulting from radial motion of the target. These two target parameters are respectively defined by

$$\tau = \frac{2\rho}{c} \qquad (6.3)$$

and

$$f_D = \frac{2\dot{\rho}}{c}, \qquad (6.4)$$

where $\rho$ is the target range, $\dot{\rho}$ is the range-rate, and $c$ is the speed of electromagnetic-
wave propagation (i.e. the speed of light); the dot in $\dot{\rho}$ denotes differentiation with
respect to time. The Doppler frequency $f_D$ represents a "shift" in the carrier frequency
$f_c$; it is positive when the target is moving toward the radar and negative otherwise. We
may then express the received signal denoted by $r(t)$, in terms of the complex envelope
$\tilde{s}(t)$, as follows:

$$r(t) = \sqrt{2E_R}\, \mathrm{Re}[\tilde{s}(t-\tau)\exp[j2\pi(f_c + f_D)t + \phi] + \tilde{n}(t)] \qquad (6.5)$$

where $E_R$ is *energy of the received signal*, the phase $\phi$ is a *random variable* incurred
from reflecting surface of the target, and $\tilde{n}(t)$ is complex envelope of the *front-end
receiver noise* centered on the carrier frequency $f_c$.

### 6.3.2      Bank of matched filters and envelope detectors

Typically, at the front end of the radar receiver we have a bank of matched filters. The
impulse response of a *matched filter* is defined as the complex conjugate of the complex
transmitted signal envelope $\tilde{s}(t)$, shifted in time and frequency by scaled versions of

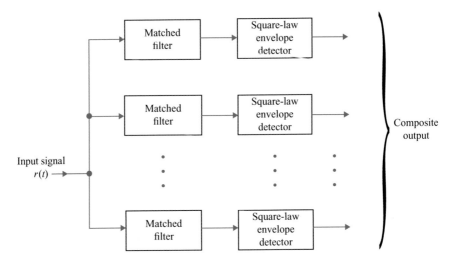

**Figure 6.2.** Radar receiver front-end, made up of bank of matched filters followed by square-law envelope detectors. In the absence of receiver noise, the composite output is the ambiguity function of the complex envelope $\tilde{s}(t)$, representing the noiseless version of the radar return (input signal) $r(t)$.

desired time- and frequency-resolutions in the time–frequency space of interest. Recognizing that a matched filter is basically equivalent to a correlator, it follows that the bank of matched filters acts as a *time–frequency correlator* of $\tilde{s}(t)$ with time- and frequency-shifted versions of itself.

In the absence of receiver noise, the squared magnitude of the time–frequency correlator output constitutes the *ambiguity function* (Woodward, 1953). To this end, we follow every matched filter by a *square-law envelope detector*, as illustrated in Figure 6.2. Using $\phi(\tau, f)$ to denote the ambiguity function and recognizing that the complex envelope $\tilde{s}(t)$ has unit energy and, then, invoking the *Schwarz inequality*,[4] we find that (Van Trees, 1971)

$$\phi(\tau, f_{\mathrm{D}}) \leq \phi(0, 0) = 1 \qquad \text{for all } \tau \text{ and } f_{\mathrm{D}}. \tag{6.6}$$

Most importantly, the resulting real-valued two-dimensional output of each envelope detector, involving time delay $\tau$ and Doppler shift $f_{\mathrm{D}}$, defines an inter-pulse vector denoted by $\mathbf{y}_n$, where the subscript $n$ denotes discrete time. The vector $\mathbf{y}_n$ performs the role of *measurement* in the state-space model of the radar target, as discussed next.

### 6.3.3    State-space model of the target

From Chapter 4, we infer that there are two equations that mathematically describe the state-space model of a radar target:

(1) *System equation*, which describes evolution of the target's *state* across time in accordance with the nonlinear equation:

$$\mathbf{x}_n = \mathbf{a}(\mathbf{x}_{n-1}) + \boldsymbol{\omega}_n, \tag{6.7}$$

where $\mathbf{x}_n$ denotes the *state* of the radar target at discrete time $n$ and $\boldsymbol{\omega}_n$ denotes the additive *system noise* accounting for environmental uncertainty about the target. In (6.7), it is assumed that the underlying physics of interactions between the transmitted signal and the target is *nonlinear* but not time varying.

(2) *Measurement equation*, which describes dependence of the measurement $\mathbf{y}_n$ on the state $\mathbf{x}_n$, as shown by

$$\mathbf{y}_n = \mathbf{b}(\mathbf{x}_n) + \mathbf{v}(\boldsymbol{\theta}_{n-1}), \qquad (6.8)$$

where the vector $\mathbf{v}(\boldsymbol{\theta}_{n-1})$ denotes the *measurement noise* that acts as the "driving force." It is by virtue of dependence of this noise on the waveform-parameter vector, denoted by $\boldsymbol{\theta}_{n-1}$, that the transmitter *controls* accuracy of state estimation in the receiver. Here again, (6.8) assumes a radar environment that exhibits nonlinear but not time-varying behavior. The rationale for using $\boldsymbol{\theta}_{n-1}$ as argument in $\mathbf{v}(\boldsymbol{\theta}_{n-1})$ rather than $\boldsymbol{\theta}_n$ is justified at the end of the section.

Application of the state-space model described in (6.7) and (6.8) hinges on four basic assumptions:

(1) The nonlinear vectorial functions $\mathbf{a}(\cdot)$ and $\mathbf{b}(\cdot)$ in (6.7) and (6.8) are both smooth and otherwise arbitrary; as already mentioned, the absence of subscript in these two functions implies time invariance.
(2) The system noise $\boldsymbol{\omega}_n$ and measurement noise $\mathbf{v}_n$ are zero-mean Gaussian distributed and statistically independent of each other; this assumption is justified for a target in space.
(3) The covariance matrix of system noise is known.
(4) The state is independent of both the system noise and measurement noise.

Examining (6.7) and (6.8), we immediately see that the state $\mathbf{x}_n$ is *hidden* from the observer (i.e. receiver). The challenge for the receiver, therefore, is to exploit dependence of the measurement vector on the state to compute a *reliable estimate* of the state and do so in a *sequential and on-line manner*.

## 6.3.4   Dependence of measurement noise on the transmitted signal

With the objective of sequential and on-line estimation of the target's state in mind, we need to determine the statistical characteristics of the measurement noise. To this end, we first recognize that the measurement noise covariance is dependent on the waveform-parameter vector $\boldsymbol{\theta}_{n-1}$ of the waveform generator in the transmitter; hence the notation $\mathbf{R}(\boldsymbol{\theta}_{n-1})$ for the covariance matrix of $\mathbf{v}(\boldsymbol{\theta}_{n-1})$. Moreover, the inverse of the *Fisher information matrix* is the CRLB[5] on the state-estimation error covariance matrix for an unbiased estimator. Denoting the Fisher information matrix by $\mathbf{J}$, we may consider the inverse matrix $\mathbf{J}^{-1}$ as a suitable characterization for "optimal waveform selection" and thus write (Kershaw and Evans, 1994):

$$\mathbf{R}(\boldsymbol{\theta}_{n-1}) = \boldsymbol{\Gamma} \mathbf{J}^{-1}(\boldsymbol{\theta}_{n-1}) \boldsymbol{\Gamma}, \qquad (6.9)$$

where, for a target in space, $\boldsymbol{\Gamma}$ is defined by the diagonal matrix

$$\boldsymbol{\Gamma} \overset{\Delta}{=} \mathrm{diag}\left[\frac{c}{2}, \ \frac{c}{4\pi f_c}\right]. \qquad (6.10)$$

As before, $c$ denotes the speed of light and $f_c$ denotes the carrier frequency. For convenience of presentation, it is desirable to separate contribution of the waveform parameter vector in the Fisher information matrix from the *received signal energy-to-noise spectral density ratio* (i.e. the SNR) defined by

$$\eta = \frac{2E_R}{N_0}, \tag{6.11}$$

where $N_0$ is the spectral density of the complex noise envelope $\tilde{n}(t)$. To this end, we may write

$$J(\boldsymbol{\theta}_{n-1}) = \eta U(\boldsymbol{\theta}_{n-1}). \tag{6.12}$$

Accordingly, we rewrite (6.9) in the desired form (Kershaw and Evans, 1994)

$$R(\boldsymbol{\theta}_{n-1}) = \frac{1}{\eta} \Gamma U^{-1}(\boldsymbol{\theta}_{n-1}) \Gamma, \tag{6.13}$$

where the matrix $U(\boldsymbol{\theta}_{n-1})$ is merely a scaled version of the Fisher information matrix $J(\boldsymbol{\theta}_{n-1})$. This new matrix is a two-by-two symmetric matrix whose elements are described as mean-square values of the following errors:

- delay estimation error,
- Doppler estimation error, and
- cross Doppler–delay estimation error.

As pointed out previously, we have chosen to work with a transmit-waveform that combines linear frequency modulation with Gaussian pulse-amplitude modulation. Let $\lambda$ denote the *pulse duration* measured in seconds and $b$ denote the *chirp rate* in hertz/second across the pulse. The two-element *waveform-parameter vector*, therefore, is defined by

$$\boldsymbol{\theta} = [\lambda, b]^\mathrm{T}. \tag{6.14}$$

Correspondingly, the complex envelope of the transmit-waveform is given by

$$\tilde{s}(t) = (\pi\lambda^2)^{-1/4} \exp\left[-\left(\frac{1}{2\lambda^2} - jb\right)t^2\right], \tag{6.15}$$

where $t$ denotes continuous time. Moreover, Kershaw and Evans (1994) showed that the covariance matrix of the measurement noise $v(\boldsymbol{\theta}_{n-1})$ is given by

$$R(\boldsymbol{\theta}_{n-1}) = \begin{bmatrix} \dfrac{c^2\lambda^2}{2\eta} & -\dfrac{c^2b\lambda^2}{2\pi f_c\eta} \\ \dfrac{c^2b\lambda^2}{2\pi f_c\eta} & \dfrac{c^2}{(2\pi f_c)^2\eta}\left(\dfrac{1}{2\lambda^2} + 2b^2\lambda^2\right) \end{bmatrix}, \tag{6.16}$$

where, for the sake of simplicity, we have omitted the dependence on time $n$ in the right-hand side of (6.16).

### 6.3.5    Closing remarks

It is important to note that the formula for $R(\boldsymbol{\theta}_{n-1})$ in (6.16) is valid so long as the assumption that the energy per transmitted waveform remains constant from one cycle of the perception–action cycle to the next. Otherwise, we would have to expand the waveform parameter vector $(\boldsymbol{\theta}_{n-1})$ by adding a new time-dependent parameter $\eta(\boldsymbol{\theta}_{n-1})$.

We close this section by justifying the rationale for using $(\boldsymbol{\theta}_{n-1})$ as the waveform-parameter vector. To account for the time delay incurred due to electromagnetic wave propagation from the receiver to the transmitter, followed by the time delays for the receiver and transmitter to complete their respective parts of perception and action in the perception–action cycle, it is proposed that if $(\boldsymbol{\theta}_{n-1})$ denotes the waveform parameter vector in the transmitter, then, without loss of generality, the measurement at the receiver input is $\mathbf{y}_n$; hence the formulation of the measurement equation as shown in (6.8).

## 6.4    System design considerations

With this background on the radar environment at hand, it is apropos that we pause to discuss the basic issues involved in designing the receiver and transmitter in the perception–action cycle of Figure 6.1 for tracking a target in space. The underlying statistics of such an environment may be assumed to be *Gaussian*; this assumption is not only justifiable, but also simplifies the system design, certainly so for a target in space.

Let us begin the discussion by considering the receiver first. With optimal performance as the goal, the ideal framework for target-state estimation is the *Bayesian filter*. We say so because the Bayesian filter propagates the *posterior* of the state-estimation error covariance matrix, which, in conceptual terms, is the best that we can find for state estimation. Unfortunately, as discussed in Chapter 4, when the state-space model is nonlinear, as it is in (6.7) and (6.8), the Bayesian filter is no longer computationally feasible; hence the practical need for its approximation. For many decades past, the EKF has been used as the method of choice to perform this approximation. From a practical perspective, however, such an approach is limited to applications where nonlinearities in the system and measurement equations are of a "mild" sort. When this requirement is violated, as is it does happen in practice, we have to look to other alternatives. Among the alternatives described in the literature, the CKF stands out as a method of choice for approximating the Bayesian filter in a nonlinear setting under the Gaussian assumption. With the state-space model of (6.7) and (6.8) as the focus of interest, we look to the CKF discussed in detail in Chapter 4 and revisited in the next section as the "central" functional block for the environment-scene analyzer in the receiver for target-state estimation.

Turning next to the transmitter, the primary function here is to optimally *control the receiver* through selection of the transmitted waveform in response to feedback information from the receiver. Provided that the waveform parameters are selected optimally, any *action* taken by the transmitter will be viewed as an optimal *response* to the environment as perceived by the receiver. With optimal control of the receiver in mind, we may look to *dynamic programming* (Bellman, 1957) as the ideal framework for optimal transmit-waveform selection, at least in a conceptual sense. In effect, we are looking for a *controller* in a nonlinear closed-loop feedback system that "tunes" the transmit-waveform parameters so as to "improve" the behavior of the receiver in an effort to minimize the target-tracking errors in a statistical sense.

Unfortunately, Bellman's dynamic programming also has a practical limitation of its own; it suffers from the curse-of-dimensionality problem when the state space, measurement space, or action space is of high dimensionality; hence the practical requirement to seek an approximation to this ideal framework in designing the transmitter. Section 6.9 addresses the issues involved in the algorithmic formulation of an approximate dynamic-programming procedure for optimal waveform selection.

## 6.5 Cubature Kalman filter for target-state estimation

### 6.5.1 Cubature rule of third degree

When the nonlinear system and measurement equations of the state-space model are both Gaussian, formulation of the Bayesian filter reduces to the problem of how to compute moment integrals whose integrands are of the following form:

$$(\text{nonlinear function}) \times (\text{Gaussian density}) \tag{6.17}$$

To compute integrals numerically whose intergrands are of the form described in this expression, we may use the *cubature rule of third degree*, discussed in detail in Chapter 4. To recap briefly on the material presented therein, consider the example of an integrand described in (6.17) that consists of a nonlinear function $\mathbf{a}(\mathbf{x})$ multiplied by a multivariate Gaussian density of the vector $\mathbf{x}$, denoted by $\mathcal{N}(\mathbf{x}; \boldsymbol{\mu}, \boldsymbol{\Sigma})$, where the vector $\boldsymbol{\mu}$ is the mean and the matrix $\boldsymbol{\Sigma}$ is the covariance. According to the third-degree cubature rule, the resulting integral may be approximated in terms of the mean $\boldsymbol{\mu}$, covariance $\boldsymbol{\Sigma}$, and a set of *weighted cubature points*, denoted by $\{\alpha_i\}_{i=1}^{N_x}$ as follows:

$$\underbrace{\int_{R_{N_x}} \underbrace{\mathbf{a}(\mathbf{x})}_{\substack{\text{Nonlinear} \\ \text{function}}} \underbrace{\mathcal{N}(\mathbf{x}; \boldsymbol{\mu}, \boldsymbol{\Sigma})}_{\substack{\text{Gaussian} \\ \text{function}}} \, d\mathbf{x}} \approx \frac{1}{2N_x} \sum_{i=1}^{2N_x} \mathbf{a}(\boldsymbol{\mu} + \sqrt{\boldsymbol{\Sigma}}\alpha_i), \tag{6.18}$$

where the subscript $N_x$ denotes *dimensionality of the state space*. The square-root factor of the state-estimation error covariance S in (6.18) satisfies the factorization $\boldsymbol{\Sigma} = \sqrt{\boldsymbol{\Sigma}} \sqrt{\boldsymbol{\Sigma}}^{\mathsf{T}}$, and the *even set of* $2N_x$ *cubature points* is given by

$$\alpha_i = \begin{cases} \sqrt{N_x}\mathbf{e}_i, & i = 1, 2, \dots N_x, \\ -\sqrt{N_x}\mathbf{e}_{i-N_x}, & i = N_x + 1, \, N_x + 2, \dots, \, 2N_x, \end{cases} \tag{6.19}$$

where $\mathbf{e}_i \in R_{N_x}$ denotes the $i$th elementary column vector; the rule of (6.19) is exact for integrands that are polynomials of degree up to three or any odd integer. In effect, the $2N_x$ weighted cubature points are distributed uniformly over the surface of an ellipsoid to provide an approximate representation of the nonlinear function $\mathbf{a}(\mathbf{x})$.

### 6.5.2 Probability-distribution flow-graph of the Bayesian filter

To develop insight into the underlying behavior of the CKF, Figure 6.3 depicts the flow-graph of probability distributions for one recursion of the Bayesian filter; such a recursin occupies a fraction of the time occupied by one cycle in the perception–action cycle. This figure has been structured in such a way that it corresponds to the steps involved in its

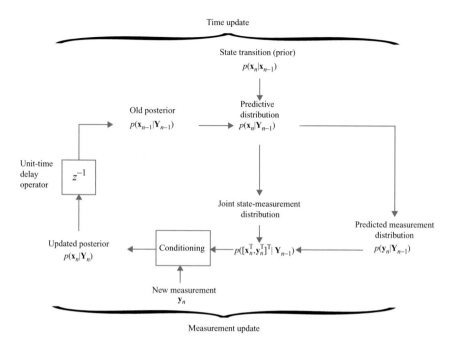

**Figure 6.3.** Probability-distribution flow-graph of the Bayesian filter under the Gaussian assumption; the figure is structured in a way that corresponds to the steps involved in deriving the CKF. The functional block, labeled conditioning, is intended to compute the updated posterior.

approximation using the cubature rule of degree three for deriving the CKF. The flow-graph of Figure 6.3 consists of two updates:

- the time update at the top and
- the measurement update at the bottom.

The time update starts with the "old" posterior, $p(\mathbf{x}_{n-1} | \mathbf{Y}_{n-1})$, where $\mathbf{x}_{n-1}$ is the state at time $n-1$ and $\mathbf{Y}_{n-1}$ sums up the past history of measurements up to and including time $n-1$. Given the *state transition matrix* or *prior*, $p(\mathbf{x}_n | \mathbf{x}_{n-1})$, the old posterior is used to compute the *predictive distribution*, $p(\mathbf{x}_n | \mathbf{Y}_{n-1})$, and with it the time update is completed.

Then, the measurement update begins with computation of the *predicted-measurement distribution* $p(\mathbf{y}_n | \mathbf{Y}_{n-1})$, which, in combination with the predictive distribution $p(\mathbf{x}_n | \mathbf{Y}_{n-1})$, yields the *joint state-measurement distribution* $p([\mathbf{x}_n^{\mathrm{T}}, \mathbf{y}_n^{\mathrm{T}}]^{\mathrm{T}} | \mathbf{Y}_{n-1})$; as usual, the superscript T denotes matrix transposition. Finally, through the application of Bayes' rule to this joint distribution, the "updated" posterior $p(\mathbf{x}_n | \mathbf{Y}_n)$ is computed. With this computation, the measurement update is completed, ready for the next recursion of the Bayesian filter.

From the distribution flow-graph of Figure 6.3, we see that the Bayesian filter propagates the posterior from one recursion to the next. Similarly, the CKF follows the footsteps of the Bayesian filter, as described next.

## 6.5.3        Time update

The first in a *two-step recursion* in the CKF is the *time-update step*. In this update, involving prediction at time $n - 1$, the CKF computes the mean $\hat{\mathbf{x}}_{n|n-1}$ and the associated covariance matrix $\mathbf{P}_{n|n-1}$ of the Gaussian predictive density, numerically using the cubature points. We write the *predicted estimate*

$$\hat{\mathbf{x}}_{n|n-1} = \mathbb{E}[\mathbf{x}_n | \mathbf{Y}_{n-1}], \tag{6.20}$$

where $\mathbb{E}[\cdot]$ is the statistical expectation operator and $\mathbf{Y}_{n-1}$ is the history of past measurements up to and including time $n - 1$. Substituting the system equation (6.7) into (6.20) yields

$$\hat{\mathbf{x}}_{n|n-1} = \mathbb{E}[(\mathbf{a}(\mathbf{x}_{n-1}) + \boldsymbol{\omega}_n) | \mathbf{Y}_{n-1}]. \tag{6.21}$$

With the system noise $\boldsymbol{\omega}_n$ assumed to be zero-mean and uncorrelated with the measurement sequence, (6.21) takes the form

$$\begin{aligned}
\hat{\mathbf{x}}_{n|n-1} &= \mathbb{E}[\mathbf{a}(\mathbf{x}_{n-1}) | \mathbf{Y}_{n-1}] \\
&= \int_{\mathbb{R}^{N_{\mathbf{x}}}} \mathbf{a}(\mathbf{x}_{n-1}) p(\mathbf{x}_{n-1} | \mathbf{Y}_{n-1}) \, d\mathbf{x}_{n-1} \\
&= \int_{\mathbb{R}^{N_{\mathbf{x}}}} \mathbf{a}(\mathbf{x}_{n-1}) \mathcal{N}(\mathbf{x}_{n-1}; \hat{\mathbf{x}}_{n-1|n-1}, \mathbf{P}_{n-1|n-1}) \, d\mathbf{x}_{n-1},
\end{aligned} \tag{6.22}$$

where, as before, $\mathcal{N}(.;.,..)$ is the conventional symbol for Gaussian distribution. Similarly, given the measurement sequence $\mathbf{y}_{1:n-1}$, which is another way of describing $\mathbf{Y}_{n-1}$, we obtain the associated error covariance matrix

$$\begin{aligned}
\mathbf{P}_{n|n-1} &= \mathbb{E}[(\mathbf{x}_n - \hat{\mathbf{x}}_{n|n-1})(\mathbf{x}_n - \hat{\mathbf{x}}_{n|n-1})^{\mathrm{T}} | \mathbf{y}_{1:n-1}] \\
&= \int_{\mathbb{R}^{N_{\mathbf{x}}}} \mathbf{a}(\mathbf{x}_{n-1}) \mathbf{a}^{\mathrm{T}}(\mathbf{x}_{n-1}) \mathcal{N}(\mathbf{x}_{n-1}; \hat{\mathbf{x}}_{n-1|n-1}, \mathbf{P}_{n-1|n-1}) \, d\mathbf{x}_{n-1} - \hat{\mathbf{x}}_{n|n-1} \hat{\mathbf{x}}_{n|n-1}^{\mathrm{T}} + \mathbf{Q}_{n-1},
\end{aligned}$$

$$\tag{6.23}$$

where $\mathbf{Q}_n$ is covariance matrix of the system noise $\boldsymbol{\omega}_n$.

## 6.5.4        Measurement update

The second update in the two-step recursion of the CKF is the *measurement-update step*. To deal with this second update, by definition, we first recognize that in a Gaussian environment the *innovations process*, defined by the prediction error $(\mathbf{y}_n - \hat{\mathbf{y}}_{n|n-1})$, is not only white but also zero-mean Gaussian. Under these conditions, the predicted measurement may be estimated in the *least-squares error sense*; that is, $\hat{\mathbf{y}}_{n|n-1}$ is the least-squares prediction of the measurement $\mathbf{y}_n$ given its past history up to and including time $n - 1$. In particular, we may express the predicted measurement distribution as follows:

$$p(\mathbf{y}_n | \mathbf{Y}_{n-1}) = \mathcal{N}(\mathbf{y}_n; \hat{\mathbf{y}}_{n|n-1}, \mathbf{P}_{\mathbf{yy},n|n-1}). \tag{6.24}$$

Using the measurement equation, we find that the predicted measurement itself and the associated covariance matrix are respectively given by

$$\hat{\mathbf{y}}_{n|n-1} = \int_{R^{N_{\mathbf{x}}}} \mathbf{b}(\mathbf{x}_n)\mathcal{N}(\mathbf{x}_n; \hat{\mathbf{x}}_{n|n-1}, \mathbf{P}_{n|n-1})\, d\mathbf{x}_n, \tag{6.25}$$

$$\mathbf{P}_{\mathbf{yy},n|n-1} = \int_{R^{N_{\mathbf{x}}}} \mathbf{b}(\mathbf{x}_n)\mathbf{b}^{\mathrm{T}}(\mathbf{x}_n)\mathcal{N}(\mathbf{x}_n; \hat{\mathbf{x}}_{n|n-1}, \mathbf{P}_{n|n-1})\, d\mathbf{x}_n - \hat{\mathbf{y}}_{n|n-1}\hat{\mathbf{y}}_{n|n-1}^{\mathrm{T}} + \mathbf{R}(\boldsymbol{\theta}_{n-1}). \tag{6.26}$$

An important point to note in (6.26) is the fact that dependence on the waveform-parameter vector $\boldsymbol{\theta}_{n-1}$ is confined to the covariance matrix $\mathbf{R}(\boldsymbol{\theta}_{n-1})$; in effect, the control action of the transmitter applied to the receiver via the environment manifests itself only by affecting the measurement noise in the state-space model.

We may now compactly express the Gaussian conditional distribution of the joint state-measurement space as

$$p([\mathbf{x}_n^{\mathrm{T}}, \mathbf{y}_n^{\mathrm{T}}]^{\mathrm{T}} \mid \mathbf{Y}_{n-1}) = \mathcal{N}\left(\begin{pmatrix} \hat{\mathbf{x}}_{n|n-1} \\ \hat{\mathbf{y}}_{n|n-1} \end{pmatrix}, \begin{pmatrix} \mathbf{P}_{n|n-1} & \mathbf{P}_{\mathbf{xy},n|n-1} \\ \mathbf{P}_{\mathbf{xy},n|n-1}^{\mathrm{T}} & \mathbf{P}_{\mathbf{yy},n|n-1} \end{pmatrix}\right), \tag{6.27}$$

where the cross-covariance

$$\mathbf{P}_{\mathbf{xy},n|n-1} = \int_{R^{N_{\mathbf{x}}}} \mathbf{x}_n \mathbf{b}^{\mathrm{T}}(\mathbf{x}_n)\mathcal{N}(\mathbf{x}_n; \hat{\mathbf{x}}_{n|n-1}, \mathbf{P}_{n|n-1})\, d\mathbf{x}_n - \hat{\mathbf{x}}_{n|n-1}\hat{\mathbf{y}}_{n|n-1}^{\mathrm{T}}. \tag{6.28}$$

On receipt of the new measurement vector $\mathbf{y}_n$, the Bayesian filter computes the updated posterior $p(\mathbf{x}_n \mid \mathbf{Y}_n)$ from (6.27), yielding

$$\begin{aligned} p(\mathbf{x}_n \mid \mathbf{Y}_n) &= \frac{p([\mathbf{x}_n^{\mathrm{T}}, \mathbf{y}_n^{\mathrm{T}}]^{\mathrm{T}} \mid \mathbf{Y}_{n-1})}{p(\mathbf{y}_n \mid \mathbf{Y}_{n-1})} \\ &= \mathcal{N}(\mathbf{x}_n; \hat{\mathbf{x}}_{n|n}, \mathbf{P}_{n|n}). \end{aligned} \tag{6.29}$$

Turning next to the *Kalman gain*, this is defined by

$$\mathbf{G}_n = \mathbf{P}_{\mathbf{xy},n|n-1} \mathbf{P}_{\mathbf{yy},n|n-1}^{-1}. \tag{6.30}$$

We may then express the filtered estimate of the state $\mathbf{x}_n$ as

$$\hat{\mathbf{x}}_{n|n} = \hat{\mathbf{x}}_{n|n-1} + \mathbf{G}_n(\mathbf{y}_n - \hat{\mathbf{y}}_{n|n-1}), \tag{6.31}$$

where, as mentioned previously, the prediction error $(\mathbf{y}_n - \hat{\mathbf{y}}_{n|n-1})$ defines the innovations process. Correspondingly, the filtered state-estimation error covariance is given by

$$\mathbf{P}_{n|n} = \mathbf{P}_{n|n-1} - \mathbf{G}_n \mathbf{P}_{\mathbf{yy},n|n-1} \mathbf{G}_n^{\mathrm{T}}. \tag{6.32}$$

Equations (6.30)–(6.32) follow from linear Kalman filter theory.

## 6.5.5   Summarizing remarks

The material presented in this section on the CKF for perception of the environment is based on two powerful tools:

(1) The third-degree cubature rule, which provides the means for numerical approximation of the Gaussian-weighted integrals that characterize the Bayesian filter.

(2) Linear Kalman filter theory, known for its mathematical elegance, the use of which completes the mathematical formulation of the CKF.

The net result is a nonlinear filter that is a method of choice for approximating the optimal Bayesian filter under the Gaussian assumption in a nonlinear setting. Moreover, derivation of this new filter is mathematically rigorous from the beginning to the end, as stressed in chapter 4.

## 6.6        Transition from perception to action

At this point in the discussion, we need to think about closing the feedback loop in the perception–action cycle of Figure 6.1 with the aim of preparing the transmitter for its role as "actuator" in the radar environment, under "perceptual guidance" of the receiver. To this end, there are two basic issues to be considered: feedback information at the receiver output, followed by cost-to-go function formulated in the transmitter; they are both discussed in what follows in this order.

### 6.6.1        Feedback information about the target

Referring back to Figure 6.1, the baseband form of "action" involves the transmitter illuminating the radar environment by transmitting the complex low-pass waveform $\tilde{s}_{n-1}$. Allowing for the time taken for "propagation delay" from the transmitter to the receiver,[6] the corresponding observable at the receiver input is $\mathbf{y}_n$ at time $n$. Recognizing that the waveform-parameter vector responsible for the generation of $\tilde{s}_{n-1}$ is $\boldsymbol{\theta}_{n-1}$ at time $n-1$, it follows that $\mathbf{y}_n$ is dependent on $\boldsymbol{\theta}_{n-1}$. This dependence shows up in (6.26) solely in the measurement noise covariance matrix $\mathbf{R}(\boldsymbol{\theta}_{n-1})$.

Now, looking ahead to the "action" to be taken by the transmitter, which has to be at time $n$ for the perception–action cycle to continue, we clearly have to think in terms of "one-step prediction into the future." To be more specific, we have to build on the filtered estimate of the state $\hat{\mathbf{x}}_{n|n}$, defined in (6.31), and formulate an expression for the one-step prediction $\hat{\mathbf{x}}_{n+1|n}$, computed at time $n$, given all the measurements up to and including time $n$. Indeed, examining (6.22), we see that this computation is perfectly feasible on replacing $n$ with $n+1$. We say so because, with this replacement, the terms under the integral in (6.22) are available, as they all pertain to the system equation at time $n$.

With the corresponding future value of the state at time $n+1$ being $\mathbf{x}_{n+1}$, we may now define the desired "predicted state-estimation error" at time $n+1$ as

$$\boldsymbol{\varepsilon}_{n+1|n} = \mathbf{x}_{n+1} - \hat{\mathbf{x}}_{n+1|n}$$

Moreover, the covariance matrix of $\boldsymbol{\varepsilon}_{n+1|n}$, namely $\mathbf{P}_{n+1|n}$, is also computable at time $n$. Here again, saying so is justified by the fact on replacing $n$ with $n+1$ in (6.23), the resulting terms on the right-hand side of the equation are available, as they also all pertain to the system equation at time $n$.

We may now make our first statement on the transition from the receiver output to the transmitter input:

Given the measurements up to and including time $n$, the one-step prediction error $\boldsymbol{\varepsilon}_{n+1|n}$ and its covariance matrix $\mathbf{P}_{n+1|n}$ are computable and, therefore, available at the receiver output at time $n$.

So, indeed, in the error-covariance matrix $\mathbf{P}_{n+1|n}$ we have the *feedback information* we need for the receiver to pass it onto the transmitter to initiate the "action" part of the perception–action cycle at time $n$. For this action to be performed accurately, accurate performance of the CKF in computing $\mathbf{P}_{n+1|n}$ is of crucial practical importance.

### 6.6.2 Posterior expected error covariance matrix

Reiterating what we have already said: in transmitting the complex waveform $\hat{s}_{n-1}$ to enable the receiver to perform its own "perception" part of the perception–action cycle at time $n$, we already have knowledge of the transmit-waveform parameter vector $\boldsymbol{\theta}_{n-1}$. What we need, therefore, is for the transmitter to compute the "updated" parameter vector $\boldsymbol{\theta}_n$ for completion of the cycle at time $n$ and thereafter continue the perception–action cycle for the next cycle, and so on.

With this objective in mind, we need to have a "cost-to-go function" for the transmitter to operate on, such that two requirements are satisfied:

(1) Formulation of the cost-to-go function utilizes the feedback information $\mathbf{P}_{n+1|n}$ computed at the receiver output, so as to link the transmitter to the receiver.
(2) The formulation is expanded to incorporate the updated parameter vector $\boldsymbol{\theta}_n$ as the only "unknown" to be determined.

To satisfy these two points, we propose to use the *posterior expected* error-covariance matrix, denoted by $\mathbf{P}_{n+1|n+1}(\boldsymbol{\theta}_n)$ as the appropriate cost-to-go function. The term "posterior" is used here to signify the fact that the cost-to-go function is to be formulated "after" delivery of the feedback information from the receiver to the transmitter. The second term "expected" is used to anticipate the filtering benefit to be gained as a result of transition from the "old" $\boldsymbol{\theta}_{n-1}$ to the "new" $\boldsymbol{\theta}_n$ for the perception–action cycle to move on.

First, examining (6.32), rewritten by substituting $n + 1$ for $n$, we see that the expected $\mathbf{P}_{n+1|n+1}$ does include $\mathbf{P}_{n+1|n}$ as its first term. Hence, point (1) is satisfied.

Second, the remaining term $\mathbf{G}_{n+1}\mathbf{P}_{\mathbf{yy},n+1|n}\mathbf{G}^{\mathrm{T}}_{n+1}$ in the formula for $\mathbf{P}_{n+1|n+1}$ involves the measurement $\mathbf{y}_{n+1}$; hence, the dependence of $\mathbf{P}_{n+1|n+1}$ on $\boldsymbol{\theta}_n$ is satisfied. But, we have to establish that $\boldsymbol{\theta}_n$ is the only unknown. To this end, we use (6.30) to write

$$\mathbf{G}_{n+1}\mathbf{P}_{\mathbf{yy},n+1|n}\mathbf{G}^{\mathrm{T}}_{n+1} = \mathbf{P}_{\mathbf{xy},n+1|n}\mathbf{P}^{-1}_{\mathbf{yy},n+1|n}\mathbf{P}_{\mathbf{xy},n+1|n}, \qquad (6.33)$$

where we have made use of the symmetric property that applies to the covariance matrices on the right-hand side of (6.33). Now, we make two observations:

- From (6.25) and (6.28), we see that $\mathbf{P}_{\mathbf{xy},n+1|n}$ is independent of $\boldsymbol{\theta}_n$.
- From (6.26), we see that $\mathbf{P}_{\mathbf{yy},n+1|n}$ is indeed dependent on $\boldsymbol{\theta}_n$ and, most importantly, this dependence is limited solely to the measurement covariance matrix $\mathbf{R}(\boldsymbol{\theta}_n)$.

In conceptualized terms, we may now sum up the discussion up to this point as follows:

$$\mathbf{P}_{n+1|n}(\boldsymbol{\theta}_{n-1}) \xrightarrow{\substack{\text{Expected transition} \\ \text{from perception to action}}} \mathbf{P}_{n+1|n+1}(\boldsymbol{\theta}_n).$$

Correspondingly, in words, we may make our second and final statement on the transition from perception in the receiver to action in the transmitter as follows:[7]

The posterior expected error-covariance matrix $\mathbf{P}_{n+1|n+1}$ is dependent on the new waveform-parameter vector $\boldsymbol{\theta}_n$ for completing the perception–action cycle at time $n$ and this dependence is confined entirely to an additive term defined by the measurement-noise covariance matrix $\mathbf{R}(\boldsymbol{\theta}_n)$.

The sole dependence of $\mathbf{P}_{n+1|n+1}$ on $\mathbf{R}(\boldsymbol{\theta}_n)$ makes the computation of $\boldsymbol{\theta}_n$ at time $n$ practically feasible.

Associating the posterior error vector $\boldsymbol{\varepsilon}_{n+1|n+1}(\boldsymbol{\theta}_n)$ with $\mathbf{P}_{n+1|n+1}(\boldsymbol{\theta}_n)$, we will, henceforth, use $g(\boldsymbol{\varepsilon}_{n+1|n+1}(\boldsymbol{\theta}_n))$ as the *cost-to-go function* for the transmitter to optimize it with respect to the unknown $\boldsymbol{\theta}_n$ for preparing the perception–action cycle to go to the next cycle at time $n$. In particular, we look upon the error $\boldsymbol{\varepsilon}_{n+1|n+1}$ as the "most logical" random vector to deliver information about the radar environment to the transmitter from the receiver. We say so in light of what we know from, first, the state-space model and, second, optimal estimation of the state given the measurements at the receiver input, past as well as the present.

## 6.7 Cost-to-go function

Our next task is to develop a formula for the cost-to-go function $g(\boldsymbol{\varepsilon}_{n+1|n+1}(\boldsymbol{\theta}_n))$ that is computationally optimizable with respect to the unknown waveform-parameter vector $\boldsymbol{\theta}_n$. With this objective in mind, we will consider two criteria: one based on the mean-square error and the other based on Shannon's information theory.

### 6.7.1 Cost-to-go function using mean-square error

The *posterior expected estimation error* $\boldsymbol{\varepsilon}_{n+1|n+1}(\boldsymbol{\theta}_n)$, corresponding to the error-covariance matrix $\mathbf{P}_{n+1|n+1}(\boldsymbol{\theta}_n)$, is defined by

$$\boldsymbol{\varepsilon}_{n+1|n+1}(\boldsymbol{\theta}_n) = \mathbf{x}_{n+1} - \hat{\mathbf{x}}_{n+1|n+1}(\boldsymbol{\theta}_n). \tag{6.34}$$

Using the mean-square error criterion, our first cost-to-go function is thus given by

$$g(\boldsymbol{\varepsilon}_{n+1|n+1}(\boldsymbol{\theta}_n)) = \mathbb{E}[\| \boldsymbol{\varepsilon}_{n+1|n+1}(\boldsymbol{\theta}_n) \|^2], \tag{6.35}$$

where the symbol $\|\cdot\|$ signifies the *Euclidean norm*. Using straightforward algebraic manipulations, we may equivalently express the cost-to-go function in (6.35) in terms of the *Fisher information metric* as follows:

$$g(\boldsymbol{\varepsilon}_{n+1|n+1}(\boldsymbol{\theta}_n)) = \text{Tr}[\mathbf{P}_{n+1|n+1}(\boldsymbol{\theta}_n)], \tag{6.36}$$

where the operator $\text{Tr}[.]$ denotes the *trace* of the enclosed matrix. By definition, the *trace* of a matrix is equal to the sum of its diagonal terms, hence the simplicity of the formula in (6.36).

However, a serious limitation of the cost-to-go function defined in (6.36) is the fact that it only accounts for diagonal terms of the posterior expected error-covariance metric $\mathbf{P}_{n+1|n+1}(\boldsymbol{\theta}_n)$, which means that valuable information about the target contained in the off-diagonal terms is ignored. In other words, the mean-square-error criterion for the cost-to-go function violates the *principle of information preservation* that was strongly emphasized in Chapter 1.

## 6.7.2    Cost-to-go function using Shannon's entropy

To overcome the information-preserving limitation of the mean-square error criterion, we next look to *Shannon's entropy* as the measure of information content of a random vector. To this end, we use the formula for the posterior estimation error vector $\varepsilon_{n+1|n+1}(\theta_n)$ to define the second formula for cost-to-go function as follows:

$$g(\varepsilon_{n+1|n+1}(\theta_n)) = \underbrace{H(\varepsilon_{n+1|n+1}(\theta_n))}_{\text{Entropy}}$$

$$= \int_{-\infty}^{\infty} \underbrace{\varepsilon_{n+1|n+1}(\theta_n)}_{\substack{\text{Random error} \\ \text{vector}}} \; \underbrace{p(\varepsilon_{n+1|n+1}(\theta_n))}_{\substack{\text{Conditional probability} \\ \text{density function}}} \, d\varepsilon_{n+1|n+1}(\theta_n),$$

(6.37)

where the integration is performed with respect to the error vector itself. The integral formula of (6.37) may be expressed in the following compact form (Cover and Thomas, 2006):

$$g(\varepsilon_{n+1|n+1}(\theta_n)) = \det[\mathbf{P}_{n+1|n+1}(\theta_n)],$$

(6.38)

where the operator det[.] denotes *determinant* of the enclosed matrix.

Now, we find that elements of the expected error-covariance matrix are all accounted for in formulating the entropy-based cost-to-go function. This second formula of (6.38) for the cost-to-go function, therefore, is superior to the first one of (6.36) in preserving all the information content of the expected-error covariance matrix $\mathbf{P}_{n+1|n+1}(\theta_n)$, but the improved information preservation is attained at the cost of increased computational complexity.

## 6.7.3    Another information-theoretic viewpoint of the entropy-based cost-to-go function

Equation (6.37) for the cost-to-go function is defined in terms of Shannon's entropy. It is illuminating to reformulate the definition in terms of another concept in Shannon's information theory, known as the *mutual information*.[8] To be specific, we would like to know the mutual information between the posterior estimation error vector $\varepsilon_{n+1|n+1}$ and the related waveform parameter vector $\theta_n$. This mutual information is equivalently expressed as the difference between two conditional entropies, as follows:

$$I(\varepsilon_{n+1|n+1}; \theta_n \mid \theta_{0:n-1}) = H(\varepsilon_{n+1|n+1} \mid \theta_{0:n-1}) - H(\varepsilon_{n+1|n+1} \mid \theta_{0:n-1}, \theta_n), \quad (6.39)$$

where

$$\begin{aligned} \theta_{0:n} &= \{\theta_0, \theta_1, \ldots, \theta_{n-1}, \theta_n\} \\ &= \{\theta_{0:n-1}, \theta_n\}. \end{aligned}$$

(6.40)

The term on the left-hand side of (6.39) denotes the mutual information between $\varepsilon_{n+1|n+1}$ and $\theta_n$, conditioned on the past history $\theta_{0:n-1}$ that extends up to and including time $n-1$. The first term on the right-hand side of the equation defines the conditional entropy of $\varepsilon_{n+1|n+1}$, given the past history $\theta_{0:n-1}$. As for the second term, it denotes the conditional entropy of $\varepsilon_{n+1|n+1}$ given the complete history $\theta_{0:n}$ up to and including time $n$ in light of (6.40).

The first conditional entropic term in (6.39) may be ignored on account of the fact that the past history $\theta_{0:n-1}$ is already known at time $n - 1$ and, therefore, there is no new information to be had. Accordingly, (6.39) reduces to the simplified form:

$$I(\varepsilon_{n+1|n+1}; \theta_n | \theta_{0:n-1}) = -H(\varepsilon_{n+1|n+1} | \theta_{0:n-1}, \theta_n)$$

$$= -H(\varepsilon_{n+1|n+1} | \theta_{0:n}), \tag{6.41}$$

where the entropic term $H(\varepsilon_{n+1|n+1} | \theta_{0:n})$ is another way of writing $H(\varepsilon_{n+1|n+1}(\theta_n))$. In light of (6.41), we may now make the following statement:[9]

Minimizing the conditional entropy $H(\varepsilon_{n+1|n+1}(\theta_{0:n}))$ is equivalent to maximizing the mutual information between the posterior estimation error vector $\varepsilon_{n+1|n+1}$ and the waveform-parameter vector $\theta_n$, given the past history $\theta_{0:n-1}$.

In expressing the cost-to-go function in terms of mutual information as in (6.41), we are, in effect, stating that there is *information gain* about the radar environment to be had in going from one cycle of the perception–action cycle to the next, and so on.

## 6.8     Cyclic directed information-flow

At this point in the discussion, in light of what we have already learned about perception of the radar environment in the receiver and action in the transmitter, it is instructive that we restructure the perception–action cycle of Figure 6.1 of the cognitive tracking radar in the form of a *cyclic directed information-flow graph*, as depicted in Figure 6.4; this graph follows from the state-space model of (6.7) and (6.8). Examination of this figure with global feedback reveals two fundamental transmission paths, one being *bottom up* (i.e. feedforward) and the other *top down* (i.e. feedback).

### 6.8.1     Bottom-up transmission path

The bottom-up (i.e. forward) transmission path in Figure 6.4 embodies the "perception" part of the perception–action cycle in the receiver. It begins with the measurement $\mathbf{y}_n$ at the receiver input, initiating the following sequence of computations all at time $n$:

- filtered estimation of the state $\mathbf{x}_n$, denoted by $\hat{\mathbf{x}}_{n|n}$, and associated error covariance matrix $\mathbf{P}_{n|n}$, followed by
- predicted estimation of the future state $\mathbf{x}_{n+1}$, denoted by $\hat{\mathbf{x}}_{n+1|n}$, and, finally,
- covariance matrix of the prediction error vector $\varepsilon_{n+1|n}$, denoted by $\mathbf{P}_{n+1|n}$.

The bottom-up transmission path, centered on the cubature Kalman filtering in the receiver, is depicted on the right-hand side of Figure 6.4. Basically, this path of directed information flow describes the *perceptual dynamics* of the receiver. Herein, it is assumed that the radar environment is nonlinear, hence justifying the use of cubature Kalman filtering for state estimation.

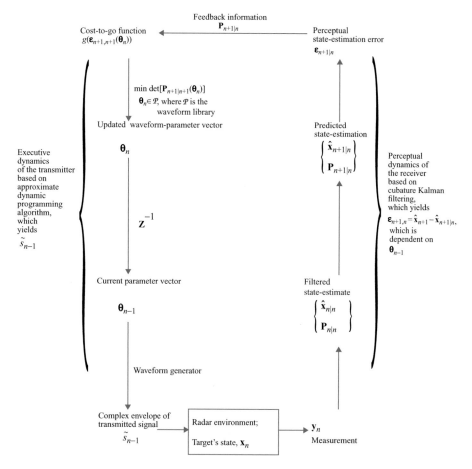

**Figure 6.4.** The basic cycle of directed information flow-graph, based on the state-space model of (6.7) and (6.8). (a) Right-hand side: perceptual dynamics of the receiver, using bottom-up processing of the measurement $\mathbf{y}_n$. Here, it is assumed that the radar environment is nonlinear; hence the use of the CKF. (b) Left-hand side: executive dynamics of the transmitter, using top-down processing of the cost-to-go function, $g(\varepsilon_{n+1,n+1}(\boldsymbol{\theta}_n))$. The symbol $\mathbf{Z}^{-1}$ represents a bank of unit-time delays.

## 6.8.2    Top-down transmission path

The top-down (i.e. feedback) transmission path in Figure 6.4 embodies the "action" part of the perception–action cycle in the transmitter. It begins with the feedback information $\mathbf{P}_{n+1|n}$ at the transmitter input delivered by the receiver, which initiates the following sequence of computations that look into the future by one time step:

- formulation of the cost-to-go function $g(\varepsilon_{n+1|n+1}(\boldsymbol{\theta}_n))$, defined in terms of the posterior state-estimation error vector $\varepsilon_{n+1|n+1}$ with dependence on the "to be updated" parameter vector $\boldsymbol{\theta}_n$, followed by
- optimization of the cost-to-go function, yielding the unknown $\boldsymbol{\theta}_n$, and, finally,
- setting the stage for a repeat of the perception–action cycle at time $n + 1$.

Most importantly, the cost-to-go function provides the transmitter with a measure of how well the receiver is doing in extracting information about the environment from the radar returns.

The top-down transmission path, centered on approximate dynamic programming, to be discussed in the next section, is depicted on the left-hand side of Figure 6.4. Basically, this path describes the *executive dynamics* of the transmitter, a primary function of which is to compute the "new" waveform-parameter vector $\boldsymbol{\theta}_n$ at time $n+1$.

## 6.9    Approximate dynamic programming for optimal control

To begin the discussion on indirect control of the receiver by the transmitter through optimal selection of the waveform-parameter vector, we assume that the condition for starting the perception–action cycle is defined by setting the measurement at the receiver input at some initial value, denoted by $\mathbf{y}_0$. With time $n$ denoting the $n$th cycle, the perception–action cycle progresses forward iteratively in accordance with the cyclic directed information flow-graph of Figure 6.4.

Before proceeding further with formulation of the procedure for transmit-waveform selection, however, we have to resolve a practical problem in the following sense: the formulation of Bellman's dynamic-programming algorithm presented in Chapter 5 not only demands that the environment be *Markovian*, but also that the algorithm has perfect knowledge of the state. In reality, however, the transmitter of a radar tracker has an *imperfect* estimate of the state reported to it by the receiver. Accordingly, we are faced with an *imperfect state-information problem* that is subject to the state-space model of the radar environment. To resolve this problem, we introduce an *information vector* that is available to the transmitter at time $n$, as discussed in Chapter 5. Specifically, the information vector is defined by

$$\mathbf{I}_n \overset{\Delta}{=} (\mathbf{y}_n, \boldsymbol{\theta}_{n-1}), \quad \text{with the initial condition } \mathbf{I}_0 = \mathbf{y}_0, \tag{6.42}$$

where, moving on beyond this initial condition, we have

$$\begin{Bmatrix} \mathbf{y}_{0:n} \\ \boldsymbol{\theta}_{0:n-1} \end{Bmatrix} = \begin{Bmatrix} \mathbf{y}_0, \ \mathbf{y}_1, \ \dots, \ \mathbf{y}_n \\ \boldsymbol{\theta}_0, \ \dots, \ \boldsymbol{\theta}_{n-1} \end{Bmatrix}. \tag{6.43}$$

From these two equations, we readily obtain the recursion

$$\mathbf{I}_n = (\mathbf{I}_{n-1}, \ \mathbf{y}_n, \ \boldsymbol{\theta}_{n-1}), \qquad n = 1, 2, \dots \tag{6.44}$$

Equation (6.44) may be viewed as the descriptor of state evolution of a "new" dynamic system with perfect-state information and, therefore, amenable to dynamic programming. According to (6.44), we have:

- $\mathbf{I}_{n-1}$ is the old (past) value of the state;
- $\boldsymbol{\theta}_{n-1}$ is the transmit-waveform parameter vector selected at time $n-1$; it is responsible for generating the transmit waveform $\hat{s}_{n-1}$,
- the measurement $\mathbf{y}_n$ is viewed as a *random disturbance* resulting from the control policy in the transmitter, attributed to the waveform-parameter $\boldsymbol{\theta}_{n-1}$; and
- $\mathbf{I}_n$ is the current state.

Note that the terminology adopted in (6.44) is, in its own way, consistent with the system equation (6.7).

At cycle time $n$, the control policy for waveform selection in the transmitter seeks to find the set of best waveform parameters, for which the cost-to-go function $g(\varepsilon_{n+1|n+1}(\boldsymbol{\theta}_n))$ is to be minimized in the time steps denoted by $n : n + L - 1$ for a rolling horizon of $L$ steps. Correspondingly, the *control policy* for this set of steps is denoted by $\pi_n = \{\mu_n, \ldots, \mu_{n+L-1}\}$ with the policy function $\mu(\mathbf{I}_n) = \boldsymbol{\theta}_{n-1} \in \mathcal{P}$, which, in effect, maps the feedback information into an action in the *waveform library* denoted by $\mathcal{P}$. The objective, therefore, is to find an optimal policy $\pi_n^*$ at time $n$ that updates the waveform selection by solving the following minimization problem:

$$\min_{\pi_n} \mathbb{E}\left[ \sum_{i=n}^{n+L-1} g(\varepsilon_{i|i}(\boldsymbol{\theta}_{i-1})) \right]. \tag{6.45}$$

For example, using the cost-to-go function of (6.35), the expected mean-square error of the radar tracker is defined by

$$g(\varepsilon_{i+1|i+1}(\boldsymbol{\theta}_i)) = \mathbb{E}_{\mathbf{x}_{i+1}, \mathbf{y}_{i+1}|\mathbf{I}_i, \boldsymbol{\theta}_i}[(\mathbf{x}_{i+1} - \hat{\mathbf{x}}_{i+1|i+1})^{\mathrm{T}}(\mathbf{x}_{i+1} - \hat{\mathbf{x}}_{i+1|i+1})], \quad i = n, \ldots, n + L - 1, \tag{6.46}$$

where it is understood that the posterior expected state estimate $\hat{\mathbf{x}}_{i+1|i+1}$ is dependent on $\mathbf{I}_i, \mathbf{x}_{i+1}, \mathbf{y}_{i+1}$, and $\boldsymbol{\theta}_i$; the parameter vector $\boldsymbol{\theta}_i$ is the only unknown to be determined. However, for an $L$-step dynamic-programming algorithm, we need to predict $L$ steps ahead, which means that accurate performance of the CKF in the receiver is of crucial importance to overall performance of the cognitive tracking radar.

Referring back to the top-down transmission path on the left-hand side in the cyclic directed information-flow graph of Figure 6.4, the function of the approximate dynamic-programming algorithm in the transmitter is to look into the future by one time step and compute the "to-be-selected" parameter vector $\boldsymbol{\theta}_n$ for the next cycle by minimizing the cost-to-go function $g(\varepsilon_{n+1|n+1}(\boldsymbol{\theta}_n))$ with respect to $\boldsymbol{\theta}_n$.

When, however, there is provision for a horizon looking a prescribed number of steps into the future, then the dynamic-programming algorithm solves the optimization problem through a recursive procedure starting at time $n$. Specifically, the dynamic-programming algorithm is made up of two parts (Haykin *et al.*, 2011):

*Terminal point*

$$\mathcal{T}(\mathbf{I}_{n+L-1}) = \min_{\boldsymbol{\theta}_{n+L-1} \in \mathcal{P}} g(\varepsilon_{n+L|n+L}(\boldsymbol{\theta}_{n+L-1})). \tag{6.47}$$

*Intermediate points*

$$\mathcal{T}(\mathbf{I}_i) = \min_{\boldsymbol{\theta}_i \in \mathcal{P}} [g(\varepsilon_{i+1|i+1}(\boldsymbol{\theta}_i)) + \mathbb{E}_{\mathbf{x}_{i+1}, \mathbf{y}_{i+1}|\mathbf{I}_i, \boldsymbol{\theta}_i}[\mathcal{T}_{i+1}]], \tag{6.48}$$

where $g(\varepsilon_{i+1|i+1}(\boldsymbol{\theta}_i))$ is a fixed cost and $i = n, \ldots, n + L - 2$. The $\mathcal{T}_{i+1}$ in (6.47) is itself defined by

$$\mathcal{T}_{i+1} = \mathcal{T}(\mathbf{I}_i, \mathbf{y}_{i+1}, \boldsymbol{\theta}_i). \tag{6.49}$$

As mentioned previously, $\mathcal{P}$ in (6.47) and (6.48) is the waveform library.

With (6.47) pertaining to the terminal point, (6.48) pertains to the intermediate points that go *backward in time* from the terminal point in $(L-1)$ steps. The desired optimal policy, $\{\mu_n^*, \ldots, \mu_{n+L-1}^*\}$ for optimal control is thus obtained in the following two-step manner:

(1) The terminal point, defined in (6.47), is minimized for every possible value of the information vector $\mathbf{I}_{n+L-1}$ to obtain $\mu_{n+L-1}^*$; meanwhile, the $\mathcal{T}_{n+L-1}$ of (6.49) is also computed.

(2) Next, $\mathcal{T}_{n+L-1}$ is substituted into calculation of the intermediate points defined in (6.48) to obtain $\mu_{n+L-2}^*$ over every possible value of $\mathbf{I}_{n+L-2}$. This second step is repeated until the optimal policy $\mu_n^*$ has been computed, at which point the controller acts by selecting the waveform-parameter vector $\boldsymbol{\theta}_n$ for signal transmission at time $n$, and with it the next cycle in the perception–action cycle begins.

Unfortunately, this two-step procedure is computationally too demanding to be feasible in practice. Therefore, we have to simplify the computational complexity of the dynamic-programming algorithm by introducing two approximations, described next.

### 6.9.1     Step 1: cost-to-go function for compressing information about the radar environment

Inclusion of the state $\mathbf{x}_{i+1}$ under the expectation on the right-hand side of (6.48) speaks for itself. As for the observable $\mathbf{y}_{i+1}$, its inclusion under the expectation is justified on the following ground: in accordance with (6.31), the expected estimate $\hat{\mathbf{x}}_{i+1|i+1}$ depends on $\mathbf{y}_{i+1}$, which itself depends nonlinearly on $\mathbf{x}_{i+1}$. Consequently, we have to get around the computational difficulty of evaluating this expectation. To do this, information about the radar environment (contained in the measurements) is compressed by looking to the cost-to-go function for minimization in the transmitter. For example, opting for the mean-square error criterion, we use (6.36) to define the cost-to-go function, as shown by

$$g(\boldsymbol{\varepsilon}_{n+1,n+1}(\boldsymbol{\theta}_n)) = \mathrm{Tr}[\mathbf{P}_{n+1|n+1}(\boldsymbol{\theta}_n)].$$

As discussed previously, $\mathbf{P}_{n+1,n+1}(\boldsymbol{\theta}_n)$ is the posterior expected estimation error-covariance matrix, which is dependent on the unknown parameter vector $\boldsymbol{\theta}_n$ that is to be determined.

### 6.9.2     Step 2: approximation in the measurement space

In a physical context, the state-space model of the radar environment across time is an infinite-dimensional continuous-valued space with respect to both the state and measurement equations. Moreover, the dimension of this model grows exponentially with depth of the optimization horizon $L$. Specifically, at each step of the dynamic-programming algorithm, we need to examine an infinite number of possibilities such that the perfect state information vector $\mathbf{I}_n$ can evolve to a new value on the next time step. To simplify this very cumbersome computation, we apply the same approximation technique used in developing the CKF in Section 6.5; that is, the expectation operation on the right-hand side of (6.48) is approximated by using the *third-degree cubature rule*. According to the CKF formulation, the predicted measurement at time $i$ is Gaussian-distributed with mean $\hat{\mathbf{y}}_{i+1|i}$ as in (6.25) and covariance matrix $\mathbf{P}_{\mathbf{yy},i+1|i}$ as in (6.26). Therefore, the expectation term on the right-hand side of (6.48) may now be redefined in terms of the measurement space, as follows:

$$\mathbb{E}_{\mathbf{x}_{i+1},\mathbf{y}_{i+1}|I_i,\mathbf{\theta}_i} \mathrm{Tr}[\mathbf{P}_{i+1|i+1}(\mathbf{\theta}_i)] = \mathrm{Tr}[\mathbb{E}_{\mathbf{x}_{i+1},\mathbf{y}_{i+1}|I_i,\mathbf{\theta}_i}\mathbf{P}_{i+1|i+1}(\mathbf{\theta}_i)]$$

$$= \mathrm{Tr}\left[ \int_{R_{N_y}} \underbrace{\mathbf{P}_{i+1|i+1}(\mathbf{\theta}_i)}_{\substack{\text{Nonlinear} \\ \text{function}}} \underbrace{\mathcal{N}(\mathbf{y}_{i+1}; \hat{\mathbf{y}}_{i+1}; \mathbf{P}_{\mathbf{yy},i+1|i}(\mathbf{\theta}_i))}_{\text{Gaussian distribution}} \, d\mathbf{y}_{i+1} \right].$$

(6.50)

Note that, in the first line on the right-hand side of the equation, we have interchanged the orders of expectation and trace operators, which is justified because they are both *linear*. The integrand on the right-hand side of the second-line in (6.50) is now in a form similar to that described in (6.17). Accordingly, using the cubature rule of (6.18) to approximate this integral, we obtain the desired approximation in the measurement space, as shown by

$$\mathbb{E}_{\mathbf{x}_{i+1},\mathbf{y}_{i+1}|I_i,\mathbf{\theta}_i}[\mathrm{Tr}(\mathbf{P}_{i+1|i+1})] \approx \mathrm{Tr}\left( \frac{1}{2N_y} \sum_{i=1}^{2N_y} \mathbf{P}_{i+1|i+1}(\hat{\mathbf{y}}_{i+1|i} + \mathbf{P}_{\mathbf{yy},i+1|i}^{1/2}(\mathbf{\theta}_i)\alpha_i^{1/2}) \right), \quad (6.51)$$

where $N_y$ is the dimensionality of the measurement space. It is important to note that, in (6.51), the posterior expected-error covariance matrix $\mathbf{P}_{i+1|i+1}$ is expressed as a *function* of the sum term $\hat{\mathbf{y}}_{i+1|i} + \mathbf{P}_{\mathbf{yy},i+1|i}^{1/2}(\mathbf{\theta}_i)\alpha_i^{1/2}$ in accordance with the cubature rule, and $\mathbf{P}_{\mathbf{yy},i+1|i}^{1/2}$ is the square root of the measurement covariance matrix $\mathbf{P}_{\mathbf{yy},i+1|i}$. As for the cubature points $\alpha_i$, they are defined in accordance with (6.19), with $2N_y$ denoting the number of cubature points, since the approximation is being performed in the measurement space.

Finally, using the approximation of (6.51) based on the cubature rule, the dynamic-programming algorithm in (6.47)–(6.49) is simplified into the following pair of equations:

*Terminal point*

$$\mathcal{T}(\mathbf{I}_{n+L-1}) \approx \min_{\mathbf{\theta}_{n+L-1} \in \mathcal{P}} \mathrm{Tr}(\mathbf{P}_{n+L|n+L}(\mathbf{\theta}_{n+L-1})). \quad (6.52)$$

*Intermediate points*

$$\mathcal{T}(\mathbf{I}_i) \approx \min_{\mathbf{\theta}_i \in \mathcal{P}} \mathrm{Tr}\left( \mathbf{P}_{i+1|i+1}(\mathbf{\theta}_i) + \frac{1}{2N_y} \sum_{i=1}^{2N_y} \mathbf{P}_{i+1|i+1}(\hat{\mathbf{y}}_{i+1|i} + \mathbf{P}_{\mathbf{yy},i+1|i}^{1/2}(\mathbf{\theta}_i)\alpha_i^{1/2}) \right), \quad (6.53)$$

for $i = n, \ldots, n + L - 2$. In both (6.52) and (6.53), we have made use of the formula in (6.36), assuming the use of the mean-square error criterion for the cost-to-go function. It should also be reemphasized that the mean $\hat{\mathbf{y}}_{i+1|i}$ is independent of the waveform parameter vector $\mathbf{\theta}_i$ but that the covariance matrix $\mathbf{P}_{\mathbf{yy},i+1|i}$ is dependent on it, in accordance with (6.26).

To summarize, the terminal point (6.52) computes the cost-to-go function looking $L$ cycles into the future, where $L$ is the prescribed depth of horizon. Then, starting with the computed cost $T(\mathbf{I}_{n+L-1})$, the approximate dynamic-programming algorithm described in (6.53) takes over by computing the sequence of cost-to-go functions by working *backward in time from the terminal point step by step* until we arrive at the cycle time $n$. At this point, the dynamic-programming computation is completed.

### 6.9.3     Special case: dynamic optimization

The approximate dynamic-programming algorithm of (6.52) and (6.53) includes *dynamic optimization* of the cognitive radar tracker as a special case, for which there is no provision of looking into the future; that is, $L = 1$. In this special case, the terminal point in (6.52) defines the *dynamic optimization algorithm*, with only a single cost-to-go function to be optimized. For example, using the mean-square error formula of (6.36) as we have done in (6.52), we write

$$T(\mathbf{I}_n) = \min_{\boldsymbol{\theta}_n \in \mathcal{P}} \mathrm{Tr}[\mathbf{P}_{n+1|n+1}(\boldsymbol{\theta}_n)]. \tag{6.54}$$

In words, the dynamic optimization algorithm encompasses the top-down (feedback) transmission path, which starts with the cost-to-go function formulated in the transmitter at the current cycle, time $n$, and moves on to the waveform-selection process from $\boldsymbol{\theta}_{n-1}$ to $\boldsymbol{\theta}_n$ to prepare for the next cycle. Most importantly, the whole computation is practically feasible in an *on-line manner* by virtue of setting the horizon depth $L = 1$.

## 6.10     The curse-of-dimensionality problem

The *curse-of-dimensionality problem* was first discussed in Chapter 4 in the context of non-linear filtering, where there are only two spaces to consider: the state space and measurement space. In cognitive radar, however, the problem is compounded in difficulty by two issues:

- the presence of three different spaces and
- predictive modeling, attributed to a horizon of depth into the future in the dynamic-programming algorithm.

The three spaces that characterize cognitive tracking radar are as follows:

(1) *State space.* The system equation for this space describes evolution of the state $\mathbf{x}_n$ over time $n$. Naturally, the state space lies in the radar environment where the target is embedded.
(2) *Measurement space.* The equation for this second space defines dependence of the measurement $\mathbf{y}_n$ on the state $\mathbf{x}_n$ at time $n$. The measurement space lies in the receiver where "perception" of the environment is carried out.
(3) *Action space.* It is in this space where "action" in the radar environment is performed in response to feedback information delivered to the transmitter by the receiver. Accordingly, the action space lies in the transmitter.

In order to put these spaces into an analytic perspective, we introduce the following three-space-related parameters:

- $N_x$, the state-space dimension;
- $N_y$, the measurement-space dimension;
- $N_g$, the waveform-parameter grid size used for the control policy; and
- $L$, the dynamic-programming horizon depth.

The last two parameters naturally relate to the action space. In general, complexity of the dynamic-programming algorithm for finding the control policy in the transmitter is on the order of

$$O\left(N_s^2(2N_y N_g)^L\right), \tag{6.55}$$

where $N_s = \max(N_y, N_x)$ and the term $N_s^2$ is the number of matrix inversions needed for computation of the posterior expected error-covariance matrix; the term $2N_y$ is for the number of cubature points in the measurement space used for computing the approximate summation term in (6.53). Herein, it is assumed that all individual optimizations in each stage of the dynamic-programming algorithm are performed over the complete set of the waveform-library grid.

We see from (6.55) that the main source of complexity lies in the dynamic-programming algorithm due to the exponential growth of computations arising from the horizon depth of $L$ steps into the future; the dimensions $N_y$ and $N_g$ are involved in this exponential growth. More specifically, at each stage of the dynamic-programming algorithm and for each cubature point in (6.53), a new search in the waveform library needs to be performed. We refer to such a complete search of the waveform library as the *global search*. As $L$ increases, the level of computation becomes exponentially unsustainable, and with it the curse-of-dimensionality problem comes into play in a serious way.

To mitigate this problem, we may have to reduce the horizon depth $L$ to try to perform the optimization by searching a *local* neighborhood of the cubature point in the immediate past cycle. Moreover, a case could be made for an *explore–exploit strategy* for waveform selection. In so doing, the dynamic-programming algorithm is constrained to search for a limited-size neighborhood in the waveform-parameter grid that is centered on the old cubature point; this search would have to be repeated at each stage of the dynamic-programming algorithm. The rationale for the explore–exploit strategy can be justified on the grounds that a change in the way in which the receiver perceives the environment is typically small from one cycle to the next. It follows, therefore, that exploring a local neighborhood of the cubature point in the immediate past makes logical sense.

## 6.11   Two-dimensional grid for waveform library

Discussion of the cognitive radar for target tracking would be incomplete without a description of how a two-dimensional grid is used for the library of modulated waveforms in the transmitter. The grid is two-dimensional because the modulated waveform has two parameters, namely the chirp rate $\lambda$ and envelope's pulse duration $b$, as indicated in (6.14).

For practical reasons, the waveform parameter vector $\boldsymbol{\theta}$ is divided into a set of *grid points*, with each grid point represented by a subset in the two-dimensional space of the waveform library $\mathcal{P}$. To elaborate, let $\lambda_{min}$ and $\lambda_{max}$ denote the minimum and maximum values for the chirp rate $\lambda$ respectively; and let $b_{min}$ and $b_{max}$ denote the minimum and maximum values of the envelope's pulse duration respectively. These minimum and maximum values are predetermined for the transmitter. Hence, the waveform library $\mathcal{P}$ is made up of a grid defined as follows:

$$\mathcal{P} = \{\lambda \in [\lambda_{min} : \Delta\lambda : \lambda_{max}], b \in [b_{min} : \Delta b : b_{max}]\}. \qquad (6.56)$$

The $\Delta\lambda$ and $\Delta b$ in (6.56) denote the step sizes of the chirp rate and envelope's pulse duration respectively.

The two-stage analysis–synthesis procedure for construction and deployment of the waveform library $\mathcal{P}$ proceeds as follows:

(1) *Analysis*

A parameter vector, denoted by $\boldsymbol{\theta}_n$ in the library $\mathcal{P}$, is assigned to that particular value of the posterior expected estimation-error covariance matrix $\mathbf{P}_{n+1|n+1}(\boldsymbol{\theta}_n)$ for which the cost-to-go function is minimized. This computation involves the use of (6.32), assisted by (6.23), (6.26), and (6.30). Here, we must reemphasize that evaluating the integrals in these four equations will require application of the cubature rule.

(2) *Synthesis*

This second stage of the procedure involves *grid searching*, where the task of the transmitter is to search over the waveform library $\mathcal{P}$ with the objective of locating the particular grid point $\boldsymbol{\theta}_n^*$, for which the trace or determinant of $\mathbf{P}_{n+1|n+1}(\boldsymbol{\theta}_n)$ attains its minimum value, depending on the criterion adopted for the cost-to-go function. With the optimum parameter vector $\boldsymbol{\theta}_n^*$ so located, the task of grid searching in the transmitter is terminated for the cycle in question.

## 6.12     Case study: tracking a falling object in space

Insofar as cognitive information processing is concerned, the primary purpose of this case study is to demonstrate that information gain is obtained through use of the perception–action cyclic mechanism of Figure 6.1 with a selectable transmit waveform. In practical radar terms, the information gain manifests itself in terms of improved tracking accuracy, compared with the corresponding traditional active radar.

To be specific, we consider an extensively studied problem in the tracking community: the problem of target reentry in space that was first described by Athans *et al.* (1968). More specifically, a ballistic target reenters the Earth's atmosphere after having traveled a long distance at high speed, which makes the remaining time to ground impact relatively short. The goal of a tracking radar in this scenario is to intercept and track

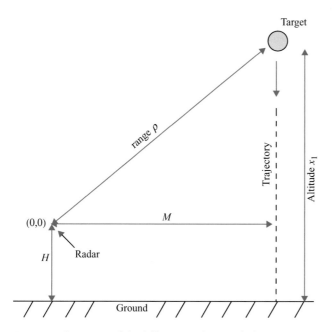

**Figure 6.5.** Geometry of the falling target's scenario in space.

the ballistic target. Figure 6.5 describes the underlying geometry of the falling target's scenario. In the reentry phase, two types of force are in effect:

- The most dominant force is *drag*, which is a function of speed and has a substantial nonlinear variation in altitude.
- The second force is due to *gravity*, which accelerates the target toward the center of the Earth. This tracking problem is particularly difficult because the target's dynamics change ever so rapidly with time.

### 6.12.1    Modeling the reentry problem

Under the combined influence of drag and gravity acting on the target, the following differential equation governs its motion (Athans *et al.*, 1968):

$$
\left.
\begin{aligned}
\dot{x}_1 &= -x_2 \\
\dot{x}_2 &= \underbrace{-\frac{\xi(x_1)gx_2^2}{2x_3}}_{\text{drag}} + g \\
\dot{x}_3 &= 0
\end{aligned}
\right\},
\tag{6.57}
$$

where $x_1$, $x_2$, and $x_3$ are respectively the *altitude*, *velocity*, and *ballistic coefficient* of the three-dimensional state **x**. The ballistic coefficient depends on the target's mass, shape, cross-sectional area, and air density. The dots in the left-hand side of (6.57) denote differentiation with respect to continuous time $t$. The term $\xi(x_1)$ in the second line of (6.57) is air density; it is modeled as an exponentially decaying function of the altitude $x_1$, given by

$$\xi(x_1) = \xi_0 \exp(-\gamma x_1) \tag{6.58}$$

with the proportionality constant $\xi_0 = 1.754$ and the exponential factor $\gamma = 1.49 \times 10^{-4}$. In the second line of (6.57), the acceleration due to gravity $g = 9.8 \text{ ms}^{-2}$.

To convert (6.57) into a form that corresponds to the state-space model, we define the three-dimensional state as

$$\mathbf{x} = [x_1, x_2, x_3]^\mathrm{T}.$$

The system equation for continuous time $t$ can now be expressed by the first-order non-linear differential equation

$$\dot{\mathbf{x}}_t = \mathbf{f}(\mathbf{x}_t),$$

where the dot in $\dot{\mathbf{x}}_t$ denotes differentiation with respect to time $t$ and $\mathbf{f}(\cdot)$ is a known non-linear function. Using the *Euler approximation* with a small integration step $\delta$, we may reformulate this differential equation in the desired discrete form:

$$\begin{aligned}\mathbf{x}_n &= \mathbf{x}_{n-1} + \delta \mathbf{f}(\mathbf{x}_{n-1}) \\ &= \mathbf{a}(\mathbf{x}_{n-1}),\end{aligned} \tag{6.59}$$

where, as before, $n$ denotes discrete time. In order to account for imperfections in the "noiseless" system equation (6.59) that are caused by the combinations of lift force, small variations in the ballistic coefficient, and spinning motion, we add the zero-mean Gaussian system noise $\boldsymbol{\omega}_n$, obtaining the expanded noisy system equation:

$$\mathbf{x}_n = \mathbf{a}(\mathbf{x}_{n-1}) + \boldsymbol{\omega}_n, \tag{6.60}$$

where we have

$$\mathbf{a}(\mathbf{x}_{n-1}) = \boldsymbol{\Phi}\mathbf{x}_{n-1} - \mathbf{G}[D(\mathbf{x}_{n-1}) - g]. \tag{6.61}$$

The two new matrices $\boldsymbol{\Phi}$ and $\mathbf{G}$ in (6.61) are defined as follows:

$$\boldsymbol{\Phi} = \begin{pmatrix} 1 & -\delta & 0 \\ 0 & 1 & 0 \\ 0 & 0 & 1 \end{pmatrix} \tag{6.62}$$

and

$$\mathbf{G} = \begin{bmatrix} 0 & \delta & 0 \end{bmatrix}^\mathrm{T},$$

where $\delta$ is the small integration step. The drag $D(\mathbf{x}_n)$, expressed in terms of the three components of the state vector $\mathbf{x}_n$, namely $x_{1,n}$, $x_{2,n}$, and $x_{3,n}$ at time $n$, is defined by

$$D(\mathbf{x}_n) = \frac{\xi(x_{1,n}) g x_{2,n}^2}{2 x_{3,n}} \tag{6.63}$$

where $\xi(x_{1,n})$ is the air density of (6.58) evaluated at time $n$. We assume that the system noise $\omega_n$ is zero-mean Gaussian with the covariance matrix

$$
\mathbf{Q} = \begin{pmatrix} q_1 \dfrac{\delta^3}{3} & q_1 \dfrac{\delta^2}{2} & 0 \\[2mm] q_1 \dfrac{\delta^2}{2} & q_1\delta & 0 \\[2mm] 0 & 0 & q_2\delta \end{pmatrix}. \tag{6.64}
$$

The parameters $q_1$ and $q_2$ control the amount of system noise in target dynamics and ballistic coefficient respectively. For the simulations to be presented, we used the values $q_1 = 0.01$, $q_2 = 0.01$, and $\delta = 1\,\mathrm{s}$. The system equation (6.60) for the object falling in space is now fully described.

## 6.12.2   Radar configurations

We use LFM with both up-sweep and down-sweep chirps and Gaussian pulse-amplitude modulation, which composes the waveform library with the parameter vector

$$
\boldsymbol{\Theta} = \{\lambda \in [10^{-6},\ 300\times 10^{-6}],\ b \in [-3\times 10^{10},\ -10^{10}] \cup [10^{10},\ 30\times 10^{10}]\}
$$

and grid step-sizes $\Delta\lambda = 10^{-5}$ and $\Delta b = 10^{10}$. The bandwidth was set to be 5 MHz. The 0 dB SNR was set at 80 km. Referring to Figure 6.5, an X-band radar is fixed at the point labeled (0, 0) and operated at a fixed carrier frequency of 10.4 GHz, with the speed of electromagnetic-wave propagation $c = 3 \times 10^8$ m/s. The depth of horizon for the cognitive tracking radar is fixed at two different values, $L = 1$ and $L = 2$.
Other points of interest:

- The traditional active radar was equipped with fixed waveform parameters: down-sweep chirp rate and pulse duration $\lambda = 20$ μs.
- The sampling rate was set to $T_s = 100$ ms and the simulations were conducted for 50 Monte Carlo runs.
- The radar was located at height $H = 30$ m measured with respect to ground; and its horizontal distance to the track (trajectory) $M = 30$ km.
- The measurement space is two-dimensional, consisting of range and range rate. Denoting these two radar parameters by $y_{1,n}$ and $y_{2,n}$ at time $n$ respectively, we write

$$
y_{1,n} = \sqrt{M^2 + (x_{1,n} - H)^2} + v_{1,n}, \tag{6.65}
$$

$$
y_{2,n} = \frac{x_{2,n}(x_{1,n} - H)}{\sqrt{M^2 + (x_{1,n} - H)^2}} + v_{2,n}, \tag{6.66}
$$

where $v_{1,n}$ and $v_{2,n}$ are the two elements of measurement noise vector $\mathbf{v}_n$. For the Gaussian-distributed measurement noise, we have

$$
\mathbf{v}_n \sim \mathcal{N}(\mathbf{0}, \mathbf{R}(\boldsymbol{\theta}_{n-1})),
$$

where the noise covariance matrix $\mathbf{R}(\boldsymbol{\theta}_{n-1})$ is dependent on the waveform parameter vector $\boldsymbol{\theta}_{n-1}$. Note also that with dimensionality of the measurement space being two, $N_y = 2$ in the approximate formula for the dynamic programming algorithm in (6.53).

With the transmitter and the receiver being collocated, the received signal energy depends inversely on the fourth power of the target range $\rho$. For this reason, the SNR of the radar returns for the target observed at range $\rho$ was modeled according to

$$\eta_\rho = \left(\frac{\rho_0}{\rho}\right)^4, \tag{6.67}$$

where $\rho_0$ denotes the range at which we have the 0 dB SNR. To reiterate, for the simulations, $\rho_0$ was set to be 80 km.

The *true* initial state of the target was defined as

$$\mathbf{x}_0 = [61 \text{ km}, 3048 \text{ m/s}, 19\,161]^T$$

and the initial state estimate and its covariance matrix were respectively assumed to be

$$\hat{\mathbf{x}}_{0|0} = [61.5 \text{ km}, 3400 \text{ m/s}, 19\,100]^T$$

and

$$\mathbf{P}_{0|0} = \text{diag}(10^6, 10^4, 10^4).$$

## 6.12.3    Performance metric

With simplified computational cost in mind, we used the ensemble-averaged root mean-square error (EA-RMSE) as the metric to evaluate performance of the fore-active tracking radar, compared with the traditional active radar of fixed transmit-waveform. The EA-RMSE is defined as

$$\text{EA-RMSE}(n) = \sqrt{\frac{1}{N}\sum_{k=1}^{N}\left(x_{1,n}^{(k)} - \hat{x}_{1,n}^{(k)}\right)^2 + \left(x_{2,n}^{(k)} - \hat{x}_{2,n}^{(k)}\right)^2}, \tag{6.68}$$

where $(x_{1,n}^{(k)}, x_{2,n}^{(k)})$ and $(\hat{x}_{1,n}^{(k)}, \hat{x}_{2,n}^{(k)})$ are respectively the true and estimated positions of the target at time $n$ in the $k$th Monte Carlo run with $n = 1, 2, \ldots$, and $N$ is the total number of simulation runs. In a similar manner, we may also define the EA-RMSE for range rate.

## 6.12.4    Simulation results

Here we show the simulation results for two different scenarios (Xue, 2010):

- The perception–action cycle of Figure 6.1 with selectable transmit-waveform. As mentioned previously, this mechanism is the first step towards radar cognition. The purpose of the case study is to demonstrate the information gain resulting from the perception–action cycle as the first stage towards cognition.
- The corresponding traditional active radar with fixed transmit-waveform, which is included so as to provide a frame of reference for the information gain.

With the curse-of-dimensionality problem as an issue of computational concern, we study two cases: $L = 1$ for the special case of dynamic optimization and $L = 2$ for the case of dynamic programming. Figures 6.6, 6.7, and 6.8 respectively plot the RMSEs for the following three elements of target's state: altitude, velocity, and ballistic coefficient, all three of which occupy the same time scale.

## 6.12.5   Comments on the simulation results

Examination of the RMSE results plotted in Figures 6.6–6.8 leads us to make three important observations:

(1) In all three figures, the perception–action cyclic of Figure 6.1 outperforms the traditional active radar with fixed transmit-waveform by a large margin.

(2) For estimates of all three target parameters, altitude, velocity, and ballistic coefficient, the trajectories computed for horizon depths $L = 1, 2$ appear to be essentially identical until we reach time $n = 3\,\mathrm{s}$. Thereafter, the use of dynamic programming for $L = 2$ begins to increasingly outperform the use of dynamic optimization for $L = 1$. Of course, this improvement in estimation accuracy is achieved at the expense of increased computational complexity.

(3) Tracking accuracies of the target altitude and target velocity follow roughly similar trajectories, which appears to indicate that the choice of linear frequency modulation combined with amplitude modulation is a good compromise for transmit-waveform selection by the cognitive radar.

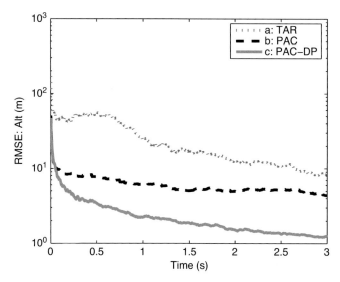

**Figure 6.6.** RMSE of altitude for: (a) traditional active radar with fixed waveform (TAR); (b) perception–action cycle with dynamic optimization (PAC); (c) perception–action cycle with dynamic programming (PAC- DP).

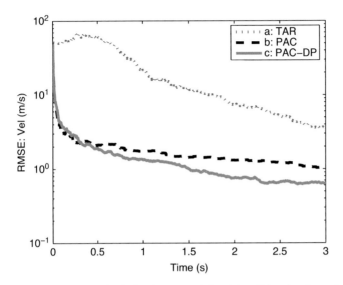

**Figure 6.7.** RMSE of velocity (range rate) for: (a) traditional active radar with fixed waveform (TAR); (b) perception–action cycle with dynamic optimization (PAC); (c) perception–action cycle with dynamic programming (PAC-DP).

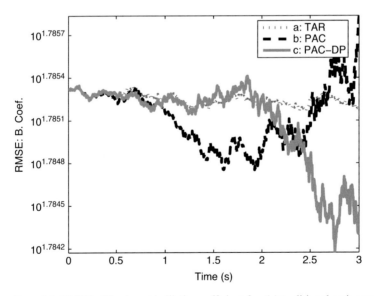

**Figure 6.8.** RMSE of the target ballistic coefficient for: (a) traditional active radar with fixed waveform (TAR); (b) perception–action cycle with dynamic optimization (PAC); (c) perception–action cycle with dynamic programming (PAC-DP).

The message to take from this case study is confirmation that there is significant information gain and accelerated rate of convergerce for range and range-rate estimietion to be had by virtue of the perception–action cycle as the first stage towards radar cognition, compared with the corresponding traditional active radar.

## 6.13     Cognitive radar with single layer of memory

Now that we have thoroughly studied the practical benefit gained from the perception–action cycle providing the first functional pillar for a radar to be cognitive, we next turn our attention on the benefit to be gained from the second functional pillar for cognition: *memory*.

To be specific, the memory to be considered consists of a single layer: perceptual memory in the receiver, executive memory in the transmitter, and working memory. Specifically, as shown in Figure 6.9, we have the following linkages:

(1) *The perceptual memory* is reciprocally coupled to the *environmental scene analyzer* in the receiver.
(2) *The executive memory is* reciprocally coupled to the environmental scene actuator in the transmitter.
(3) *The working memory* reciprocally couples the perceptual memory and executive memory together.

With the addition of memory, the cognitive radar of Figure 6.9 now satisfies all four basic properties of cognition, as outlined in Section 6.1 and reiterated here for convenience of presentation:

• perception in the receiver followed by action in the transmitter through global feedback from the receiver to the transmitter;

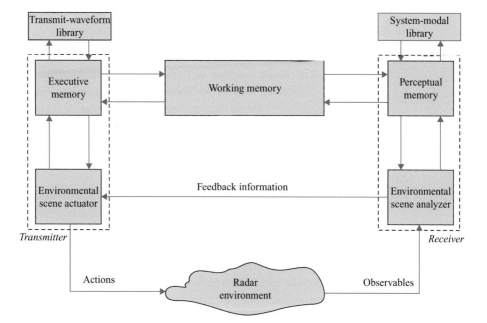

**Figure 6.9.** Cognitive radar with single level (layer) of memory.

- memory configured to predict the consequences of action (i.e. transmit-waveform selection taken by the transmitter and state-equation's parameter selection made by the receiver);
- memory-based attentional mechanism for the prioritized allocation of available resources; and
- decision-making mechanism for selecting intelligent choices in the face of environmental uncertainties.

As emphasized previously, intelligence builds on the perception–action cycle, memory and attention; moreover, it relies on the many local and global feedback loops in the cognitive radar for its facilitation.

Most importantly, we expect that the addition of memory to the perception–action cycle of Figure 6.1 will contribute its own enhancement to the information-processing power of the cognitive radar, which will be demonstrated experimentally later on in the section.

## 6.13.1　Cyclic directed information flow in cognitive radar with single layer of memory

Although the structure of the cognitive radar with one layer of memory in Figure 6.9 is much more elaborate than that of the perception–action cycle pictured in Figure 6.1, it is logically intuitive to find that, insofar as evaluation of the feedback information is concerned, these two structures are both governed by the same cyclic directed information flow of Figure 6.5 with a fundamental difference:

The successive information-processing cycles carried out by the triple-memory system, within their own local and global feedback loops, have the resultant effect of always reducing the amount of feedback information passed to the transmitter by the receiver, continually from one cycle to the next.

To justify this profound statement, we make two related observations:

(1) *The perception–action cycle acting alone*

For prescribed nonlinear function **a**(.) and system noise covariance matrix in the receiver and prescribed waveform library in the transmitter, the information-processing capability of the perception–action cycle of Figure 6.1 is powerful enough to select the optimal transmit-waveform to illuminate the environment. However, this optimum selection "may not be the best" using prescribed parameters, as the system equation could be misaligned with respect to the actual state of the target.

(2) *The perception–action cycle with added memory*

With the addition of a library of nonlinear state-transition functions and system noise variances assigned to the perceptual memory in the receiver in Figure 6.9, we have *balanced* the resources (libraries) available to the transmitter and receiver of the cognitive radar. Accordingly, through bottom-up and top-down adaptations performed between the environmental scene analyzer and perceptual memory in the receiver, and the corresponding bottom-up and top-down adaptations performed between the environmental scene actuator and executive memory in the transmitter, both being augmented by the reciprocal coupling between the perceptual memory and executive memory through the working memory, the triple-memory system just described impacts the information-processing cycle of Figure 6.5 through two successive optimized selections:

- First, the CKF is enabled to optimize the filtered estimate of the state, given the nonlinear function $\mathbf{a}(.)$ and system noise variance retrieved by the perceptual memory, so as to match the incoming measurements.
- Second, the approximate dynamic programming optimizes retrieval of the transmit-waveform for the next cycle, given the subset of waveform library selected by the executive memory.

Consequently, at each cycle in time in Figure 6.9, we find that the amount of feedback information sent from the receiver to the transmitter is successively reduced as we proceed from one cycle to the next. In other words, the use of memory, an inherent structural ingredient of cognitive radar, will always result in improved tracking accuracy, albeit at the expense of increased system complexity.

## 6.13.2    Communication among subsystems in cognitive radar

To understand how the subsystems constituting the cognitive radar communicate with each other, we may refer to Figure 6.10. The communication process in this figure starts with the measurements received at the radar input from the environment, thereafter proceeding along the following six steps:

*Step 1*. Through the bottom-up link, the receiver sends the current measurements to the perceptual memory; the goal here is to retrieve the particular nonlinear function $\mathbf{a}(.)$ and associated system noise variance that are the best match to the current measurements.

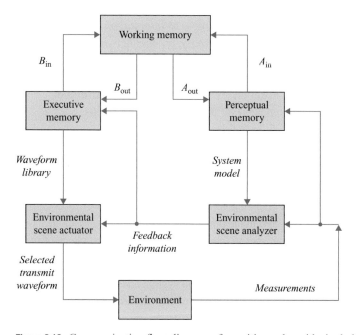

**Figure 6.10.** Communication flow-diagram of cognitive radar with single level of memory.

*Step 2.* Through the top-down link, the new parameters of the system model are fed back to the environmental scene analyzer, thereby updating the system equation (6.7) and, with it, the state-space model of the radar environment. Meanwhile, the new features about the environment learned in the hidden layer of the perceptual memory are sent as *Information* $A_{in}$ to the working memory to be temporarily stored for future use in Step 6.

*Step 3.* With the updated state-space model at hand in the environmental scene analyzer, the CKF is now ready to process the received measurements for computing the predicted state-estimation error vector one step into the future. Step 3 is completed by computing the feedback information, based on this error vector, which is sent to the environmental scene actuator in the transmitter by the environmental scene analyzer.

*Step 4.* Through the bottom-up link, the feedback information is sent to the executive memory; the goal here is to retrieve that particular subset of the transmit-waveform library that is the best fit to the radar environment represented by the feedback information. Meanwhile, the features about the feedback information and, therefore, the environment learned in the hidden layer of the executive memory are sent as *Information* $B_{in}$ to be temporarily stored in the working memory for Step 6.

*Step 5.* Through the top-down link, the new subset of waveform library is fed back to the environmental scene actuator. Therein, the cost-to-go function is formulated, at which point in time the optimum transmit waveform is selected from the retrieved set of waveform library by the dynamic-programming algorithm for action in the environment.

*Step 6.* Finally, with the features obtained from the hidden layer of the perceptual memory as *Information* $A_{in}$ and the corresponding features obtained from the hidden layer of the executive memory as *Information* $B_{in}$, these two sets of features are processed in the working memory, producing two outputs: the output $A_{out}$, corresponding to *Information* $A_{in}$, is sent to the perceptual memory; and the other output $B_{out}$, corresponding to *Information* $B_{in}$, is sent to the executive memory as the consequences of action taken in the receiver or transmitter.

Then, Steps 1–6 are repeated for the next cycle, and so on.

### 6.13.3    Communications between scene analyzer and perceptual memory

Irrespective of the application of interest, a cognitive radar's perception of the environment is continually affected by the current measurements as well as cognitive information about the environment stored in the perceptual memory. Stated in another way, we may say that every percept (i.e. snapshot of the perception process) is made up of two components:

- The first component of the percept refers to *recognition* and, therefore, retrieval of a set of nonlinear functions and associated system noise variances, which are stored in the perceptual memory for the purpose of representing past measurements.
- The second component of the percept refers to *categorization* (classification) of features in the new measurements that are *matched* with the memory.

The end result of these two components is the updated selection of nonlinear state-transition function and system noise variance that match the current measurements. In both components, the processing is performed in a self-organized and synchronous manner.

### 6.13.4   Communications between environmental scene actuator and executive memory

Just as the receiver is continually influenced by incoming measurements, so it is with the transmitter being continually influenced by the incoming feedback information about the environment from the receiver. Stated in another way, every execution (i.e. snapshot of the decision-making process) is made up of two components:

- The first component refers to *recognition* and, therefore, retrieval of a particular set in the waveform library, which is stored in the executive memory for the purpose of representing past values of feedback information from the receiver.
- The second component refers to *categorization* of the current feedback information that is *matched* with the memory.

The end result is selection of that updated transmit-waveform in the waveform library that matches the current value of feedback information. Here again, the processing in both components is carried out in a self-organized and synchronous manner.

### 6.13.5   Communications within the triple-memory system

Finally, we come to the communication process carried out between the perceptual memory in the receiver and the executive memory in the transmitter. This communication is carried out through the *working memory*, which is reciprocally coupled to the perceptual memory on the one side and reciprocally coupled to the executive memory on the other side. In effect, the working memory acts as on the "mediator" between the two memories through a *matching process*.

Referring to Figure 6.10, we have *Information* $A_{in}$ that is input into the working memory from the perceptual memory. In light of what was described previously under Step 2, this information represents the features learned in the hidden layer of the perceptual memory. In other words, $A_{in}$ provides the working memory with information about the environment, viewed *directly* through perception by the environmental scene analyzer.

Moreover, we also have *Information* $B_{in}$ that is input into the working memory from the executive memory. In light of what was described under Step 4, this second information represents the features learned in the hidden layer of the executive memory. In other words, $B_{in}$ provides the working memory with information about the radar environment, viewed *indirectly through the feedback link* from the environmental scene analyzer.

It follows, therefore, that there will be *matching* between the two inputs: $A_{in}$ and $B_{in}$. Accordingly, it is the function of the working memory to exploit this matching, thereby producing the respective outputs, $A_{out}$ and $B_{out}$. The outputs $A_{out}$ and $B_{out}$ are fed back to the perceptual memory and the executive memory respectively, supplying them both the *relationships discovered*: first, between the particular nonlinear function

**a**(.) and associated system noise variance selected in the receiver for *perception* of the environment and, second, between the particular waveform selected in the transmitter for action in the environment. In so doing, the working memory reports a pair of *consequences* for actions taken:

- One consequence for the perceptual memory is to have selected the function, **a**(.), and system-noise variance as the possible match for the state equation (6.7), given the current measurements.
- The other consequence for the executive memory is to have selected a particular transmit waveform as the possible match for the incoming feedback information.

This twofold statement is in accord with the previous statement first made in Chapter 2: the function of memory is to predict the "consequences of selection/action taken" by a cognitive dynamic system and do so in a self-organized and synchronous manner.

Although the radar environment is *common* to both the receiver and transmitter, they view it differently: the receiver perceives the environment *directly* through the incoming measurements, whereas the transmitter views it *indirectly* through feedback information from the receiver. Obviously, then, there must be matching between the environmental modeling process carried out in the perceptual memory and the waveform-categorization process carried out in the executive memory. Through its predictive capability, the function of the working memory is to maximize this matching process on each cycle, so that action taken in the environment by the transmitter matches perception of the environment by the receiver in a way better than it did on the previous cycle.

## 6.13.6    Experimental validation of improved tracking performance attributed to memory

To validate the improvement in target-tracking accuracy resulting from addition of the triple-memory system in Figure 6.9, we present the results herein of a difficult computer experiment, in which a target performs a "coordinated turn" in a three-dimensional Cartesian space. The dimensionality of the state-space model is seven, made up of range and velocity in each of the three Cartesian coordinates plus the curvature of the coordinated turn itself; this radar scenario was discussed previously in Chapter 4. Three scenarios are considered in the computer experiment (Xue, 2010):

- a traditional active radar with fixed transmit waveform;
- the perception–action cycle pictured in Figure 6.1 with selectable transmit-waveform, representing the first step towards radar cognition; and
- the cognitive radar pictured in Figure 6.9.

The target parameters are as follows:

- the initial range across the three Cartesian coordinates is [10 km, 10 km, 5 km]
- the corresponding initial range-rate is; and [100 km/s, 150 km/s, 0 m/s]
- the corner turning rate: 20°.

The results of the experiment are presented in Figure 6.11.

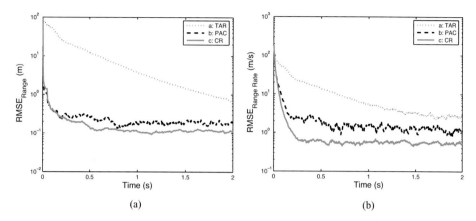

**Figure 6.11.** RMSE for: (a) range; (b) range rate. TAR: traditional active radar with fixed transmit waveform; PAC: perception–action cycle as the first stage towards radar cognition; CR: cognitive radar.

Figure 6.11a plots the RMSE for range estimation for the three scenarios: Figure 6.11b plots the corresponding RMSE for range-rate (velocity) estimation. Examination of the figure leads to the performance results summarized in Table 6.1.

From the graphical plots of Figure 6.11 and the summarizing results presented in Table 6.1, we now make three observations:

(1) We see successive improvements in estimating the target range and velocity as we proceed through the perception–action cycle of Figure 6.1 with selectable transmit-waveform, and then the cognitive radar of Figure 6.9, both compared with the corresponding traditional active radar with fixed transmit-waveform.

(2) The perception–action cycle improves the accuracy of range estimation more so than velocity estimation.

(3) On the other hand, the cognitive radar improves the accuracy of velocity estimation more profoundly than range estimation; that is, the opposite to that under point (2).

Moreover, convergence rates of estimating both range and range rate are accelerated considerably, compared with the traditional active radar.

**Table 6.1** Comparison of the three scenarios for tracking accuracy

|  | $\text{RMSE}_{\text{range}}$ (m) | $\text{RMSE}_{\text{velocity}}$ (m/s) |
|---|---|---|
| (i) Traditional active radar with fixed transmit-waveform | 0.7 | 1.2 |
| (ii) Perception–action cycle with selectable transmit waveform | 0.2 | 1.0 |
| (iii) Cognitive radar | 0.1 | 0.5 |

## 6.14    Intelligence for dealing with environmental uncertainties

In Section 6.1, following Chapter 2, we made the statement that to be cognitive a radar has to embody four processes: the perception–action cycle, memory, attention, and intelligence. Among these four processes, intelligence is the most complex process involved in cognition and the hardest one to define. The difficulty in pinning down the role of intelligence in cognition is attributed to the fact that the other three processes, perception, memory, and attention, make their individual contributions to intelligence in varying degrees. Nevertheless, in Chapter 2, insofar as intelligence is concerned, we did make the following statement, reiterated here for convenience of presentation:

Intelligence is the ability of a cognitive dynamic system to continually adjust itself through an adaptive process by making the receiver respond to new changes in the environment so as to create new forms of action and behavior in the transmitter.

This is a very profound statement; unfortunately, it cannot be proven mathematically, not yet. An informative way to develop insight into the role of intelligence in cognitive dynamic systems is through experimental means. In this context, the cognitive radar with one level of memory is well suited for such a study.

### 6.14.1    Experimental demonstration of the information-processing power of cognition in the presence of environmental disturbance

To explore the importance of cognition experimentally, we revisit the simple computer experiment presented in Section 1.4 of the introductory chapter. As in that section, the state-space model of the radar is described by the following pair of linear equations:

(1) *System equation*

$$\mathbf{x}(n) = \mathbf{A}\mathbf{x}(n-1) + \boldsymbol{\omega}(n). \tag{6.69}$$

(2) *Measurement equation*

$$\mathbf{y}(n) = \mathbf{B}\mathbf{x}(n) + \mathbf{v}(\boldsymbol{\theta}_{n-1}). \tag{6.70}$$

Compared to the experiment in Section 1.4, we have made only one "significant" change: whereas the system noise $\omega_n$ was white and Gaussian in Section 1.4, its power spectrum is now no longer constant; rather, its intensity changes with time. Despite this difference, $\omega_n$ is still Gaussian distributed, which, in turn, means that we are still dealing with a state-space model that remains to be linear and Gaussian. Accordingly, use of the Kalman filter satisfies the perceptual needs of the receiver, as was the case in Section 1.4.

With the state consisting of target range and range-rate, the transition matrix $\mathbf{A}$ in (6.69) is a two-by-two matrix defined by

$$\mathbf{A} = \begin{bmatrix} 1 & T \\ 0 & 1 \end{bmatrix},$$

where $T$ is the sampling period. Correspondingly, the covariance matrix of the system noise $\boldsymbol{\omega}_n$ is defined by

$$\mathbf{Q} = \sigma^2 \begin{bmatrix} \frac{1}{3}T^3 & \frac{1}{2}T^2 \\ \frac{1}{2}T^2 & T \end{bmatrix},$$

where $\sigma^2$ characterizes the "intensity" of system noise $\boldsymbol{\omega}_n$, which may vary with time; its dimension is measured in $(\text{length})^2/(\text{time})^3$; see Bar-Shalom *et al.* (2001). Figure 6.12a and b plots a sample waveform of the system noise and its bell-shaped amplitude spectrum, which are respectively characterized as follows:

(1) The amplitude waveform includes a *disturbance*, whose duration extends from 2 to 4 s with intensity $\sigma^2 = 16$; outside this duration, the system noise is essentially white with variance $\sigma^2 = 0.49$.

(2) The disturbance was generated by passing white Gaussian noise through a linear filter with a Gaussian-shaped transfer function, thereby yielding the bell-shaped amplitude spectrum plotted in Figure 6.11b.

Thus, as already mentioned, insofar as the distribution of the overall system noise is concerned, it remains to be Gaussian.

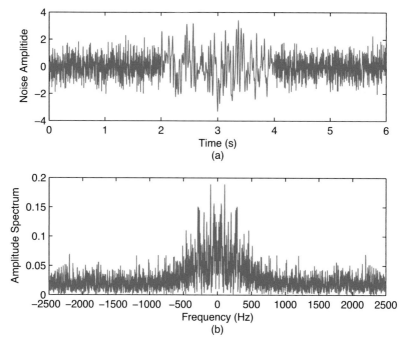

**Figure 6.12.** (a) Waveform of system noise $\boldsymbol{\omega}_n$ that includes a disturbance extending from 2 to 4 s. (b) Amplitude spectrum of the system noise $\boldsymbol{\omega}_n$.

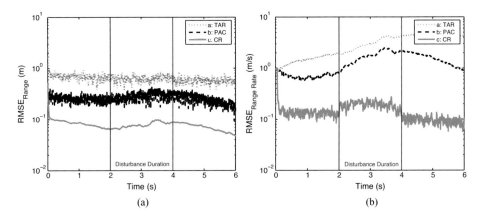

**Figure 6.13.** Illustrating the impact of unexpected disturbance arising in three different radar scenarios: traditional active radar (TAR), perception–action cycle (PAC), and cognitive radar. The results presented in the figure pertain to: (a) range estimation; (b) range-rate estimation.

It is assumed that the target is traveling towards the radar along a straight line from a distance of 3 km and with a speed of 200 m/s. Figure 6.13a and b plots the RMSE results for range and range-rate respectively for all three scenarios as usual.

Examining the experimental results presented in Figure 6.13b, where the case of range-rate estimation appears to be most sensitive to environmental uncertainties, three observations are made:

(1) Tracking performance of the traditional active radar, with a prescribed system model and a fixed transmit waveform, is severely sensitive to the environmental uncertainties occurring in the system noise $\omega_n$; this deficiency in tracking performance is attributed to a lack of intelligence in the radar system.

(2) The perception–action cycle, the first step towards radar cognition, appears to exhibit two undesirable deficiencies of its own:

  • First, tracking performance remains sensitive to environmental uncertainties, albeit to a smaller extent than in the traditional active radar; this slightly reduced sensitivity is attributed to the "limited form of intelligence" facilitated by the embodiment of global feedback around the environment.

  • Second, once the disturbance comes to an end, the radar tracker is unable to recover its trajectory prior to occurrence of the disturbance.

(3) Overall, the cognitive radar handles the presence of environmental uncertainties in the system noise $\omega_n$ reasonably well; most importantly, the radar's tracking trajectry provides a clear timing indication of where the disturbance actually occurred.

The conclusion to be drawn from these experimental observations can be summarized as follows:

The intelligent decision-making capability of cognitive radar in dealing with environmental uncertainties is enhanced significantly by two processes built into the system design:

(1) The use of an integrated memory system consisting of perceptual memory, executive memory, and working memory.

(2) The increased number of local and global feedback loops distributed throughout the cognitive radar, with the end result that the cognitive radar is capable of adjusting itself through an adaptive process to withstand the presence of environmental uncertainties.

To elaborate on this statement, from feedback theory it is well known that feedback is a *double-edged sword* in the following sense: feedback can work for you or it can work against you. In the context of this statement just made on intelligence, the distributed feedback loops under point (2) provide the basis for improved tracking performance, but, at the same time, they may raise the potential for instability (i.e. oscillatory behavior). This is where the integrated use of memory appears to come to the rescue by acting as the *stabilizer*. Although, as of yet, we have no mathematical proof for this statement, the simulations presented in Figure 6.13b do justify the practical importance of memory and intelligence in a cognitive radar, which is intuitively satisfying.

## 6.15   New phenomenon in cognitive radar: chattering

In the course of the computer experiment just described, a highly interesting phenomenon, called the *chattering phenomenon*, was discovered (Xue, 2010). This new phenomenon refers to "intensity enhancement" in random fluctuations in the measurement noise waveforms observed in the estimation of range as well as range-rate, and the observation applies to two scenarios: the cognitive radar and the first step to radar cognition (i.e. the perception–action cycle), both of which involve the use of feedback. The interesting point is that, despite this enhancement in the measurement noise in both of these scenarios, the tracking performance is improved over that of the traditional active radar as it is chatter free due to the lack of global feedback. The observations just made are substantiated by examining the measurement noise processes for the usual three scenarios (Xue, 2010): traditional active radar, the perception–action cycle of Figure 6.1, and cognitive radar of Figure 6.9. Sample waveforms of the experimental results are presented in Figure 6.14 for range estimation.

To explain the reasons behind the chattering phenomenon, we offer the following statements:

(1) The measurement noise in the receiver is indirectly *controlled* by the transmitted signal through the radar returns from the target, which is a fact regardless of whether the radar is cognitive or not. Indeed, it is because of this fact that the measurement noise in the state-space model of the radar is denoted by $v(\theta_{n-1})$, where $\theta_{n-1}$ is the waveform-parameter vector responsible for generating the transmitted signal at time $n - 1$.
(2) The single feature that is common to the cognitive radar and its perception–action cycle mechanism acting alone, and which distinguishes both of them from the traditional active radar, is global feedback from the receiver to the transmitter.
(3) The reason for a significantly more intensive chattering in cognitive radar (shown in Figure 6.14c) than that in the corresponding perception–action cycle is attributed to the fact that cognitive radar has more global feedback loops (embodying the environment) than the perception–action cycle.

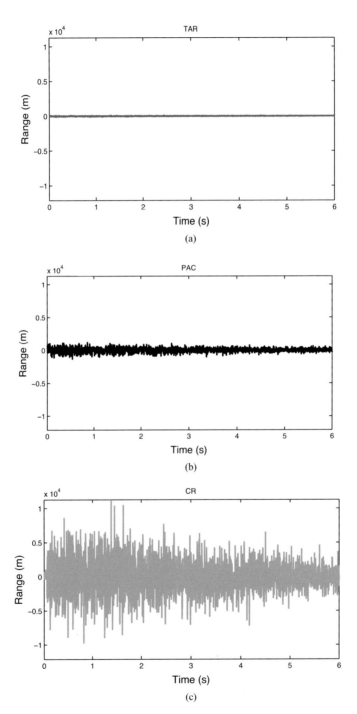

**Figure 6.14.** The chattering phenomenon experienced in the estimation of range for (a) traditional active radar (TAR), (b) perception–action cycle (PAC) for first stage towards radar cognition, and (c) cognitive radar (CR).

So, the conclusion to be drawn from these two points prompts us to make the following statement:

The chattering phenomenon is an inherent property of "all" radars with global feedback, be they cognitive or not.

Interestingly enough, the chattering phenomenon is also known to occur in ordinary nonlinear feedback control systems (Utkin, 1992; Khalil, 2002; Habibi and Burton, 2003; Utkin *et al.*, 2009). However, the chattering phenomenon appears to operate differently in cognitive radars than in ordinary closed-loop feedback control systems, as summarized here:

(1) In cognitive radar, the chattering phenomenon is attributed to measurement noise observed in the receiver, which is indirectly affected by the controlling action of the transmitter; most importantly, the chattering appears to be "harmless."
(2) In ordinary nonlinear feedback control systems, on the other hand, the chattering phenomenon is attributed to system noise; the phenomenon, therefore, is considered to be "harmful," as it leads to a low control accuracy.

One other point that should be emphasized, which is common to cognitive radar and ordinary nonlinear feedback control systems: the actions taken in both cases must be *discontinuous*.

So, we conclude this discussion on the chattering phenomenon with the follow-up statement (Xue, 2010):

The chattering effect is a physical phenomenon characterized by high-frequency and finite-amplitude oscillations that appear in a closed-loop feedback control system, where actions are applied discontinuously.

In the case of a cognitive radar, the action taken by the transmitter is certainly discontinuous by virtue of the fact that the transmit waveform is selected in a discontinuous manner as the controller (i.e. approximate dynamic-programming algorithm) switches from one grid point to another in the waveform library.

## 6.16   Cognitive radar with multiscale memory

In light of what we now know about the information-processing power of cognition applied to radar, albeit confined to a single layer of memory as in Figure 6.9, we may further expand the *structural complexity* of cognitive radar by introducing *multiscale memory* into the perception–action cycle building on the single layer of memory along the lines depicted in Figure 6.9.

The multiscale distributed memory of this new cognitive radar is structured in Figure 6.15 as follows:

(1) *Multiscale perceptual memory*, which is reciprocally coupled with the environmental scene analyzer in the receiver.

(2) *Multiscale executive memory*, which is reciprocally coupled with the environmental scene actuator in the transmitter.

(3) *Working memory*, which couples the supervised layer of the perceptual memory on the right-hand side of Figure 6.15 with the corresponding supervised layer of the executive memory on the left-hand side of the figure. In so doing, the perceptual memory and executive memory are enabled to operate in *synchrony* with each other from one cycle to the next. In implementational terms, we may think of the working memory as a *heterogeneous associative memory* that is activated for a relatively short period of time; this kind of memory was discussed in Chapter 2.

Also, as explained in Chapter 2 on the perception–action cycle, a "deep" neural architecture for the distributed memory system permits the representation of a wide family of signal-processing functions in a more compact fashion than is possible with a single-layered "shallow" architecture. Simply stated:

The use of a multiscale memory makes it possible to trade off space for time, wherein features extracted by the individual layers of memory become increasingly more abstract and, therefore, easier to recognize as we go up in the hierarchy.

Just as importantly, as a result of using the complex multiscale memory system, the numbers of global and local feedback loops distributed throughout the entire cognitive radar tend to increase *exponentially*. In a corresponding way, the radar acquires *multiple levels of cyclic directed information flows*, which, in turn, strengthen the radar's capabilities of perception, memory, attention, and intelligence. In other words, the information-processing power of a cognitive radar is enhanced with the increased *depth* of the multiscale memory in Figure 6.15, up to a certain point.

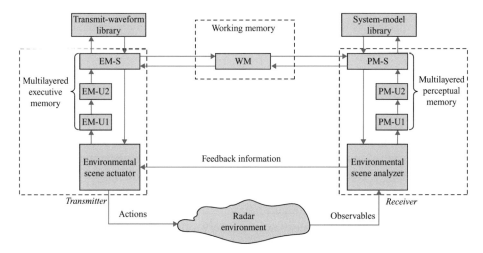

**Figure 6.15.** Cognitive radar with multiscale memory. Acronyms: (1) Perceptual memory: PM; unsupervised learning: PM-U1, and PM-U2; supervised learning: PM-S. (2) Executive memory: EM; unsupervised learning: EM-U1, and EM-U2; supervised learning: EM-S.

### 6.16.1  Cyclic directed information flow in cognitive radar with multiscale memory

Following on from what we said previously in Section 6.8, the cyclic directed information flow-graph of Figure 6.4 applies equally well to cognitive radar with multiscale memory. Of course, we should bear in mind the fundamental difference covered under points (1) and (2) on page 200 between the cognitive radar with single-layer memory in Figure 6.9 on the one hand and its own perception–action cyclic mechanism in Figure 6.1 acting alone on the other hand. Those differences become even more pronounced in the case of cognitive radar with multiscale memory. Accordingly, we may go on to state that, in terms of radar-tracking performance:

The estimation accuracy of range and range-rate will progressively improve with increasing depth of the multiscale memory.

However, we may eventually reach a point in the memory expansion beyond which we end up with diminishing returns.

### 6.16.2  Other potential benefits of multiscale memory

Two other practical benefits, which we may *expect* to be gained from the increased information-processing power of a cognitive radar with multiscale memory, are summarized here.

(1) *Environmental disturbances.* The occurrence of environmental uncertainties has a direct effect on the system equation (6.7), thereby affecting the measurement equation (6.8) on account of the dependence of the measurements on the state. When the cognitive radar is embodied with multiple layers of memory as structured in Figure 6.15, the cognitive radar is expected to deal progressively better with these uncertainties.

(2) *Interrupted transmission.* If the incoming measurements happen to suffer a *temporal discontinuity* for some unknown reason, then the perceptive capability of the radar receiver is correspondingly interrupted. In a difficult situation of this kind, it is possible for the working memory to step in and close the perception–action cycle, thereby bridging the temporal gap in the measurements at the receiver input.

### 6.16.3  Features of features: distinctive characteristic of multiscale memory

Up to this point in the discussion on multiscale memory, an issue that has not received the detailed attention it deserves is the following:

Why do we need multiscale perceptual and executive memories?

The simple answer to this fundamental question is the fact that the *complexity* of real-life radar returns (signals) is typically too difficult for a perceptual memory to capture it in a single-layered neural network. To elaborate, radar returns are characterized by an underlying *sparse composition*, which means that any given set of radar returns can be represented by a relatively small number of descriptors, called *features*.

In the extraction of inherent features of radar returns, it turns out that the most effective and computationally efficient perceptual memory structure is one based on a *deep-belief neural network*. To be specific, network *depth* is considered to be highly desirable because it provides the basis for *trading off space for time* (Bengio and LeCun, 2007). Accordingly, the *bottom-up* processing of radar returns in the perceptual memory proceeds as follows:

- The first layer of the memory captures a "coarse" set of features characterizing the incoming radar returns.
- The second layer is stimulated by the features captured in the first layer, thereby producing a "finer" representation of the radar returns.
- This procedure of extracting *features of features* (Selfridge, 1958) is continued in the memory's successive layers for as long as required.

In an information-theoretic sense, the objective behind the features-of-features strategy is to find a *low-entropy representation* of the radar returns, and the lower the better. To elaborate on this statement, we know from the definition of entropy as a measure of information that for a random event to have low entropy, its probability of occurrence must be close to unity (Shannon, 1948). It follows therefore that an output with low entropy is easier to recognize then the original radar returns. Just as importantly, the features-of-features strategy described herein can be realized by using an unsupervised learning process, which is accounted for by the two so-called unsupervised layers in Figure 6.15.

From a practical perspective, however, it is important that the *matching process* just described be *optimal*. To achieve this optimality, this requires the use of *supervised learning*, which is the last stage in the features-of-features strategy as depicted in Figure 6.15. Most importantly, the particular set of system-model parameters residing in the perceptual memory's library and picked as a result of the optimal matching process is passed on to the environmental scene analyzer in a top-down manner for the purpose of state estimation; the job of the perceptual memory is then finished for the particular cycle under test.

Much of what has been said thus far on the multiscale perceptual memory in the receiver applies equally well to the multiscale executive memory in the transmitter, except for two minor modifications in conceptual terms:

(1) Feedback information from the receiver to the transmitter is substituted for the role of radar returns in the receiver.
(2) Matching the transmit-waveform in the executive memory's library to information about the environment contained in the feedback process is substituted for matching the system-model in the perceptual memory's library to information about the target-state contained in the radar returns.

## 6.17    The explore–exploit strategy defined

Earlier in Section 6.10 dealing with the curse-of-dimensionality problem, we mentioned the *explore–exploit strategy* as a way of easing the computational burden that arises usually when dynamic programming is employed in the transmitter for waveform selection.

But this strategy was mentioned therein in the case of a fore-active radar employing the perception–action cycle all on its own (i.e. in the absence of memory). Now, in a cognitive radar, memory is an integral part of the system. Building on the discussion presented in the preceding section, the exploration part of the strategy involves selection of a cognit (i.e., piece of knowledge), representing a small set of waveform vectors in the transmitter's library, so as to "closely" match the feedback information about the environment on a Euclidean basis; this selection is made in immediate neighborhood of the previously transmitted waveform. Then, this particular set of waveform vectors is downloaded to the controller, where "optimality" is performed by picking the one waveform vector that minimizes the cost-to-go function; this second part of the strategy defines exploitation. In this way, a "global search" of the entire waveform library is replaced with a "local search" confined to a subset of the whole library, thereby easing the computational burden. With changes in execution being relatively slow, we expect that evolution of the local search to be correspondingly smooth.

Clearly, the explore–exploit strategy can be equally applied to the matching process aimed at the optimal selection of system-model parameters in the receiver.

Summing up:

The explore–exploit strategy, intended for easing the computational burden incurred in addressing the matching needs of both the transmitter and receiver in cognitive radar, provides the algorithmic mechanisms for top-down attention based on memory in both the transmitter and receiver.

## 6.18    Sparse coding

Another important issue emerging from the material presented in Section 6.16 on cognitive radar is how to design a multiscale memory. To this end, we may follow the traditional approach of extracting feature representations by reducing the dimensionality of the incoming radar returns as we successively move from one layer to the next. Such an approach is exemplified by *identity mapping* (also referred to as *autoencoding*), which was discussed in Chapter 2. The motivation behind this approach is *redundancy reduction*.[10]

However, several findings on sensory information processing, reported recently in the literature, advocate an entirely different approach, namely *sparse coding*. In this new way of thinking, the preference is to go for *overcomplete representations* of natural images. In reality, the radar returns from an object illuminated by the transmitter are just as natural as ordinary natural images. It follows, therefore, that there is much to be gained from sparse coding in the design of multiscale memory. In so doing, the analogy between cognitive radar and the visual brain becomes that much closer.

To be specific on what is meant by sparse coding in sensory information processing, we may make the following statement (Olshausen and Field, 1996):

The principle of sparse coding refers to a neural code, in which each sensory input of a nervous system is represented by the strong activation of a relatively small number of neurons out of a large population of neurons in the system.

From a practical perspective, the advantages of sparse coding include the following (Olshausen and Field, 1996):

- increased storage capacity in memories;
- representation of complex signals in a manner easier to recognize at higher levels of memory by virtue of the fact that the features are more likely to be linearly separable and more robust to noise; and
- of course, saving of energy.

## 6.18.1    Mathematical basis of sparseness

Before delving into a detailed account of sparse coding, it is instructive that we begin with a brief exposé of the mathematical basics of sparseness (Donoho and Elad, 2003).

Suppose we are given an input signal $\mathbf{x} \in \mathcal{R}_N$ and the issue of interest is to find a *code* $\mathbf{z} \in \mathcal{R}_M$ such that we have

$$\mathbf{W}\mathbf{z} = \mathbf{x}. \tag{6.71}$$

The $\mathbf{W}$ in (6.71) is a rectangular $N$-by-$M$ matrix, denoting a *dictionary* of generating elements defined by the set $\{\mathbf{w}_k\}_{k=1}^{M}$, where $\mathbf{w}_k \in \mathcal{R}_N$. These generating elements (i.e. columns of the matrix $\mathbf{W}$) are *normalized*; that is, $\mathbf{w}_k^\mathsf{T}\mathbf{w}_k = 1$ for all $k$; this normalization is done in order to prevent trivial solutions. However, there is no fixed relationship between the two dimensions $M$ and $N$, except for $M$ being larger than $N$. In other words, the number of generating elements is larger than the dimensionality of the input signal $\mathbf{x}$. It is for this reason that the dictionary is said to be *overcomplete*, in the sense that some of the generating elements are *linearly dependent*. As such, it would be improper to refer to the generating elements as basis vectors; rather, they are referred to simply as *atoms*.

With sparseness as the issue of interest, we would like to find the most sparse code $\mathbf{z}$ for solving the following constrained optimization problem (Donoho and Elad, 2003):

$$\text{Minimize the norm } \|\mathbf{z}\|_0 \text{ subject to } \mathbf{x} = \mathbf{W}\mathbf{z}, \tag{6.72}$$

where $\|\mathbf{z}\|_0$ is the $l^0$ norm, highlighting nonzero entries in the code $\mathbf{z}$. Unfortunately, the solution to this problem requires a *combinatorial search*, which is computationally intractable in high-dimensional spaces. To get around this difficulty, the problem statement in (6.72) is reformulated as an unconstrained optimization problem (Kavukcuoglu *et al.*, 2008):

$$F(\mathbf{x}, \mathbf{z}; \mathbf{W}) = \underbrace{\frac{1}{2} \| \mathbf{x} - \mathbf{W}\mathbf{z} \|^2}_{\substack{\text{Decoding-}\\\text{error energy}}} + \underbrace{\lambda_z \| \mathbf{z} \|_1}_{\substack{\text{Regularization}\\\text{term}}}, \tag{6.73}$$

where $\|\mathbf{z}\|_1$ denotes the $l^1$ norm of the code $\mathbf{z}$. In a mathematical sense, (6.73) may be viewed as the closest *convexication* of the statement in (6.72). Convex optimization is well studied in the literature (Boyd and Vandenberghe, 2004), hence the practical importance of solving (6.73) over (6.72). In any event, given the tractability of (6.73) and intractability of (6.72), what is rather surprising is the following statement (Donoho and Elad, 2003):

For certain dictionaries that are sufficiently sparse, solutions to (6.73) are the same as the solutions to (6.72).

This statement makes (6.73) as a basis for sparse coding all that more important.

To elaborate on (6.73): the function $F(\mathbf{x}, \mathbf{z}; \mathbf{W})$ on the left-hand side of this equation is the cost function to be minimized. This cost function is made up of two terms:

(1) With coding in mind, we may view the matrix product $\mathbf{Wz}$ as an approximate reconstruction, that is, a *decoded* version of the input signal $\mathbf{x}$. Hence, the first term $\frac{1}{2}\|\mathbf{x} - \mathbf{Wz}\|^2$, represents the energy of the decoding error, measuring how well the code $\mathbf{z}$ describes the signal vector $\mathbf{x}$.

(2) Explanation of the second term, $\lambda_z\|\mathbf{z}\|_1$, is more subtle. It is introduced to ensure *sparsity* of the code $\mathbf{z}$ by penalizing nonzero values of the code units. To be more specific, this second term may be viewed as a *regularization* term, with $\lambda_z$ denoting the *regularization parameter*, and the $l^1$ norm $\|\mathbf{z}\|_1$ performing the role of regularizer. As such, assigning a small value to the regularization parameter $\lambda_z$ implies that we have high confidence in the dictionary $\mathbf{W}$. On the other hand, assigning a high value to $\lambda_z$ implies that we have high confidence in the code $\mathbf{z}$ that makes up for low confidence in the dictionary $\mathbf{W}$. Typically, the regularization parameter $\lambda_z$ is assigned a value somewhere between these two limiting conditions.

One last comment is in order. The learning algorithm obtained by minimizing the cost function of (6.73) is essentially similar to the sparse coding algorithm described by Olshausen and Field (1996).

## 6.18.2 Sparse encoding symmetric machine

With this brief mathematical introduction to sparse coding, our next task is to formulate a neural network structure for its implementation. The first approach we have chosen to follow is based on the *principle of the encoder–decoder*, which we are encouraged to do so by virtue of what we know about *autoencoding* (identity mapping), discussed in Section 2.8.2. Thus, recognizing that the cost function $F(\mathbf{x}, \mathbf{z}; \mathbf{W})$ in (6.73) only accounts for decoding error, we need to expand it so as to also account for encoding error. To shed light on how this expansion could be accommodated, we refer to Figure 6.16, which is motivated by bottom-up and top-down processing that are a distinctive characteristic of perceptual and executive memories.

In Figure 6.16, referred to as SESM, we clearly see the following two operations:

(1) *Bottom-up* processing of the input signal $\mathbf{x}$, which leads to the following formula for the *encoding-error energy*:

$$E_{\text{enc}}(\mathbf{x}, \mathbf{z}; \mathbf{W}^{\mathrm{T}}) = \frac{1}{2}\|\mathbf{z} - \mathbf{W}^{\mathrm{T}}\mathbf{x}\|^2. \tag{6.74}$$

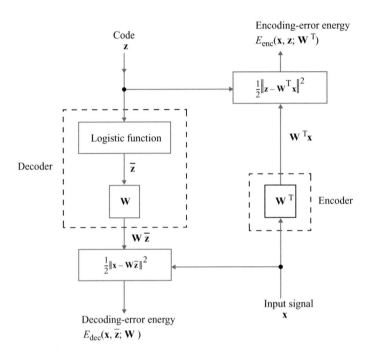

**Figure 6.16.** The sparse encoding symmetric machine (SESM), embodying bottom-up processing of the input signal **x** and top-down processing of the code **z**.

(2) *Top-down* processing of the code **z**, which yields the *decoding-error energy*

$$E_{dec}(\mathbf{x}, \overline{\mathbf{z}}; \mathbf{W}) = \frac{1}{2} \| \mathbf{x} - \mathbf{W}\overline{\mathbf{z}} \|^2, \tag{6.75}$$

where $\overline{\mathbf{z}}$ is a nonlinearly transformed version of the code **z**. For example, the transformation may be achieved by using the *point-wise logistic function*:

$$\overline{z}_i = \varphi(z_i) = \frac{1}{1 + \exp(-az_i)}, \qquad i = 1, 2, \ldots, M,$$

for some fixed gain $a$. This nonlinear transformation has the effect of limiting the amplitude of each unit in the code **z** to lie inside the range [0, 1].

Putting the two pieces of bottom-up and top-down processing described in (6.74) and (6.75) together, we may now formulate the *overall cost function for sparse coding* as follows (Ranzato *et al.*, 2007):

$$L(\mathbf{W}, z) = \eta \underbrace{E_{enc}(\mathbf{x}, \mathbf{z}; \mathbf{W}^T)}_{\substack{\text{Encoding-error} \\ \text{energy}}} + \underbrace{E_{dec}(\mathbf{x}, \overline{\mathbf{z}}; \mathbf{W})}_{\substack{\text{Decoding-error} \\ \text{energy}}} + \underbrace{\lambda_z \| \overline{\mathbf{z}} \|_1}_{\substack{\text{Regularization} \\ \text{for transformed} \\ \text{code, } \overline{\mathbf{z}}}} + \underbrace{\lambda_w \| \mathbf{W} \|_1}_{\substack{\text{Regularization} \\ \text{for dictionary, } \mathbf{W}}}. \tag{6.76}$$

To describe the overall cost function $L(\mathbf{W}, \mathbf{z})$, we may identify the following:

- The sum of the first two terms in (6.76) represents a *linear combination* of the two energy terms of encoding and decoding errors, with $\eta$ as a *learning-rate parameter*.

- The third term in (6.76) is a rewrite of the regularization term introduced previously in (6.76), with one difference: the code **z** in (6.73) is replaced by its nonlinearly transformed version $\bar{\mathbf{z}}$, defined in accordance with the activation function of (6.76). Specifically, the regularizer in (6.76) is defined by[11]

$$
\begin{aligned}
\|\bar{\mathbf{z}}\|_1 &= \sum_{i=1}^{M} \log(1 + \bar{z}_i^2) \\
&= \sum_{i=1}^{M} \log(1 + \varphi^2(z_i))^2,
\end{aligned} \tag{6.77}
$$

which assesses sparseness of the code **z** for a given input signal **x** by assigning a penalty that depends on how activation is distributed across units of the code **z**. These representations in which activation is spread over many of the code units should be penalized to a greater extent than those in which only a few code units carry the load (Olshausen and Field, 1996).

- The fourth term in (6.76) is an $l^1$ *regularization* on the set of linear filters in the encoder, so as to suppress noise in the signal vector **x** and push for individual localization of the filters.

The encoding- and decoding-error energy terms in (6.76) are respectively defined as follows:

$$
E_{\text{enc}} = \frac{1}{2}\|\mathbf{x} - \mathbf{W}^{\mathrm{T}}\mathbf{z}\|^2 \tag{6.78}
$$

and

$$
E_{\text{dec}} = \frac{1}{2}\|\bar{\mathbf{z}} - \mathbf{W}\mathbf{x}\|^2, \tag{6.79}
$$

both of which couple the unknowns, **W** and **z** (through $\bar{\mathbf{z}}$), in their own individual ways. It follows, therefore, that it is not possible to minimize the overall cost function $L(\mathbf{W}, \bar{\mathbf{z}})$ with respect to both **W** and **z** simultaneously. If the code **z** is known, then minimization of this cost function is a convex optimization problem. On the other hand, if **W** is fixed, then the optimum code **z**\* may be computed by using gradient descent to minimize the cost function with respect to **z**. Building on the twofold rationale just described, the learning algorithm for computing the matrix **W** and code **z** is as follows (Ranzato *et al.*, 2007):

(1) For a given input signal **x** and encoder-parameter setting **W**, the method of gradient descent is used to obtain the optimal code **z**\*.

(2) With both the input signal **x** and optimal code **z**\* clamped using the values computed in step 1, *one step* of gradient descent is enough to update the matrix **W**, as shown by

$$
\mathbf{W}^+ = \mathbf{W} - \alpha\frac{\partial L}{\partial \mathbf{W}},
$$

where $\partial L/\partial \mathbf{W}$ is the gradient and $\alpha$ is the step-size parameter.

After this two-step training is completed, the sparse coding algorithm converges to a state where we find the following twofold condition:

- the decoder produces a good reconstruction of the input signal **x** from a sparse (rectified) code; and
- the encoder predicts the optimal code through a simple feed-forward propagation process.

The last issue to be considered is that of building a multiscale memory, using the SESM of Figure 6.16 as the basic functional module. To do so, we may use unsupervised learning and proceed along the features-of-features strategy that is descriptive of autoencoding. Specifically, the features captured by the encoder of the first module are applied as input to the second layer of the memory. This feedforward procedure is continued for as many layers as required. Construction of the multiscale memory is completed by adding a supervised classifier as the top layer of the hierarchy, using a multilayer perceptron trained with the back-propagation algorithm.

From the discussion presented herein on the SESM, it is apparent that although there are three unknowns to be considered (namely, the sparse code and the encoder and decoder matrices), the SESM reduces them to two by virtue of the fact that the encoder and decoder matrices are the transpose of each other in accordance with the encoding–decoding principle. Unfortunately, this computational advantage of the SESM is the source of its computational weakness, in that it is sensitive to the way in which the input data stream is scaled. To illustrate, suppose that the input is *scaled up* by the factor $K$. Then, in order to maintain the sparse code unchanged, the encoder matrix would like to be *scaled down* by the factor $K$. Correspondingly, the decoder matrix would like to be *scaled up* by the same factor $K$ in order to perform the data reconstruction correctly. The net result is that the SESM is *non-robust* with respect to scaling the input data stream and learns very slowly.

To overcome this limitation of the SESM, we look to another closely related sparse coding model, described next.

## 6.18.3  Predictive sparse decomposition

The predictive sparse decomposition (PSD) model, first described by Kavukcuoglu *et al.* (2008), overcomes the *scaling problem* by untying the algebraic relationship between the encoder and decoder matrices. To be specific, whereas these two matrices in the SESM are respectively denoted by $\mathbf{W}$ and $\mathbf{W}^T$, in the PSD model they are respectively denoted by $\mathbf{W}$ and $\mathbf{U}$. Thus, the encoder matrix $\mathbf{U}$ of dimensions $M \times N$ is now independent of the decoder matrix $\mathbf{W}$ of dimensions $N \times M$.

In algorithmic terms, the PSD model follows the same mathematical steps as the SESM up to and including (6.73). In particular, to make the learning process computationally efficient, the PSD model does the following:

A nonlinear regressor is trained to map the input vector $\mathbf{x}$ to sparse representations defined by the code $\mathbf{z}$.

To this end, consider the following nonlinear mapping vector:

$$\mu(\mathbf{x}; \mathbf{G}, \mathbf{U}, \mathbf{b}) = \mathbf{G} \tanh (\mathbf{Ux} + \mathbf{b}), \qquad (6.80)$$

where $\mathbf{b}$ is a bias vector. The hyperbolic tangent function on the right-hand side of the equation is responsible for the nonlinearity, operating on its argument, element by element. As for the new matrix $\mathbf{G}$, it is a diagonal matrix whose *gain* coefficients are adjustable so as to permit the mapping outputs in (6.80) to compensate for input-scaling variations, given that data reconstruction is performed in the decoder with a filter matrix $\mathbf{W}$ whose generating elements are all normalized.

To simplify matters, let

$$\mathbf{P}_f = \{\mathbf{G}, \mathbf{U}, \mathbf{b}\} \tag{6.81}$$

denote the parameters that are to be learned by the encoder in a *predictive* manner. Let the optimal code $\mathbf{z}^*$ be defined by

$$\mathbf{z}^* = \arg \min_{\mathbf{z}} F(\mathbf{x}, \mathbf{z}; \mathbf{W}), \tag{6.82}$$

where the function $F(\mathbf{x}, \mathbf{z}; \mathbf{W})$ is defined in (6.73). We may then formulate the algorithmic goal of the PSD model as follows:

Make the prediction of the nonlinear regressor $\mu(\mathbf{x}; \mathbf{P}_f)$ as close as possible to the optimal code $\mathbf{z}^*$.

A straightforward approach to this optimization is to carry it out *after* the optimal code $\mathbf{z}^*$ has been computed by solving (6.82). However, for a computationally efficient learning process, it is preferable to optimize the encoder's $\mathbf{P}_f$ and the decoder matrix $\mathbf{W}$ *jointly*. To set the stage for this joint optimization, the cost function in (6.73) is expanded by including an additional term to represent the encoder, as shown by

$$F(\mathbf{x}, \mathbf{z}; \mathbf{W}, \mathbf{P}_f) = \underbrace{\frac{1}{2} \| \mathbf{x} - \mathbf{W}\mathbf{z} \|^2}_{\substack{\text{Decoding-}\\ \text{error energy}}} + \underbrace{\lambda_z \| \mathbf{z} \|_1}_{\substack{\text{Regularization}\\ \text{term}}} + \underbrace{\frac{1}{2}\eta \| \mathbf{z} - \mu(\mathbf{x}, \mathbf{P}_f) \|^2}_{\substack{\text{Regression-error}\\ \text{energy}}}, \tag{6.83}$$

where $\eta$ is the learning-rate parameter. In effect, the additional term representing the regression-error energy enforces the code $\mathbf{z}$ to be as close as possible to the nonlinear regressor $\mu(\mathbf{x}, \mathbf{P}_f)$.

With the cost function of (6.83) at hand, the learning process proceeds on the basis of an on-line coordinate gradient-descent algorithm that alternates between two steps for each training sample $\mathbf{x}$ as follows (Kavukcuoglu *et al.*, 2008):

*Step 1.* Keeping the parameters $\mathbf{P}_f$ and $\mathbf{W}$ fixed in the encoder and decoder respectively, minimize the cost function $F(\mathbf{x}, \mathbf{z}; \mathbf{W}, \mathbf{P}_f)$ of (6.83) with respect to the code $\mathbf{z}$, starting from the initial condition supplied by the nonlinear regressor $\mu(\mathbf{x}, \mathbf{P}_f)$.

*Step 2.* Using the optimal value of the code $\mathbf{z}$ computed in Step 1, update the parameters $\mathbf{P}_f$ and $\mathbf{W}$ using the one-step gradient-descent method.

It appears that, after training the PSD model using this two-step learning algorithm, the model provides a computationally fast and smooth approximation to the optimal sparse representation, and yet it achieves better accuracy than exact sparse coding algorithms on visual object recognition tasks (Kavukcuoglu *et al.*, 2008). Therefore, it may be said that the PSD model is a method of choice for implementing multiscale memory.

One other issue of interest is how to infer the code $\mathbf{z}$, once the learning process has been completed. For an *approximate inference*, a single matrix-vector multiplication suffices. To be specific: in light of (6.80) and (6.81), computation of the approximate code is obtained by passing the input vector $\mathbf{x}$ through the nonlinear regressor denoted by $\mu(\mathbf{x}, \mathbf{P}_f)$. For a more *accurate inference*, we may use (6.83), which, however, is computationally more demanding.

## 6.19    Summary and discussion

### 6.19.1    Cognitive radar defined

For a radar to be *cognitive* in the full sense of the word as we know it in human cognitive terms, the radar has to embody four fundamental processes (putting language aside):

(1) *Perception* of the environment by the environmental scene analyzer in the receiver, followed by *action* taken in the environment by the environmental scene actuator in the transmitter through global feedback from the receiver to the transmitter. The resulting perception–action cycle manifests itself in *information gain* about the target state, which will increase from one cycle to the next.

(2) *Memory* with one layer or more, distributed across the entire radar system. The function of memory is to predict the consequences of actions taken by the radar and do so in a self-organized and adaptive manner; action examples include the selection of waveform-parameter vector in the transmitter and the selection of parameters characterizing the system equation in the receiver.

(3) *Attention* driven by memory. Its function is to prioritize the allocation of available resources in accordance with the incoming streams of radar returns from the target through memory-based algorithmic attentional mechanisms.

(4) *Intelligence* driven by perception, memory, and attention. Its function is to enable the algorithmic control and decision-making mechanism in the transmitter to identify intelligent choices in the face of environmental uncertainties. As such, intelligence is by far the most profound of all the four fundamental processes involved in cognition.

### 6.19.2    Cognitive functional integration across time

One of the objectives of the perception–action cycle in cognitive radar is to *bridge time*. In this context, time plays three key roles in the operation of a cognitive radar, regardless of how many layers of memory are built into its design:

(1) Time separates the incoming stimuli (i.e., radar returns) so as to guide the overall radar system behavior.

(2) Time separates the stimuli responsible for perception of the environment by the receiver from corresponding actions taken in the environment by the radar transmitter in response to feedback information from the receiver.

(3) Time separates the occurrence of sensory feedback in the receiver from further action taken in the transmitter.

To elaborate further, temporal organization of the radar system behavior hinges on integration of the percepts and actions across time, as follows:

- percepts with other percepts;
- actions with other actions; and
- percepts with other actions.

This temporal way in which the information-processing steps is performed in cognitive radar is termed the *functional integration across time*, which is a cardinal property of cognitive radar.

### 6.19.3 State estimation and control

Given the nonlinear state-space model of (6.7) and (6.8), to satisfy optimality of perception in the receiver as well as optimality of control in the transmitter, we may consider the following pair of tools:

(1) *CKF for perception* (i.e. environmental scene analysis). In a nonlinear environment and under the Gaussian assumption, the CKF provides the "best" known approximation to the optimal Bayesian filter in the radar receiver. Stated another way, when we are confronted with nonlinear state estimation in a Gaussian environment, the CKF provides a method of choice. Here, we are assuming that the nonlinearity is beyond the reach of the EKF that is known for its computational simplicity. Simply put, the CKF improves the state-estimation process over the EKF at the cost of increased computational complexity, particularly when the state space is of high dimensionality.

(2) *Approximate dynamic programming for control*. Here again, when it comes to action in the environment, *approximate dynamic programming* provides a method of choice for optimal control of the receiver by the transmitter indirectly via the environment. To elaborate more, for optimal waveform selection in the transmitter in response to feedback information from the receiver, we may not be able to do better than approximate dynamic programming.

Thus, when it comes to designing a cognitive radar in a nonlinear and Gaussian environment, there is much to be said for cubature Kalman filtering for state estimation in the receiver and approximate dynamic programming for control in the transmitter. For a non-Gaussian environment, we may consider the use of Gaussian-sum approximation for nonlinear filtering and look to parallel processing for implementing the filter.

### 6.19.4 Cyclic directed information flow

Continuing on the concluding statements just made on state estimation and control that are the *dual* of each other, we first recall that the cyclic directed information flow-graph of Figure 6.4 was originally formulated for the perception–action cycle of Figure 6.1, representing the first step towards radar cognition. As stressed along the way, this same baseline flow graph applies to a cognitive radar, bearing in mind the following statement:

The amount of feedback information (i.e., entropy), measured in accordance with Shannon's information theory, is progressively reduced as the number of layers in the memory of the cognitive radar is correspondingly increased.

In other words, in a tracking application, the tracking accuracy of the radar is progressively improved with increasing layers of memory in the cognitive radar. Equally well, other information-processing capabilities of the radar are also progressively improved.[12]

### 6.19.5    Analogy between exploring radar environment via feedback information and saccadic eye movements in the visual brain

Feedback information computed in the receiver of a cognitive radar plays a key role in how the transmitter is enabled to explore a dynamic environment for tracking target movements. To be more precise, the environmental scene actuator in the transmitter sees the radar environment indirectly through the feedback information passed onto it by the environmental scene analyzer in the receiver that has direct access to the observable data. Most importantly, despite the fact that the feedback information link resides inside a closed feedback loop, the cognitive radar remains stable.

The ability of the environmental scene analyzer to explore the environment via the feedback information may be likened to the phenomenon of saccadic eye movements in the visual brain.

To elaborate, as explained by Morrone and Burr (2009), the discrete ballistic movements of the eyes, from one point of interest to another in a quick manner, direct our eyes toward a visual target; yet, we are typically unaware of the actual eye movements. Just as importantly, despite the continually saccadic eye movements, the vision is always clear and stable.

From these realities, therefore, it follows that here is another physical phenomenon to justify the analogy between cognitive radar and the visual brain.

### 6.19.6    Sparse coding

For the effective and computationally efficient implementation of perceptual memory for cognitive radar, we need only to look at *human vision* and find that sparse coding plays a key role in perceiving the outside world. In recent years, sparse coding has attracted the attention of researchers on many fronts: vision, neural computation, mathematics, and signal processing. The net result is that we now have a plethora of sparse-coding procedures. In this chapter, we focused on two models, SESM and PSD, both of which are rooted in signal processing and, therefore, easy to understand.

The SESM is simple to implement by virtue of the fact that the encoder's filter matrix is the transpose of the decoder's filter matrix. However, this encoding–decoding property is also responsible for the lack of robustness to scaling variations in the input data.

The PSD model overcomes this limitation of the SESM by untying the algebraic relationship between filter matrices in the encoder and decoder. Its algorithmic implementation is based on the minimization of a cost function that consists of two parts:

• the first part is essentially the cost function employed in Olshausen–Field's sparse coding algorithm;

- the second part is defined in terms of a nonlinear regressor that maps a patch of the input data stream to sparse representation of that patch.

The net result is an algorithm that is computationally fast and accurate in performance.

### 6.19.7  Closing statements

We close the discussion on cognitive radar by viewing it in two different ways:

(1) Cognitive radar is a *complex correlative-learning system*, in which *control-based perception* is performed in the receiver, *perception-based control* is performed in the transmitter, *feedback information* links them together, and memory acts as the *stabilizer* guarding against the cumulative destabilizing tendency of the local and global feedback loops distributed across the cognitive radar.
(2) Cognitive radar is an *intelligent decision-making system*, capable of *risk management* by selecting intelligent choices in both the transmitter and receiver in the face of environmental uncertainties.

Putting these two viewpoints together, we conclude with a final statement that sums up the information-processing power of cognition (Wicks, 2010):

If a radar system can be made to work by traditional methods, then cognition can make it work better.

Most importantly, cognitive information processing is *software centric*. As such, it can be applied not only to a new radar system, but equally well to an old one.

### Notes and practical references

1. *Historical note on fore-active radar*
   In Kershaw and Evans's (1994) classic paper, a theory was described for optimal transmit-waveform selection for illuminating the environment by linking the receiver to the transmitter. In so doing, a closed loop around the environment was formed, whereby it became possible for the transmitter to exercise indirect control over the receiver via the environment. Although the computation described in that paper was performed in an *off-line* manner, it may be justifiably argued that, in mathematical terms, the Kershaw–Evans paper was the first formulation of a fore-active radar; that is, the first step towards radar cognition. However, no such statement was originally made in the paper, as cognition was not known in the radar literature at that time.
   Another related publication that should be noted is the book by Guerci (2010). This book, consisting of five chapters, covers material on two classes of radar: multiple-input multiple-output (MIMO) radar and fully adaptive knowledge-based radar. With regard to the latter, recognizing that knowledge is no substitute for memory as emphasized in Chapter 2, the fully adaptive knowledge-based radar in figures 5.1 of the book is basically a sophisticated fore-active radar; that is, it is the

first step towards cognition. In the final analysis, the fact of the matter is that, no matter how fully adaptive a radar is, adaptivity, be it full or otherwise, will not result in cognition.

For a radar to be cognitive, there has to be *learning* through continued interactions with the environment; the learning process is embodied in memory, which is an essential ingredient for the radar to be cognitive. Simply put, cognition implies adaptivity, but not the other way around.

2. *Monostatic versus bistatic radar*

In a monostatic radar, the transmitter and receiver are collocated. On the other hand, in a bistatic radar, they are located in different places; accordingly, the use of global feedback in a bistatic radar requires the use of a physical link.

3. *Controlled oscillator for waveform generator*

Adaptive waveform generation in the transmitter may be accomplished in an *on-line* manner through the use of a *controlled oscillator*. However, implementation of such an approach may well be more complicated than the use of a waveform library that is discrete by nature.

4. *The Schwarz inequality*

Consider two functions of the variable $x$, denoted by $\phi_1(x)$ and $\phi_2(x)$, both of which satisfy the conditions

$$\int_{-\infty}^{\infty} \phi_i^2(x) \; dx < \infty \qquad \text{for } i = 1, 2.$$

*Schwarz's inequality* states that

$$\left| \int_{-\infty}^{\infty} \phi_1(x)\phi_2(x) \; dx \right|^2 \leq \int_{-\infty}^{\infty} \phi_1^2(x) \; dx \int_{-\infty}^{\infty} \phi_2^2(x) \; dx,$$

where the equality holds if, and only if, we have

$$\phi_1(x) = c\phi_2(x),$$

where $c$ is an arbitrary constant.

5. *The Fisher information matrix and Cramér–Rao lower bound*

In the estimation of a parameter vector $\boldsymbol{\theta}$, we are usually interested in two properties:

• the mean of the estimator and

• the covariance matrix of the estimation error.

Denoting the estimator by $\hat{\boldsymbol{\theta}}$, the mean is simply the expectation of $\hat{\boldsymbol{\theta}}$. The covariance matrix is defined by the expectation of the outer product $(\boldsymbol{\theta} - \hat{\boldsymbol{\theta}})(\boldsymbol{\theta} - \hat{\boldsymbol{\theta}})^{\mathrm{T}}$, where the superscript denotes matrix transposition. Provided that the estimator is unbiased, the *Cramér–Rao bound* establishes a lower bound on the error covariance matrix.

To elaborate, let $p(\mathbf{x}|\boldsymbol{\theta})$ denote the conditional probability density function of a random vector $\mathbf{X}$ represented by the observed sample vector $\mathbf{x}$ given the parameter vector $\boldsymbol{\theta}$. The issue of interest is to estimate $\boldsymbol{\theta}$ given $\mathbf{x}$. From Chapter 4, we recall that this estimation can be carried out using the likelihood function

$$l(\boldsymbol{\theta}|\mathbf{x}) = p(\mathbf{x}|\boldsymbol{\theta}).$$

Specifically, the *maximum likelihood estimate* of $\boldsymbol{\theta}$, denoted by $\hat{\boldsymbol{\theta}}$, is given by

$$\hat{\boldsymbol{\theta}}_{ML} = \max_{\boldsymbol{\theta}} \log l(\boldsymbol{\theta} | \mathbf{x}),$$

where log denotes the natural logarithm. Taking the partial derivative of the log-likelihood function $\log l(\boldsymbol{\theta} | \mathbf{x})$ with respect to $\boldsymbol{\theta}$, we get the *score function*:

$$\mathbf{s}(\boldsymbol{\theta}) = \frac{\partial}{\partial \boldsymbol{\theta}} \log l(\boldsymbol{\theta} | \mathbf{x}).$$

A score function is said to be "good" if it is near zero; otherwise, it is said to be "bad." The covariance matrix of the score function is called the *Fisher information matrix*:

$$J(\boldsymbol{\theta}) = \mathbb{E}[\mathbf{s}(\boldsymbol{\theta})\mathbf{s}^\mathsf{T}(\boldsymbol{\theta})],$$

which, after some manipulation, leads to the following formula (Van Trees, 1968):

$$J(\boldsymbol{\theta}) = -\mathbb{E}\left[\frac{\partial}{\partial \boldsymbol{\theta}}\left(\frac{\partial}{\partial \boldsymbol{\theta}} \log l\boldsymbol{\theta}(x)\right)\right],$$

where $\mathbb{E}$ is the expectation operator. The information matrix $\mathbf{J}$ is named in honor of Fisher, who originated the idea of maximum likelihood estimation.

Returning to the basic issue at hand, it turns out the error covariance matrix $\mathbf{C}$ of the estimator $\hat{\boldsymbol{\theta}}$ is lower bounded as follows:

$$\mathbf{C} = \mathbb{E}[(\boldsymbol{\theta} - \hat{\boldsymbol{\theta}})(\boldsymbol{\theta} - \hat{\boldsymbol{\theta}})^\mathsf{T}] \geq \mathbf{J}^{-1},$$

where $\mathbf{J}^{-1}$ is the inverse of the Fisher information matrix. Correspondingly, the mean-square error of $\hat{\theta}_i$, the $i$th element of the estimator $\hat{\boldsymbol{\theta}}$, is lower bounded as follows:

$$c_{ii} = \mathbb{E}[(\theta_i - \hat{\theta}_i)^2] \geq (\mathbf{J}^{-1})_{ii},$$

where $(\mathbf{J}^{-1})_{ii}$ is the $ii$th element of the inverse of the Fisher information matrix. This bound defines the Cramér–Rao lower bound.

6. *Duration of cognition cycle*

The time taken for the completion of a single perception–action cycle is made up essentially of three terms: $\tau_{target}$, the round trip to and from the unknown radar target; $\tau_{rec}$, the time taken for the state-estimation algorithm in the environmental scene analyzer (e.g. CKF) to reach a "steady-state value" for the state estimate; $\tau_{trans}$, the time taken for the environmental scene actuator (e.g. approximate dynamic-programming algorithm) to complete the updating of the transmit-waveform parameter vector.

For convenience of mathematical presentation, two assumptions are made:

(1) Time is normalized by considering the cognition-cycle duration as one unit of time.

(2) If the transmitter illuminates the radar environment at time $n - 1$, then the corresponding measurement is made at the receiver input at time $n$, assuming a delay of one time unit. In effect, the waveform-parameter vector $\boldsymbol{\theta}_{n-1}$, generated in the transmitter at time $n - 1$, reaches the receiver input at time $n$.

Note that the sum term, $\tau_{\text{target}} + \tau_{\text{rec}} + \tau_{\text{trans}}$, is upper bounded by the pulse-repetition period of the radar.

7. *Statement on the transition from perception to action*

The statement made on p. 182, on the transition from perception of the environment carried out in the receiver to action performed in the transmitter in response to feedback information from the receiver, follows an insightful observation first made by Kershaw and Evans (1994). In mathematical terms, the observation is traced to (6.26), referring to the covariance matrix, $\mathbf{P}_{\text{yy},n|n-1}$, the dependence of which on the waveform-parameter vector $\boldsymbol{\theta}_{n-1}$ is confined entirely to the measurement-noise covariance matrix $\mathbf{R}(\boldsymbol{\theta}_{n-1})$.

8. *Mutual information*

Consider a pair of random variables $X$ and $Y$. In the Shannon theory, the *mutual information* between these two random variables is defined by

$$I(X;Y) = \int_{-\infty}^{\infty} \int_{-\infty}^{\infty} p_{X,Y}(x,y) \left( \log \frac{p_{X,Y}(x,y)}{p_X(x) p_Y(y)} \right) \mathrm{d}x \, \mathrm{d}y,$$

where $p_{X,Y}(x,y)$ is the joint probability density function of $X$ and $Y$, and $p_X(x)$ and $p_Y(y)$ are its two marginals. If the random variables $X$ and $Y$ are statistically independent, then

$$p_{X,Y}(x,y) = p_X(x) p_Y(y),$$

in which case the mutual information $I(X;Y)$ reduces to zero. In general, we have

$$I(X;Y) \geq 0.$$

The mutual information is related to entropy as follows:

$$I(X;Y) = H(X) - H(X\,|\,Y),$$

where $H(X)$ is the entropy of $X$ and $H(X\,|\,Y)$ is the conditional entropy of $X$ given $Y$. Equivalently, we may write

$$I(X;Y) = H(Y) - H(Y|X).$$

The entropy of $X$ is defined in terms of its probability density function $p_X(x)$ as follows:

$$H(X) = -\int_{-\infty}^{\infty} p_X(x) \log p_X(x) \, \mathrm{d}x,$$

where log denotes the logarithm to base 2, with information being measured in *bits*. Moreover, entropy provides a *measure of information*.

9. *Another information-theoretic metric*

Williams *et al.* (2007) describe another information-theoretic metric. This new metric, termed the *state conditional entropy*, provides a measure of "uncertainty" about the current state $\mathbf{x}_n$, given all the measurements up to and including time $n$. Following through the steps described therein, it is shown that minimizing the conditional state entropy given past history of the measurements, $\mathbf{y}_{0:n}$, is equivalent to maximizing the conditional mutual information between the state $\mathbf{x}_n$ and measurement $\mathbf{y}_n$ at time $n$, given past history of the measurements up to and including $n-1$, that is, $\mathbf{y}_{0:n-1}$.

10. *Redundancy reduction*

The concept of *source coding based on redundancy* originated in the 1948 classic paper by Claude Shannon, which has formed the basis for information theory as we know it today. That paper addressed many of the fundamental concepts that are of profound importance to communication engineering.

In the years following Shannon's paper, information theory aroused the research interests of a few neuroscientists, including Horace Barlow. In a classic publication (Barlow, 1961), it was observed that, at later stages of sensory information processing, neurons were less active than those at early stages. Influenced by Shannon's information theory on source coding, this phenomenon was referred to in the 1961 paper as "redundancy reduction."

However, recognizing that the brain has to decide upon actions in the face of uncertainties, there is now a clear indication in neuroscience that probability theory and statistical inference, basic to the Bayesian framework, are of profound importance, more so than Shannon's information theory. It is in light of this new way of thinking that in the follow-up paper (Barlow, 2001), entitled "Redundancy reduction revisited," the original hypothesis of redundancy reduction is said to be wrong, in that compressed representations of stimuli are unsuitable for the brain, but right in drawing attention to the fact that redundancy is still important in sensory information processing.

11. *Regularizer for the sparse code*

Referring to the decoding cost function described in (6.73), we see that the regularization term pertaining to the code $\mathbf{z}$ is defined in terms of its $l^1$ *norm*, as shown by

$$\| \mathbf{z} \| = \sum_{i=1}^{N} |z_i|.$$

From an applied mathematics perspective, use of the $l^1$ norm is favored, as it lends itself to convex optimization. However, from a sensory information-processing perspective, the preferred choice for the sparse-code regularizer is one that mimics the prior distribution of linear-filter coefficients in the encoder applied to natural images; such distributions are *heavy-tailed*.

Olshausen and Field (1996) chose the logarithmic expression $\log(1 + z^2)$ because it induces sparsity and lends itself to optimization. Later, it was discovered that this distribution is the well-known *Student-t distribution* (Olshausen, 2011).

12. *Cognitive radar with multilayer memory*

For experimental results demonstrating the information-processing power of cognitive radar with multilayer memory using simulations, the reader is referred to Haykin and Xue (2012).

# 7    Cognitive radio

Interest in a new generation of engineering systems enabled with cognition, started with *cognitive radio*, a term that was coined by Mitola and McGuire (1999). In that article, the idea of cognitive radio was introduced within the software-defined radio (SDR) community. Subsequently, Mitola (2000) elaborated on a so-called "radio knowledge representation language" in his own doctoral dissertation. Furthermore, in a short section entitled "Research issues" at the end of his doctoral dissertation, Mitola went on to say the following:

'How do cognitive radios learn best? merits attention'. The exploration of learning in cognitive radio includes the internal tuning of parameters and the external structuring of the environment to enhance machine learning. Since many aspects of wireless networks are artificial, they may be adjusted to enhance machine learning. This thesis did not attempt to answer these questions, but it frames them for future research.

Then, in Haykin (2005a), the first journal paper on cognitive radio, detailed expositions of signal processing, control, learning and adaptive processes, and game-theoretic ideas that lie at the heart of cognitive radio were presented for the first time. Three fundamental cognitive tasks, embodying the perception–action cycle of cognitive radio, were identified in that 2005 paper:

- radio-scene analysis of the radio environment performed in the receiver;
- transmit-power control and dynamic spectrum management, both performed in the transmitter; and
- global feedback, enabling the transmitter to act and, therefore, control data transmission across the forward wireless (data) channel in light of information about the radio environment fed back to it by the receiver.

In effect, emphasis in that 2005 paper was placed on cognitive radio as a "closed-loop feedback control system" with practical benefits and the need for precautionary measures, recognizing that feedback is a "double-edged sword."

Since its inception about a decade ago, interest in cognitive radio, its theory and applications, has grown exponentially. The driving force behind this exponential growth is summed up as follows:

Cognitive radio has the potential to mitigate the radio-spectrum underutilization problem in today's wireless communications.

With this engineering challenge in mind, we begin the study of cognitive radio with this critical issue as our starting point.

## 7.1       The spectrum-underutilization problem

The electromagnetic *radio spectrum* is a natural resource, the use of which by transmitters and receivers is licensed by governments. In November 2002, the Federal Communications Commission (FCC) published a report prepared by the Spectrum-Policy Task Force aimed at improving the way in which this precious resource is managed in the USA (FCC, 2002). The task force was made up of a team of high-level, multidisciplinary professional FCC staff – economists, engineers, and attorneys – from across the commission's bureaus and offices. Among the task force's major findings and recommendations, the second finding on page 3 of the report is rather revealing in the context of spectrum utilization:

In many bands, spectrum access is a more significant problem than physical scarcity of spectrum, in large part due to legacy command-and-control regulation that limits the ability of potential spectrum users to obtain such access.

Indeed, if we were to scan portions of the radio spectrum including the revenue-rich urban areas, we would find that:

(1) some frequency bands in the spectrum are largely unoccupied most of the time;
(2) some other frequency bands are only partially occupied; and
(3) the remaining frequency bands are heavily used.

The underutilization of the electromagnetic spectrum leads us to think of a new term commonly called *spectrum holes*, for which we offer the following definition:

A spectrum hole is a band of frequencies assigned to a primary (licensed) user, but, at a particular time and specific geographic location, the band is not being utilized by that user.

Spectrum utilization can be improved significantly by making it possible for a *secondary* (*cognitive radio*) *user* (who is not being serviced) to access a spectrum hole unoccupied by the primary (legacy) user at the right location and time in question. *Cognitive radio* offers a new way of thinking on how to promote efficient use of the radio spectrum by exploiting the existence of spectrum holes.

   The *spectrum-utilization efficiency* of cognitive radio is assessed in the context of four practical system issues:

(1) *Accuracy and reliability*, with which the spectrum holes are identified (detected).
(2) *Computational speed*, with which the spectrum-hole identification is accomplished.
(3) *Management of resources*, which involves the allocation of spectrum holes among competing secondary users in the cognitive radio network.
(4) *Coexistence of the cognitive radio network alongside the legacy radio network*, which will have to be accomplished in a harmonious manner for the good of all users, both secondary and primary.

Requirements (1) and (2) are responsibilities of receivers in the cognitive radio network, while the transmitters in the network are responsible for requirement (3). As for requirement (4), with the legacy radio network having paid for the use of the spectrum and legally approved by regulatory agencies, the responsibility for this last requirement rests with the cognitive radio network, viewed as a *system of systems*.

## 7.2    Directed information flow in cognitive radio

With every cognitive dynamic system having its own characteristic perception–action cycle, so it is with cognitive radio. However, before proceeding to address this issue, it is instructive to come up with a definition for what we mean by a "user" in a radio network. To this end, we first recognize that at each end of a wireless communication channel we have a *transceiver*, which embodies a transmitter and a receiver combined together as one whole unit. So, when we speak of a radio user, we offer the following definition:

A user is a communication link that connects the transmitter of a transceiver at one end of the link that is in communication with the receiver of another transceiver at the other end of the link. Moreover, the term "secondary user" is adopted for a cognitive radio, so as to distinguish it from the term "primary user", which is reserved for a legacy (i.e. licensed) radio unit.

Note the terms "primary user" and "secondary user" were used in the previous section, ahead of this definition.

Referring to Figure 7.1a, we see that on the right-hand side of the figure we have an RX-CR unit whose cognitive function is *radio scene analysis* (RSA), where CR stands for cognitive radio. On the left-hand side of the figure, at some remote location we have a TX-CR unit whose cognitive function is *dynamic spectrum management* (DSM) and *transmit power control* (TPC). The RSA of the RX-CR unit *senses* the radio environment with the objective of identifying spectrum holes (i.e. underutilized subbands of the radio spectrum). This information is passed onto the TX-CR unit via the *feedback channel*. At the same time, through its own RSA the TX-CR unit will have identified the spectrum holes in its own specific neighborhood. The combined function of the DSM and TPC in the TX-CR unit is to *identify* a spectrum hole that is common to it as well as the RX-CR unit, through which transmission over a *data channel* in the radio environment can be carried out. In this way, directed information flow across the cognitive radio is established on a cycle-by-cycle basis.

### 7.2.1    The perception–action cycle in cognitive radio

What we have just described is the very essence of the perception–action cycle for a communication link in cognitive radio; this cycle is depicted in Figure 7.1b, representing a subset of Figure 7.1a with a minor difference: the functional block labeled "nonparametric spectrum estimation" in the receiver has been used in Figure 7.1b so as to add more specificity to the notion of radio scene analysis.

From Chapter 2, we recall that, for a dynamic system to be cognitive, it has to embody four distinct processes: perception, memory, attention, and intelligence. On the basis of Figure 7.1b, we now address how these four processes are indeed satisfied, one by one. In so doing, we will have not only justified the rationale for radio cognition, but also paved the way for the material to be covered in subsequent sections of the chapter.

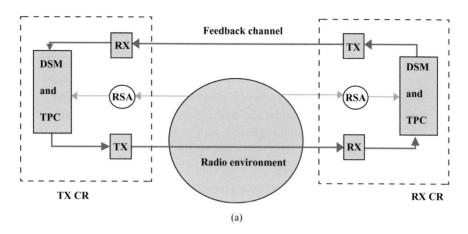

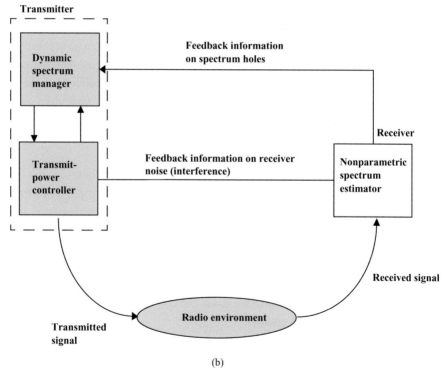

**Figure 7.1.** (a) Directed-information flow in cognitive radio. DSM: dynamic spectrum manager; TPC: transmit-power controller; RSA: radio-scene analyzer; RX: receiver; TX: transmitter; TX CR: transmitter unit in the transceiver of cognitive radio; RX CR: receiver unit in the transceiver of cognitive radio. (b) Perception–action cycle of cognitive radio unit.

(1) *Perception*

By using nonparametric spectrum estimation for perception in the receiver, the task of finding spectrum holes is achieved without having to formulate a "model" of the radio environment; hence, we may bypass the possible need for perceptual memory. In Chapter 3 we pointed out that spectrum estimation is an ill-posed inverse problem, which, therefore, requires the use of *regularization*. In Chapter 3, we also established that the MTM satisfies this requirement. Moreover, through the use of *time–space processing*, MTM provides the means for identifying the spectrum holes at a particular point in time as well as location in space. It is for these two reasons and a few others to be elaborated on in Section 7.5 that we view the MTM as a method of choice for perception (i.e. spectrum sensing) of the radio environment.

(2) *Learning and memory*

The task of dynamic spectrum management, to be discussed in Section 7.16, relies on the use of a learning process called *Hebbian learning*, inspired by the human brain; Hebbian learning was discussed in some detail in Chapter 2. An important characteristic of Hebbian learning is the inherent capability of *self-organization*. Thus, the dynamic spectrum manager has the practical means to dynamically choose and assign a set of appropriate links for communication to each cognitive-radio unit by learning the underlying environmental communication patterns. Knowledge thus learned about the communication patterns of the primary users in a radio network and, to some extent, those of other secondary users in the local neighborhood is stored in *memory*. Moreover, the "synaptic" weights of a *self-organized feature map* are *adaptively* updated in response to new inputs, on a cycle-by-cycle basis.

Looking at the block diagram of Figure 7.1b, we see that the dynamic spectrum manager is *reciprocally coupled* to the transmit-power controller. Specifically, through the use of *game-theoretic ideas*, to be discussed in Section 7.9, and by virtue of information received from the nonparametric spectrum estimator through the feedback channel about interference levels in chosen feed-forward communication channels, the transmit-power controller is enabled to *adaptively* adjust the transmitted radio signal, subject to prescribed constraints. In this *resource-allocation game*, the cognitive radio acquires the ability to reach equilibrium fast enough.

(3) *Attention*

To illustrate how the process of attention manifests itself in cognitive radio, consider the following example. A serious accident has occurred at some particular point in time and specific location in space, thereby resulting in a "surge" in wireless-communication traffic. By virtue of built-in space–time processing, the nonparametric spectrum estimator sends information to the dynamic spectrum manager in Figure 7.1b, identifying which particular subbands of the radio spectrum have become congested due to the accident. Furthermore, the dynamic spectrum manager itself focuses its attention on those remaining subbands with lower interference levels. In so doing, communication over the newly found cognitive radio link with users at both ends is maintained, bypassing the congested subbands.

Moreover, through its own self-organized learning process, the dynamic spectrum manager builds a *predictive model* of the radio environment. Using this model, the cognitive radio is enabled to do the following:

- pay attention to communication patterns and
- predict the availability-duration of spectrum holes, which, in turn, determines the predicted horizon of the transmit-power control mechanism.

(4) *Intelligence*

As it is with human cognition, intelligence in cognitive radio builds itself on the processes of perception, memory, and attention, just described under points (1), (2), and (3) respectively. To appreciate the importance of intelligence, consider a cognitive radio network with multiple secondary users whose communication needs would have to be accommodated in a satisfactory manner. Accordingly, the perception–action cycle of Figure 7.1b would have to be expanded in a corresponding way, such that the secondary users share:

- the radio environment for their individual forward communication needs and
- separate wireless channel for their individual feedback requirements.

In such a scenario, intelligence manifests itself as follows:

Through a decision-making mechanism involving intelligent choices in the transmitter, the available resources (i.e. spectrum holes and battery power) are equitably assigned to the secondary users in accordance with a prescribed protocol in the face of environmental uncertainties, and in such a way that the transmit-power of each user does not exceed a prescribed limit.

The environmental uncertainties include the reality that spectrum holes come and go in some stochastic manner, which may, therefore, mandate *robustification* of the transmit-power controller, an issue that is discussed in Section 7.13.

From the rationale just presented under points (1)–(4), it is now apparent that the four cognitive processes involved in radio for communication are satisfied quite differently from those in radar for remote sensing. This difference should not be surprising, bearing in mind the following two differences:

- cognitive radio and cognitive radar are intended for entirely different applications and
- the transmitter and receiver are located in different places in cognitive radio, whereas they are commonly collocated in cognitive radar.

# 7.3    Cognitive radio networks

As noted above, a *cognitive radio network* has to accommodate the individual communication needs of multiple secondary users. In so doing, significant complexity is appended to the functional operation of the network over and above that of cognitive radio. To be more precise, the cognitive radio network is no longer a system; rather, it assumes the form of a *system of systems*, demanding brand new practical considerations that will be addressed in forthcoming sections of the chapter. The purpose of this section is also to pave the way forward for those considerations.

In a cognitive radio network, there will naturally be a multiplicity of perception–action cycles that go on simultaneously across a common radio environment, with one cycle for each secondary user (i.e. communication link connecting a pair of cognitive radio transceivers). We may, therefore, think of a cognitive radio network as an *ensemble* of perception–action cycles, with the term "ensemble" being used here in a *stochastic* sense. What makes the study of such an ensemble not only fascinating in theoretical terms, but also challenging in practical terms, is the following list of practiced issues:

(1) *The spectrum holes come and go.* This issue arises, because, in reality, a spectrum hole is a subband that is actually owned by a licensed primary user who happens not to need that particular subband for its own employment at a certain point in time and location in space. Legally speaking, therefore, the subband in question must be given up by the secondary user in the cognitive radio network if, and when, the primary user demands it back for its own specific need.
(2) *Time delay that is incurred in the feedback channel.* This second issue arises because information extracted about the environment in the *RX-CR unit* naturally requires some time for transmission across the feedback channel to the *TX-CR unit*. Moreover, there is variability in this time delay from one perception–action cycle to another.
(3) *Security of the network.* This third issue is needed to protect the communication privacy of each secondary user in the face of attacks applied to the network by malicious users at random points in place and time.

Over and above these practical issues, there are, of course, propagation effects that also arise in the forward transmission across the data channel. From an analytic point of view, we may argue (with some justification) that this latter issue can be taken care of through the use of well-known diversity techniques (Molisch, 2005).

With this kind of provision in place, in this chapter we focus on issues (1) and (2). Issue (3) on security is beyond the scope of this chapter; nevertheless, Chapter 8 includes a brief discussion of security under "Unexplored issues."

Previously, we referred to a cognitive radio network as a system of systems. To be more precise, we may think of it as a *complex, stochastic, multiloop feedback control system* in which:

- *complexity* is inherited from the network itself,
- *stochasticity* is due to issues (1), (2), and (3) addressed above, and
- *multiloop feedback control* is attributed to the feedback channel from the RX-CR user to the TX-CR user for each communication link in the network.

From this brief characterization of a cognitive radio network, it is apparent that *control* plays a dominant role in the study of such networks. In this context, therefore, we may look to control as a fundamental process that is common not only to cognitive radio, but also cognitive radar that was considered in Chapter 6. Indeed, we may generalize this profound statement by going one step further:

Control is at the heart of all cognitive dynamic systems.

But, then, recognizing that control (applied in the transmitter) implies perception (performed in the receiver), we are simply reemphasizing the critical role of the perception–action cycle.

Summing up, the overall objective of a cognitive radio network is:

Reliable communication among secondary users in the cognitive radio network, whenever and wherever the need arises for it.

Moreover, recognizing the temporary availability of spectrum holes, the network must have the *agility* to realize this important objective as fast as practically feasible.

## 7.4   Where do we find the spectrum holes?

With spectrum holes playing a critical role in the underlying theory and design of cognitive radio networks, the key question is: where are they to be found in the radio spectrum? To set the stage for addressing this basic question, it is instructive to categorize the subbands of the radio spectrum in terms of occupancy:

(1) *White spaces*, which are free of RF interferers, except for *noise* due to natural and/ or artificial sources.
(2) *Gray spaces*, which are partially occupied by interferers and noise.
(3) *Black spaces*, the contents of which are completely full due to the combined presence of communication among primary users of the legacy radio network and (possibly) interfering signals plus noise.

The first place where we may find spectrum holes is the *television (TV) band*. The transition of all terrestrial television broadcasting from analog to digital, using the ATSC (Advanced Television Systems Committee) Standard, was accomplished in 2009 in North America. Moreover, in November 2008, the FCC in the USA ruled that access to the *ATSC-digital television (DTV)* band be permitted for wireless devices. Thus, for the first time ever, the way was opened in 2009 for the creation of "white spaces" for use by low-power cognitive radios. The availability of these white spaces will naturally vary across time and from one geographic location to another. In reality, however, noise is not likely to be the sole occupant of the ATSC-DTV band when a TV broadcasting station is switched off. Rather, interfering signals of widely varying power levels do exist below the DTV pilot. In other words, some of the subbands constituting the ATSC-DTV band may indeed be actually "gray," not "white".[1]

Consider next the commercial cellular networks deployed all over the world. In the current licensing regime, only *primary users* have exclusive rights to transmit. However, it is highly likely to find small spatial footprints in large cells where there are no primary users. Currently, opportunistic low-power usage of the cellular spectrum is *not* allowed in these areas, even though such usage by cognitive radios in a *femtocell* with a small base station is not detrimental to the *primary user* (Buddhiko, 2007). Thus, spectrum holes may also be found in commercial cellular bands; naturally, spread of the spectrum holes varies over time and space. In any event, account has to be taken of *interference* arising from conflict relationships between transmitters (base stations) of various radio infrastructure providers that coexist in a region (Subramanian and Gupta, 2007). Consequently, the spectrum holes found in cellular bands may also *not* all be white spaces.

The important point to take from this discussion is that, regardless of where the spectrum holes exist, be they in the ATSC-DTV band or cellular band, we are confronted with the practical reality that the spectrum holes may be made up of white and gray spaces. This possibility may, therefore, complicate applicability of a simple hypothesis-testing procedure that designates each subband as black (blocked space) or white (exploitable) space, using energy detection or cyclostationarity characterization.

## 7.4.1     Signal fading

The detection of spectrum holes is a difficult signal-processing problem. Unfortunately, this problem is made that much more difficult due to *signal fading* that manifests itself in two ways: multipath and shadowing (Molisch, *et al.* 2009).

Over a short distance scale (separating the receiver from the transmitter) comparable to one wavelength, we find significant fluctuations in the received signal power, which are attributed to the *multipath phenomenon*. This phenomenon arises due to the fact that the receiver's antenna picks up multiple reflections of the transmitted signal, with each reflected component having its own amplitude, phase, and delay. These components tend to interfere with each other in a constructive or destructive manner, depending on how the antenna is positioned. This first source of fading is referred to as *small-scale fading*. In statistical terms, small-scale fading is described by a *Rayleigh distribution* if there is no line of sight from the receiver to the transmitter and by a *Rician distribution* if a line of sight does exist (Molisch, 2005). The special case of Rayleigh distribution is also referred to as *flat fading* in the literature.

The *shadowing phenomenon* arises due to the presence of major obstacles (e.g. hills, buildings, and foliage) along the radio propagation between the transmitter and receiver. On account of the physical size of these obstacles, there can be significant variations in the received signal power, with the distance scale being in the order of hundreds of wavelengths. For this reason, shadowing is referred to as *large-scale fading*. In statistical terms, large-scale fading is described by a *log-normal distribution* (Molisch, 2005).

## 7.4.2     Wireless microphone sensing requirements

To illustrate how challenging the detection of spectrum holes can be in practice, we consider the spectrum-sensing requirements involved in using cognitive radio for wireless microphones (Shellhammer, 2010).

The wireless microphones that operate in the TV white space typically operate in the ultrahigh frequency (UHF) band and can be tuned over a range of frequencies. They must operate on a multiple of 25 kHz from the edge of a TV channel; but with a local oscillator (LO) frequency offset, they can be anywhere within the TV channel in practice. To ensure that a TV white space does not cause harmful interference to the wireless microphone receiver, the TV white-space-*sensing* device must cope with very *weak signals*. In the FCC (2008) report and order, the spectrum-sensing threshold was set at –114 dBm, where dBm stands for average power expressed in decibels with 1 mW as the frame of reference. However, recently the FCC issued a second memorandum opinion and order (FCC, 2010) in which the sensing threshold was changed to –107 dBm. This power level is referenced to a 0 dBi sensing antenna (i.e. an *isotropic antenna*). If the gain of the antenna is less than 0 dBi, then the sensing threshold must be lowered accordingly.

The noise power in a 6 MHz TV channel is –106 dBm plus the noise figure of the receiver, which, for low-cost devices, could be as high as 10 dB. For such a low-cost consumer device with a 10 dB noise figure, the noise floor is approximately –96 dBm. In that case the SNR of the wireless microphone signal is –11 dBm, and can be lower if the antenna gain is less than 0 dBi. In addition to thermal noise at the receiver input, there is also man-made noise in the UHF band due to out-of-band spurious emissions; this man-made noise adds to thermal noise of the receiver. The termal noise is well modeled as additive white Gaussian noise (AWGN); this mathematically tractable model is not true for the case of man-made noise, which can make sensing for weak wireless microphones particularly difficult.

Wireless microphone signals are narrowband signals with a maximum bandwidth of 200 kHz as specified by the FCC and with a typical bandwidth of 100 kHz or less. These narrowband signals are subject to multipath fading. Over a small bandwidth the fading is typically *flat*; so, the fading phenomenon experienced by the wireless microphone can be considered to be Rayleigh fading. In any event, the FCC has not specified if TV white-space-sensing devices are required to operate at an average power level of –107 dBm or at an instantaneous power level of –107 dBm. If sensing is only required at the instantaneous power level of –107 dBm, and when the wireless microphone signal fades below –107 dBm, then the TV white-space-sensing device is not required to detect the wireless microphone signal. However, if the FCC is going to require sensing at an average power level of –107 dBm, then when the wireless microphone fades due to multipath, the need for sensing would still be required; future FCC testing plans will clarify this issue. In any case, sensing techniques that are robust in the presence of multipath fading are highly desirable and may ultimately be required by the FCC.

### 7.4.3   Attributes of reliable spectrum sensing

In light of these practical realities we discussed above, we may now identify the desirable attributes of a spectrum sensor for cognitive-radio applications:

(1) *Detection of spectrum holes* and their *reliable* classification into white and gray spaces; this classification may require an *accurate estimation of the power spectrum*, particularly when the spectrum hole is of a gray-space kind.
(2) *Accurate spectral resolution* of spectrum holes, which is needed for efficient utilization of the radio spectrum; after all, this efficient utilization is the driving force behind cognitive radio.

(3) *Estimation of direction-of-arrival (DoA) of interferers*, which provides the cognitive radio a *sense of spatial direction*.

(4) *Time–frequency analysis* for highlighting *cyclostationarity*, which could be used as an additional method for the reinforcement of spectrum-hole detection and also *modulated-signal classification* when the subband of interest is occupied by a primary user.

What choice do we have for spectrum sensing in cognitive radio, which collectively satisfies this list of desirable attributes? One choice could be to opt for model-dependent *energy detection*, which is known for its simplicity. Unfortunately, energy detection is not only limited in scope but, most seriously, it lacks robustness. A recommended choice is the multitaper method considered next.

## 7.5    Multitaper method for spectrum sensing

The subject of spectrum sensing of the radio environment was discussed at some length in Chapter 3. For reasons explained in that chapter, we stressed the adoption of a *nonparametric* approach as it avoids the need for modeling the RF data. Moreover, we argued in favor of the MTM as a method of choice for designing the coherent multifunction spectrum sensor. To recap on the rationale for this choice, it is apropos that we remind ourselves of the desirable features of the MTM:

(1) In multitaper spectral estimation, the *bias* is decomposed into two quantifiable components:
   - *local bias*, due to frequency components residing inside the user-selectable band from $f - W$ to $f + W$, and
   - *broadband bias*, due to frequency components found outside this band.

(2) The *resolution* of a multitaper spectral estimator is naturally defined by the bandwidth of the passband, namely $2W$.

(3) Multitaper spectral estimators offer an easy-to-quantify *tradeoff between bias and variance*; accordingly, the bias–variance dilemma is replaced by the variance–resolution dilemma

(4) Direct spectrum estimation can be performed with more than just two *DoFs*; typically, the DoFs vary from 6 to 10, depending on the time–bandwidth product.

(5) The MTM has built-in regularization, in that it provides an analytic basis for computing the best approximation to a desired power spectrum, which is not possible from observable data alone. Moreover, the MTM is robust not only in estimating the power spectrum but also in hypothesis testing for the detection of spectrum holes (F. Zhang, 2011).

(6) Through the inclusion of a multiple array of antennas on receive, the MTM acquires a space–time processing capability, whereby information on the state of the radio environment (i.e. spectrum holes) in time as well as space can be computed; as discussed previously in Section 7.2, this kind of information can be of practical importance to the cognitive process on attention in cognitive radio.

(7) The MTM provides a rigorous mathematical basis for computing the cyclostationary characteristic of incoming radio signals, which can be exploited for identifying the legacy user responsible for occupying a subband of the radio spectrum of specific interest; as such, cyclostationarity can provide another way of further enhancing the detection of spectrum holes.

(8) Last, but by no means least, multitaper spectral estimates can be used to distinguish spectral line components within the band $(f - W, f + W)$ by including the *harmonic F-test*, which was discussed in Chapter 3.

## 7.5.1   Reliability, efficiency and computational issues

Cognitive radio applications require *reliable* identification of spectrum holes that are observable directly from the incoming RF data. The term "reliable" can be interpreted in many ways, but we use it to mean that *the estimate should have a low and quantifiable bias plus low variance*. The term "reliable" can also be interpreted as "robust," in the sense that the estimate should not depend on whether the data have a Gaussian, Laplacian, or any other possible distribution, it should not matter too much if the data are "random" or contain deterministic components, and the estimate is relatively tolerant of outliers. Simultaneously, identification of spectrum holes should also make *efficient* use of the incoming RF data: there is little point in attempting to estimate a spectrum to check what bands are available if it requires such a large sample size that the spectrum changes significantly while the data are being acquired. Finally, the computations involved in identifying the spectrum holes should be performed *rapidly*, recognizing that the time-scale of operations in the cognitive radio network is relatively short compared with that of the legacy radio network (Haykin *et al.*, 2009).

What we have just described here is that reliable, efficient, and rapid identification of spectrum holes is a tough and challenging task. The MTM satisfies the requirements of reliability and efficiency by virtue of properties (1)–(6). Moreover, from a computational perspective, multitaper power spectrum estimators can be fast:[2]

By precomputing the Slepian tapers and using the state-of-the-art FFT algorithm, computation of the MTM for spectrum sensing can be accomplished in a matter of 5–20 µs, which is relatively fast for cognitive-radio applications.

As discussed in Chapter 3, the Slepian tapers are at the heart of the MTM. We may, therefore, go on to say that the MTM is well suited for cognitive radio applications on the following grounds:

(1) In practical terms, the MTM is a reliable, efficient, and computationally fast method for spectrum-hole identification.

(2) In theoretical terms, the MTM provides a mathematical basis not only for spectrum estimation, but also space–time processing, time–frequency analysis, and cyclostationarity.

The issues, just mentioned under point (2), were all discussed in Chapter 3.

## 7.6        Case study I: wideband ATSC-DTV signal[3]

In cognitive radio applications, the requirement is for the receiver of a cognitive radio transceiver to make estimates of the average power of incoming RF data *rapidly*; hence, estimating the spectrum on short data blocks (each consisting of $N$ samples, say) is a *priority*. Moreover, reliable spectrum sensing for cognitive radio applications is *complicated* not only by the many characteristics of wireless communication signals, but also by interfering signals, channel fading, and receiver noise, all of which make the task of spectrum sensing a highly demanding task.

To illustrate the applicability of the MTM to cognitive radio, we consider the DTV band in this first case study of the chapter.

### 7.6.1        Spectral characteristics of ATSC-DTV signal

The ATSC-DTV is an eight-level *vestigial sideband (VSB) modulated signal* used for terrestrial broadcasting that can deliver a Moving Picture Expert Group-2 Transport Stream (MPEG-2-TS) of up to 19.39 megabits per second (Mbps) in a 6 MHz channel with 10.72 mega-symbols per second (over-the-air medium). A pilot is inserted into the signal at 2.69 MHz below the center of the signal (i.e., one-quarter of the symbol rate from the signal center). Typically, there are also several other narrowband signals found in the analysis bandwidth, from different sources of interference, which complicate the spectrum-sensing problem even more.

Using the MTM, we may generate a reasonable estimate of the available spectrum with about 100 µs of data.

To this end, Figure 7.2 shows the minimum, average, and maximum of 20 multitaper estimates of the ATSC-DTV power spectrum. Here, the block length used was $N = 2200$

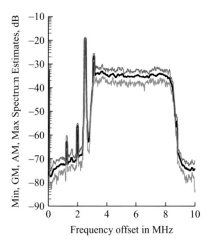

**Figure 7.2.** Multitaper estimates of the spectrum of the ATSC-DTV signal. Each estimate was made using $C_o = 6.5$ with $K = 11$ tapers. In this example, $N = 2200$, or 110 µs. The lower and upper blue curves represent the minimum and maximum estimates over 20 sections. When the figure is expanded sufficiently it can be seen to be two closely overlapping curves: the arithmetic (black) and geometric (lower blue) means. Reproduced with permission from "Spectra sensing for cognitive radio" S. Haykin *et al.* (2009). *Proc. IEEE*, **97**, 849–877.

samples, or 110 µs in duration. The blocks were offset 100% from each other. The pilot carrier is clearly visible in the figure at the lower band-edge of the MTM spectrum. In particular, there are considerable white spaces in the spectrum to fit in additional signals; hence the interest in the use of white spaces in the TV band for cognitive radio applications.[4]

The ATS-DTV signal is clearly seen to be 40 dB or more above the noise and the sharp narrowband interfering components. Both below and above the main ATSC-DTV signal, the slowly varying shapes of the baseline spectrum estimates are reminiscent of filter skirts, which make it improbable that the true thermal noise level is reached anywhere in this band.

Note also that, as anticipated, there are several *narrowband interfering signals* visible in the MTM spectrum of Figure 7.2:

- four strong signals (one nearly hidden by the vertical axis) and three others on the low side;
- two weak signals, one at 1.6 MHz (maximum level of about $3 \times 10^{-7}$, about halfway between the peaks at 1.245 and 1.982 MHz) and one halfway down the upper skirt at 8.7 MHz.

If we were to use a conventional periodogram or its tapered (windowed) version to estimate the ATS-DTV spectrum, then the presence of interfering narrowband signals such as those in Figure 7.2 would be obscured. Accordingly, in using the white spaces of the TV band, careful attention has to be given to the identification of interfering narrowband signals.

The overall message to take from the experimental results plotted in Figure 7.2 is summarized as follows:

(1) *Interfering signals do exist* in the ATSC-DTV, regardless of whether the TV station is switched on or off, as clearly illustrated in Figure 7.2.
(2) Although, indeed, it is well recognized that the ATSC-DTV band is open for cognitive radio applications when the pertinent TV station is switched off, the results presented in Figure 7.2 are intended to demonstrate the high spectral resolution that is afforded by using the MTM for spectrum sensing. Obviously, an immediate benefit of such high spectral resolution is improved radio-spectrum utilization, which, after all, is the driving force behind the use of cognitive radio, a point we have made over and over again because of its practical importance.
(3) The current FCC ruling on the use of white spaces in the ATSC-DTV band for cognitive radio applications permits such use by low-power cognitive radios only when the TV station is actually switched off. Nevertheless, the experimental results displayed in Figure 7.2 clearly show the feasibility of using low-power cognitive radios in white spaces that exist below the pilot; moreover, this is done without interference to the TV station when it is actually on, thereby offering the potential for additional improvement in radio-spectrum utilization.

## 7.7 Spectrum sensing in the IEEE 802.22 standard

In the article by Stevenson *et al.* (2009), a high-level overview of the *IEEE 802.22 standard* was described for cognitive wireless regional area networks (WRANs). For convenience of presentation, this standard is referred to simply as the IEEE 802.22 hereafter. In this standard, provision is made to define a single air-interface based on 2048-carrier *orthogonal frequency-division multiplexing* (OFDM) for a reliable end-to-end link for near line-of-sight (NLOS). Application of the IEEE 802.22 was aimed at a rural area of typically 17–30 km or more in radius (up to a maximum of 100 km) from a wireless network base station. The OFDM is described later, in Section 7.11.

Several issues relating to cognitive radio are covered in the IEEE 802.22. Two topics were given special attention:

(1) *Spectrum sensing*, which is included as a mandatory feature. The 802.22 network consists of a *base station* (BS) and a number of client stations, referred to as *customer premise equipment* (CPE). The BS controls when spectrum sensing and all such results are reported to it, which is where the final decision is made on the availability of a television channel for use by cognitive radio. As such, the spectrum sensing may be viewed as a signal processing and reporting function (Shellhammer, 2008). In the IEEE 802.22, three different licensed transmissions were specifically identified:
  - analog television (e.g. NTSC) signals;
  - DTV (e.g. ATSC) signals; and
  - wireless microphone signals.
(2) *Dynamic spectrum access*, the function of which is to dynamically identify and use the so-called "white spaces" (i.e. those subbands of the radio spectrum which are not being employed by legacy users).

   Expanding on wireless microphones as a potentially useful medium for the application of cognitive radio, their employment by secondary users is considered to be more challenging and difficult than television bands, be they analog or digital. To this end, in 2010 the Qualcomm company, San Diego, CA, conducted an open contest applicable to wireless microphones. Out of the many submissions received, three of them were selected by Qualcomm as the top ones; details of these three submissions are described at the end of the chapter.[5] Suffice it to say, although the procedures for spectrum sensing adopted by the top three winning contestants were quite different, they did have one feature in common: all three submissions used power-spectrum estimation, in one form or another, as the basis for their procedures.

## 7.8 Noncooperative and cooperative classes of cognitive radio networks

In a conventional radio network, the network is built around BS that are designed to accommodate the allocation of primary resources, namely *power* and *bandwidth*, among multiple users in some optimal manner. On the other hand, for a cognitive radio network to be broadly applicable, it is highly desirable for the network to operate in a *decentralized* manner. Insofar as the issue of *multiple access* is concerned, there are two basic mechanisms for accommodating communication needs of the secondary users, depending on how the users' *utility functions* are defined:

(1) *Noncooperative mechanism.* In this first approach, each user operates in a "greedy" manner so as to optimize its own utility function without paying attention to other users in the network.

(2) *Cooperative mechanism.* As the name implies, users of the network optimize their individual utility functions by agreeing to cooperate with each other.

Although, indeed, these two mechanisms are radically different and aimed at different applications, they do share a common prescription:

All the secondary users of a cognitive radio network follow the same *protocol*, the formulation of which involves the imposition of prescribed procedures to ensure that they work alongside the primary users of the legacy radio network in a harmonious manner.

Moreover, for ideas needed to formulate the protocol, we may look to the following three fields:

(1) *Game theory.* Here, the secondary users of a cognitive radio network are viewed as *players* or *agents*, who are collectively engaged in a *game*. Selection of the appropriate game is determined by whether the resource-allocation problem follows a cooperative or noncooperative approach. Typically, the selected game-theoretic strategy is characterized by an *equilibrium stationary point*, to which the evolution of each player's trajectory in the game converges. However, game theory does not tell us how to get there; rather, we look to information theory for guidance.

(2) *Information theory.* Earlier in this section, we mentioned there are two primary resources on which we have to focus attention: power and bandwidth. The tradeoff between these two resources is well described by Shannon's *information capacity formula*, which, in a remarkable way, embodies the following three parameters in a single formula (Cover and Thomas, 2006):
   - average value of the transmitted power,
   - channel bandwidth, and
   - power spectral density of additive noise at the channel output (i.e. receiver input).

   The formula assumes an ideal channel model, which is lossless and the channel (link) noise is Gaussian. The lossless assumption is taken care of by accounting for the path loss from the transmitter to the receiver across a wireless communication link. As for the Gaussian assumption, it is justified by invoking the *central limit theorem* of probability theory, applied to the multitude of interfering signals attributed to users in the network other than the particular user under scrutiny; the central limit theorem was discussed in Note 11 of Chapter 4.

(3) *Optimization theory.* Here, we look to a *constrained* optimization procedure, the utility function for which embodies the information-capacity formula and the constraints imposed on it in accordance with the noncooperative or cooperative mechanism. Application of the optimization procedure leads to an algorithmic solution to the resource-allocation problem, desirably in *iterative* form.

The algorithm so obtained serves the purpose of a *cognitive controller*, for the experimental evaluation of which we look to principles of *control theory*.

With this overview, we proceed, first, by studying noncooperative cognitive radio networks, the current status of which is well developed. Under the noncooperative approach, the Nash equilibrium in game theory stands out as a logical choice. This study occupies the material presented in Sections 7.9–7.15.

Then, in Section 7.16, we consider cooperative cognitive radio networks. In particular, we focus the discussion on a new class of cooperative networks known as coalitional networks.

## 7.9     Nash equilibrium in game theory

In a game-theoretic framework,[6] involving a competitive multi-agent environment with limited resources, the actions of all players in the environment are coupled with each other via the available resources. Accordingly, the task of finding a *global optimum* for the resource-allocation problem can be not only time consuming, but also computationally intractable. Moreover, the optimization would require large amounts of information being exchanged between different agents, and thereby consume precious resources. To get around these difficulties, we have to be content with a *suboptimal solution*. Prominent among the game-theoretic ideas, embodying such solutions for *finite games* with each agent having only a finite number of alternative courses of action, is that of a *Nash equilibrium*,[7] for which we offer the following definition:

A Nash equilibrium is an action profile (i.e. vector of players' actions) in a noncooperative game, in which each action is a best response to the actions of all the other players in the game.

According to this definition, the Nash equilibrium is a *stable operating (i.e. equilibrium) point* in the sense that there is no incentive for any *rational* player involved in a finite noncooperative game to change strategy given that all the other players continue to follow the equilibrium policy. The important point to note here is that the Nash-equilibrium approach provides a powerful tool for modeling nonstationary processes. Simply put, it has had an enormous influence on the evolution of game theory by shifting its emphasis toward the study of equilibria as a *predictive concept*.

With the learning process modeled as a *repeated stochastic game* (i.e. repeated version of a one-shot game), each player gets to know the past behavior of the other players, which may influence the current decision to be made. In such a game, the task of a player is to select the best mixed strategy, given information on the mixed strategies of all other players in the game; hereafter, other players are referred to simply as "opponents." A *mixed strategy* is defined as *a continuous randomization by a player of its own actions*, in which the actions (i.e. pure strategies) are selected in a deterministic manner. Stated in another way, *the mixed strategy of a player is a random variable whose values are the pure strategies of that player*.

To explain what we mean by a mixed strategy, let $a_{j,k}$ denote the $k$th action of player $j$ with $j = 1, 2, \ldots, m$, and $k = 1, 2, \ldots, K$, where $m$ is the total number of players and

$K$ is the total number of possible actions available to each user in the game. The mixed strategy of player $j$, denoted by the set of probabilities $\{p_{j,k}\}_{k=1}^{K}$, is an integral part of the *linear combination*

$$q_j = \sum_{k=1}^{K} p_{j,k} a_{j,k} \qquad \text{for } j = 1, 2, \ldots, m. \qquad (7.1)$$

Equivalently, we may express $q_j$ as the *inner product*

$$q_j = \mathbf{p}_j^{\mathrm{T}} \mathbf{a}_j \qquad \text{for } j = 1, 2, \ldots, m, \qquad (7.2)$$

where

$$\mathbf{p}_j = [p_{j,1},\ p_{j,2},\ \ldots,\ p_{j,K}]^{\mathrm{T}}$$

is the *mixed strategy vector* and

$$\mathbf{a}_j = [a_{j,1},\ a_{j,2},\ \ldots,\ a_{j,K}]^{\mathrm{T}}$$

is the *deterministic action vector*. The superscript T denotes matrix transposition. For all $j$, the elements of the mixed strategy vector $\mathbf{p}_j$ satisfy the two basic conditions of probability theory:

$$0 \le p_{j,k} \le 1 \qquad \text{for all } j \text{ and } k; \qquad (7.3)$$

$$\sum_{k=1}^{K} p_{j,k} = 1 \qquad \text{for each } j. \qquad (7.4)$$

Note also that the mixed strategies for the different players are *statistically independent*.

The motivation for permitting the use of mixed strategies is the well-known fact that every stochastic game has *at least one Nash equilibrium* in the space of mixed strategies, but not necessarily in the space of pure strategies; hence the preferred use of mixed strategies over pure strategies. The purpose of a learning algorithm is that of computing a mixed strategy, namely a sequence $\{q^{(1)}, q^{(2)}, \ldots, q^{(t)}\}$ for each player over the course of time $t$.

It is also noteworthy that the implication of (7.1)–(7.4) is that the entire set of mixed strategies lies inside a *convex simplex* or *convex hull*, whose dimension is $K - 1$ and whose $K$ vertices are the $a_{j,k}$. Such a geometric configuration makes the selection of the best mixed strategy in a multiple-player game a more difficult proposition to tackle than the selection of the best base action in a single-player environment.

To sum up, the Nash equilibrium is considered to be a concept of fundamental importance in game theory. This equilibrium point is the solution to a game-theoretic problem in which none of the players in the game has the incentive to deviate from it unilaterally. In other words, at a Nash-equilibrium point, each player's chosen strategy is the "best response" to strategies of the other players. From a practical perspective, even though a Nash-equilibrium solution is not always the best solution to the resource-allocation

problem, it is a reasonable candidate for solving this problem in a game-theoretic framework; hence its practical relevance to addressing the resource-allocation problem in
cognitive radio networks.

## 7.10     Water-filling in information theory for cognitive control

With this brief exposition of the Nash equilibrium, we may now resume our discussion
of cognitive radio networks, in which a user, referring to a communication link, plays
the role of a player in a finite-game theoretic problem. The issue of interest is that of
developing a *cognitive controller for resource allocation* with particular emphasis on
transmit-power control. To this end, adoption of the Nash equilibrium is well suited for
the issue at hand, bearing in mind the following attributes:

- decentralized implementation of the resource-allocation algorithm for achieving a
  Nash equilibrium,
- low computational complexity, and
- fast convergence to a reasonably good solution represented by a Nash-equilibrium point.

To achieve these attributes, unfortunately, game theory cannot do it all by itself. Rather,
as pointed out previously in Section 7.8, we may look to information theory for direction.
More specifically, the integration of the Nash equilibrium with the *information-theoretic
water-filling model* works nicely for solving the resource-allocation problem. For a statement of the water-filling model, we say (Cover and Thomas, 2006):

Given a set of parallel Gaussian communication links (channels), application of Shannon's
channel capacity formula provides the theoretical basis for a procedure by means of which the
available power is distributed among the various links in a manner identical to the way in which
water distributes itself in a vessel.

It is because of this analogy that we speak of a water-filling model in information theory.
Correspondingly, formulation of the controller is referred to as a *water-filling controller*.

### 7.10.1     Statement of the transmit-power control problem

Consider a cognitive radio network involving $m$ secondary users. Communication
among the users is to be *asynchronous*, with the communication process being viewed
as a *noncooperative game*. The objective of each user in the network is to maximize
its own *data transmission rate* in a greedy manner. We may then state the essence of
transmit-power control for this noncooperative scenario as follows (Haykin, 2005a):

Given a limited number of spectrum holes, select the transmit-power levels of $m$ unserviced secondary
users in a cognitive radio network so as to jointly maximize their data-transmission rates, subject to
the constraint that the interference-power limit imposed on each link in the network is not violated.

It may be tempting to suggest that the solution of this problem lies in simply increasing
the transmit-power level of each unserviced transmitter. However, increasing the
transmit-power level of any one transmitter has the undesirable effect of also increasing

the level of interference to which the receivers of all the other transmitters are subjected. The conclusion to be drawn from this reality is that it is not possible to represent the overall system performance with a single index of performance. Rather, we have to adopt a *tradeoff* among the data rates of all unserviced users in some computationally tractable fashion.

Ideally, we would like to find a *global solution* to the constrained optimization of the joint set of data-transmission rates in the network. Unfortunately, finding this global solution requires an exhaustive search through the space of all possible power allocations scenarios, in which case we find that the computational complexity needed for attaining the global solution assumes a prohibitively high level.

To overcome this computational difficulty, we use the *competitive optimality criterion*[8] (Yu, 2002) for solving the transmit-power control problem, which may now be restated as follows:

Considering a cognitive radio network viewed as a noncooperative game, maximize the performance of each secondary user, regardless of what all the other users in the network do, but subject to the constraint that a prescribed interference-power limit for each user is not violated.

This formulation of the distributed transmit-power control problem leads to a solution that is of a *local* nature; although the solution is suboptimal, it is not only insightful, but also practical.

## 7.10.2    Example: two-user noncooperative scenario

Consider the simple scenario of Figure 7.3 involving two users communicating across a flat-fading channel. The complex-valued baseband channel matrix of the network is denoted by

$$\mathbf{H} = \begin{bmatrix} h_{11} & h_{12} \\ h_{21} & h_{22} \end{bmatrix}. \tag{7.5}$$

Viewing this scenario as a noncooperative game, we may describe the two players (users) of the game as follows:

- The two *players* are represented by *communication links* 1 and 2.
- The *pure strategies* (i.e. deterministic actions) of the two players are defined by the *power spectral densities* $S_1(f)$ *and* $S_2(f)$ that respectively pertain to the transmitted signals radiated by the transmitters of communication links 1 and 2.
- The *payoffs* (i.e. rewards) to the two players are defined by the *data-transmission rates* $R_1$ and $R_2$, which correspond to communication links 1 and 2 respectively.

To proceed with a solution to this two-user transmit-control problem, we characterize the problem as follows:

(1) The noise floor of the RF radio environment is characterized by a frequency-dependent parameter: the power spectral density $S_N(f)$. In effect, $S_N(f)$ defines the "noise floor" above which the transmit-power controller must fit the transmission-data requirements of both communication links 1 and 2.

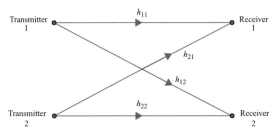

**Figure 7.3.** Signal-flow graph of a two-user communication scenario.

(2) In order to assure *reliable communication* under varying operating conditions, an *SNR gap* is included in calculating the transmission data of each user in the network.

On this basis, we may now define the *cross-coupling* between the two communication links in Figure 7.3 in terms of two *normalized interference gains* $\alpha_1$ and $\alpha_2$ by writing

$$\alpha_1 = \frac{\Gamma|h_{12}|^2}{|h_{22}|^2} \qquad (7.6)$$

and

$$\alpha_2 = \frac{\Gamma|h_{21}|^2}{|h_{11}|^2}, \qquad (7.7)$$

where $\Gamma$ is the SNR gap. Assuming that receivers of the two communication links do *not* perform any form of interference cancellation regardless of the received signal strengths, we may respectively formulate the achievable data-transmission rates $R_1$ and $R_2$ as two definite integrals involving spectrum holes 1 and 2, as shown by

$$R_1 = \int_{\text{hole 1}} \log_2\left(1 + \frac{S_1(f)}{N_1(f) + \alpha_2 S_2(f)}\right) df \qquad (7.8)$$

and

$$R_2 = \int_{\text{hole 2}} \log_2\left(1 + \frac{S_2(f)}{N_2(f) + \alpha_1 S_1(f)}\right) df. \qquad (7.9)$$

The integrands in (7.8) and (7.9) follow from the application of *Shannon's information-capacity formula*[9] to links 1 and 2 respectively. The term $\alpha_2 S_2(f)$ in the denominator of (7.8) and the term $\alpha_1 S_1(f)$ in the denominator of (7.9) are due to the *cross-coupling* between the transmitters and receivers of communication links 1 and 2; in short, they are both the effects of interfering signals. The remaining two terms $N_1(f)$ and $N_2(f)$ are *noise* terms, which are respectively defined by

$$N_1(f) = \frac{\Gamma S_{N,1}(f)}{|h_{11}|^2} \qquad (7.10)$$

and

$$N_2(f) = \frac{\Gamma S_{N,2}(f)}{|h_{22}|^2},$$  (7.11)

where $S_{N,1}(f)$ and $S_{N,2}(f)$ are respectively parts of the noise-floor's power spectral density $S_N(f)$ that define the spectral contents of spectrum holes 1 and 2.

We are now ready to formally state the competitive optimization problem for the two-user scenario of Figure 7.3 as follows:

Given that the power spectral density $S_2(f)$ *of the transmitter of communication* link 2 is fixed, maximize the transmission-data rate $R_1$ of link 1, subject to the constraint

$$\int_{\text{hole 1}} [S_1(f) + N_1(f) + \alpha_2 S_2(f)] \, df \leq P_{I,\max},$$  (7.12)

where $P_{I,\max}$ is the maximum allowable interference power.

A similar statement applies to the competitive optimization of the transmitter of communication link 2. Of course, it is understood that both $S_1(f)$ and $S_2(f)$ remain nonnegative for all $f$, as they must be in accordance with the definition of power spectral density.

Solution to the optimization problem just described follows the allocation of transmit power to users 1 and 2 in accordance with the iterative water-filling procedure. However, before discussing this new procedure in Section 7.12, we digress briefly in the next section to present a brief review of OFDM, on which the air-interface in the IEEE 802.22 is based; hence its practical importance for cognitive radio networks–see Section 7.7.

## 7.11 Orthogonal frequency-division multiplexing

A multicarrier modulation scheme that commends itself for deployment in cognitive radio networks is *OFDM*. In this scheme, closely spaced *orthogonal subcarriers* (e.g. 2048 in number in the IEEE 802.22) are used to transmit narrowband data segments simultaneously across the forward wireless link connecting the transmitter at one end of the link in the network to the receiver at the other end of the link. In effect, the forward wireless link, which commonly suffers from the *frequency-selective fading phenomenon*, is divided into a number of narrowband "flat-fading" subchannels, with each subchannel having its own subcarrier.

OFDM has many practical advantages over single-carrier transmission, as summarized here (Li and Stüber, 2006; Hanzo *et al.*, 2010):

(1) OFDM improves the efficiency of spectrum utilization by simultaneous use of multiple orthogonal subcarriers, which are densely packed.
(2) First of all, the OFDM waveform is generated in the frequency domain and then it is transformed into the time domain, thereby providing flexible bandwidth allocation.
(3) *Interleaving* (i.e., randomly shuffling) the information over different OFDM symbols (before transmission) provides against the loss of information caused by flat-fading and noise effects, and then *de-interleaving* at the receiver to recover the original information.

(4) Although the spectrum tails of subcarriers overlap with each other, at the center frequency of each subcarrier all other subcarriers are zero in accordance with the property of orthogonality in the frequency domain. Theoretically, this property prevents *intercarrier interference* (ICI). However, time- and frequency-synchronizations are critical for ICI prevention, as well as for correct demodulation; therefore, it is a major challenge in the physical layer design of cognitive radio.

(5) Since a narrowband signal has a longer symbol duration than a wideband signal, OFDM takes care of *intersymbol interference* (ISI) caused by multipath delay of wireless channels. However, guard-time intervals, which are longer than the channel impulse response, are ordinarily introduced in practice between OFDM symbols to eliminate the ISI by giving enough time for each transmitted OFDM symbol to dissipate to negligible levels.

(6) Owing to relatively low ISI, the complexity of equalization at the receiver of each user in the cognitive radio network is reduced considerably, which, in turn, leads to simplified receiver structures.

In summary, OFDM enables the provision of *frequency diversity* for combating the multipath phenomenon in wireless communications, provides for the following practical advantages compared with single-carrier data transmission schemes:

- the accomplishment of higher data rates;
- more flexibility in the selection of transmit-waveform characteristics; and
- greater robustness against channel noise and fading.

Just as importantly, application of the *FFT algorithm* provides a powerful and well-studied tool for a *computationally efficient implementation* of the OFDM and, therefore, the cognitive radio network. In this context, it is noteworthy that the choice of 2048 (an integer multiple of 2) subcarriers in the IEEE 802.22 is recommended by virtue of the FFT algorithm in mind.

## 7.12     Iterative water-filling controller for cognitive radio networks

At long last, we are now ready to formulate the cognitive controller for resource allocation in cognitive radio networks. To be precise, there are two primary resources to be considered:

(1) the transmit-power level permissible to each secondary user in the network and
(2) the number of subcarriers available in the OFDM.

The subcarriers provide frequency-domain representations of spectrum holes (i.e. unused subbands of the radio spectrum) identified by the receiver of each secondary user in the network, with a set of $2^l$ subcarriers for each spectrum hole, where $l$ is an integer.

For dynamic spectrum management of the network, the number of possible moves is *finite*. This is because the manager can only choose a subset of subcarriers from the total set of available subcarriers per spectrum hole, which is finite since the OFDM is a finite scheme. On the other hand, power is a *continuous variable* by its very nature; the transmit-power controller, therefore, has infinitely many options available to it in adjusting the power levels assigned to the subcarriers by the spectrum manager. Stated in another way: although the *action space* of the controller is of finite dimensionality, the number of possible actions is infinitely large.

To proceed, then, with mathematical formulation of the cognitive controller, let $m$ denote the number of secondary users in the cognitive radio network and let $n$ denote the number of subcarriers that can be potentially available for secondary communication in a noncontiguous OFDM scheme. Since *spectral efficiency* is a driving force behind cognitive radio, the *utility function* of each secondary user in the network of cognitive radios to be maximized is the *data transmission rate*. Thus, adoption of the water-filling model requires that each user in the network solve the following *constrained optimization problem* (Setoodeh and Haykin, 2009):

$$\max_{\mathbf{p}^i} f^i(\mathbf{p}^1, \dots, \mathbf{p}^m) = \sum_{k=1}^{n} \log_2 \left( 1 + \frac{p_k^i}{I_k^i} \right) \tag{7.13}$$

subject to

$$\left. \begin{aligned} &\sum_{k=1}^{n} p_k^i \leq p_{\max}^i \\ &p_k^i + I_k^i \leq \mathrm{CAP}_k \quad \forall k \notin \mathrm{PS} \\ &p_k^i = 0 \qquad\qquad \forall k \in \mathrm{PS} \\ &p_k^i \geq 0 \end{aligned} \right\}, \tag{7.14}$$

where $p_k^i$ denotes user $i$'s transmit power over subcarrier $k$, CAP denotes the maximum allowable level of interference power, and PS stands for a set of subcarriers employed by primary users. The noise plus interference experienced by user $i$ at subcarrier $k$, because of transmissions from other users, is given by

$$I_k^i = \sigma_k^i + \sum_{j \neq i} \alpha_k^{ij} p_k^j. \tag{7.15}$$

Recognizing that cognitive radio is *receiver centric*, in that it is at the receiver of each communication link in the network where the radio environment is sensed, $I_k^i$ is measured at receiver $i$. Note also that, with the OFDM integrated into the cognitive radio design, the logarithmic term $\log_2 \left[ 1 + \frac{(p_k^i)}{(I_k^i)} \right]$ in (7.13) is a *discrete* representation of the information capacity formula that was invoked previously in (7.8) and (7.9). To be more precise, the discrete representation of SNR in (7.13) is formulated in terms of OFDM subcarriers. As such, the following point should be carefully noted. Whereas in the two-user example described in Section 7.10 the formulation of data rates was based on spectrum holes, in (7.13) the formulation is based on subcarriers. As such, it is permissible for the subcarriers assigned to a spectrum hole to be employed by more than one secondary user in the cognitive radio network, thereby providing for a more efficient utilization of the OFDM scheme and broader applicability of the water-filling model. Yet another advantage of basing the water-filling model on OFDM is *discretization*, whereby integration in the information capacity formula is replaced by a summation of subcarrier powers. There is, therefore, much to be said on the use of (7.13) as the algorithmic basis of the *iterative water-filling controller* (IWFC). However, it should be noted that, for subcarriers to be employable by more than one secondary user, provisions must be made for the avoidance of cross-modulation between the subcarriers of two or more secondary users sharing the same subcarrier. This provision can be satisfied by assigning a specific code to each secondary user, as is done in code-division multiplexing.

The positive parameter $\sigma_k^i$ in (7.15) is the *normalized* background noise power at the receiver input of user $i$ on the $k$th subcarrier. Here again it should be noted the normalized noise parameter $\sigma$ should *not* be confused with the variance of a noise source.

The nonnegative parameter $\alpha_k^{ij}$ is the normalized *interference gain* from transmitter $j$ to receiver $i$ at subcarrier $k$; of course, we have $\alpha_k^{ii} = 1$. The term $\alpha_k^{ij}$ is the combined effect of two factors:

- propagation path-loss from transmitter $j$ to receiver $i$ at subcarrier $k$;
- subcarrier amplitude reduction due to the *frequency offset* $\Delta f$, which is inherent to OFDM.

Mathematically, $\alpha_k^{ij}$ is defined as

$$\alpha_k^{ij} = \frac{\Gamma |h_k^{ij}|^2}{|h_k^{ii}|^2},\tag{7.16}$$

where $\Gamma$ is the *SNR gap* and $h_k^{ij}$ is the *channel gain* from transmitter $j$ to receiver $i$ over the flat-fading subchannel associated with subcarrier $k$. Regarding the empirical formula for the path loss (Haykin, 2005a), we have

$$|h_k^{ij}|^2 = \frac{\beta_k}{(d_{ij})^r},\tag{7.17}$$

where $d_{ij}$ is the distance from transmitter $j$ to receiver $i$. The *path-loss exponent* $r$ varies from two to five, depending on the environment, and the *attenuation parameter* $\beta_k$ is frequency dependent. Therefore, we may write

$$\alpha_k^{ij} \propto \left(\frac{d_{ii}}{d_{ij}}\right)\tag{7.18}$$

In general, it should be noted that

$$\alpha_k^{ij} \neq \alpha_k^{ji}.\tag{7.19}$$

In any event, if user $i$'s receiver is closer to its transmitter compared with other active transmitters in the network, we will have $\alpha_k^{ij} \leq 1$.

Referring to lines 1 and 2 of constraint inequalities in (7.14), $p_{\text{max}}^i$ is user $i$'s maximum power and $\text{CAP}_k$ is the maximum allowable interference power at subcarrier $k$. $\text{CAP}_k$ is determined in a way to make sure that the permissible interference power level limit will not be violated at the primary users' receivers.

In the IWFC, user $i$ assumes that $p_k^j$ is fixed for $j \neq i$. Therefore, the optimization problem in (7.13) is a *concave maximization problem* in the power vector $\mathbf{p}^i = [p_1^i, \ldots, p_n^i]^{\text{T}}$, which can be converted to a *convex minimization problem* by introducing a minus sign and considering $-f^i$ as the objective function to be minimized. The first constraint in (7.14) states that the total transmit power of user $i$ at all subcarriers should not exceed its maximum power (power budget). The second constraint in (7.14) guarantees that the interference caused by all secondary users in the cognitive radio network at each subcarrier will be less than the maximum allowed interference at that subcarrier. If primary users in the legacy radio network do not let the secondary users employ the nonidle subcarriers in their frequency bands, then secondary users should not use those subcarriers for transmission. The third constraint in (7.14) guarantees this requirement by forcing the related components of the secondary user $i$'s power vector to be zero. If primary users allow the coexistence of cognitive radio users at nonidle subcarriers on condition that they do not violate the permissible interference power level, then the third constraint in (7.14) can be relaxed, in which case the second constraint in the set described therein suffices.

As mentioned previously, the IWFC is implemented in a decentralized manner. In order to solve the optimization problem (7.13), it is not necessary for user $i$ to know the value of $p_k^j$ for all $j \neq i$. The $I_k^i$ defined in (7.15) is measured by user $i$'s receiver rather than calculated; accordingly, secondary users do not need to exchange information. Furthermore, it is not necessary for user $i$ to know the number of other users in the network. Therefore, changing the number of users in the network does not affect the complexity of the optimization problem that should be solved by each user. Hence, there is no scaling problem, which is another attribute of the IWFC.

While the action of user $i$ is denoted by its power vector $\mathbf{p}^i$, following the notation in the game-theory literature, the joint actions of the other $m - 1$ secondary users in the cognitive radio network are denoted by $\mathbf{p}^{-i}$, where a minus sign is introduced in the superscript. Three major types of adjustment scheme $S$ can be used by the cognitive radio users to update their actions (Setoodeh and Haykin, 2009):

(1) Iterative-water-filling users update their actions in a *predetermined order*, as described by

$$\mathbf{p}^{-i}(S_t) = [\mathbf{p}^1(t+1), \ldots, \mathbf{p}^{i-1}(t+1), \mathbf{p}^{i+1}(t), \ldots, \mathbf{p}^m(t)]. \qquad (7.20)$$

According to (7:20), user $i$ updates its action by having access to the updated information from users $1, \ldots, i-1$ but not from users $i+1, \ldots, m$.

(2) *Simultaneous iterative-water-filling.* Users update their actions *simultaneously* by respecting the most recent actions of the others, as shown by

$$\mathbf{p}^{-i}(S_t) = \mathbf{p}^{-i}(t). \qquad (7.21)$$

(3) *Asynchronous iterative-water-filling*, which is an instance of an adjustment scheme that user $i$ receives update information from user $j$ at random times with some delay, as described by

$$\mathbf{p}^{-i}(S_t) = [\mathbf{p}^1(\tau_t^{i,1}), \ldots, \mathbf{p}^{i-1}(\tau_t^{i,i-1}), \mathbf{p}^{i+1}(\tau_t^{i,i+1}), \ldots, \mathbf{p}^m(\tau_t^{i,m})], \tag{7.22}$$

where $\tau_t^{i,j}$ is an integer-valued random variable satisfying the condition

$$\max_{i,j \in N}(0, t-d) \le \tau_t^{i,j} \le t+1 \qquad j \neq i, \tag{7.23}$$

which means that the delay does not exceed $d$ time units.

Recognizing the desire for avoiding both the use of central scheduling and the need for synchronization between different users in a cognitive radio network, the asynchronous adjustment scheme of (7.22) and (7.23) is the most realistic one when network simplicity is at a premium.

## 7.13    Stochastic versus robust optimization

The cognitive radio network has a dynamic nature, in which users are on the move all the time: they can leave the network and new users can join the network in a stochastic manner. Because of these stochastic phenomena, the interference-plus-noise term (7.15) in the objective function and the second constraint in (7.14) are both time varying. The IWFC, therefore, assumes the form of an optimization problem under *uncertainty*. Also, the appearance and disappearance of the spectrum holes, which depend on the activities of primary users, will change the third constraint in (7.14). The behavior of primary users and, therefore, the availability and duration of the availability of spectrum holes, can be predicted to some extent using a *predictive model*, on which more will be said in Section 7.16. During the time intervals that the activity of primary users does not change and the available spectrum holes are fixed, two approaches can be taken to deal with the uncertainty caused by the joining and leaving of other cognitive radio users and their mobility: *stochastic optimization and robust optimization* (Fukushima, 2007). The pros and cons of these two approaches are discussed in what follows.

Let the noise-plus-interference term be the summation of two components, a nominal term $\bar{I}$ and a pertubation term $\Delta I$, as follows:

$$I_k^i = \bar{I}_k^i + \Delta I_k^i. \tag{7.24}$$

In the following, the objective functions for both stochastic and robust versions of the optimization problem (7.13) are presented.

If there is good knowledge about the probability distribution of the uncertainty term $\Delta I$, then the uncertainty can be dealt with by means of probability theory and related concepts. In this case, calculation of the expected value will not be an obstacle;

therefore, the iterative water-filling algorithm problem (7.13) can be formulated as a stochastic optimization problem with the following objective function:

$$\max_{\mathbf{p}^i} \left[ \mathbb{E}_{\Delta \mathbf{I}^i} \sum_{k=1}^{n} \log \left( 1 + \frac{p_k^i}{\overline{I}_k^i + \Delta I_k^i} \right) \right], \tag{7.25}$$

where $\mathbb{E}$ denotes the statistical expectation operator with respect to

$$\Delta \mathbf{I}^i = \left[ \Delta I_1^i, \, ..., \, \Delta I_n^i \right]^{\mathrm{T}}. \tag{7.26}$$

In practice, however, little may be known about the probability distribution $\Delta I$, which means that the stochastic optimization approach utilizing the expected value is not a suitable practical approach. In stochastic situations of the kind described here, robust optimization techniques, based on the *worst-case analysis* and thereby bypassing the need for probability theory, are more appropriate. The price paid for this alternative approach is an overly conservative one, in the sense that suboptimality in performance is traded in favor of *robustness*. In spite of this shortcoming, robust optimization is in the spirit of what successful engineering network designs require in practice; it is a hallmark of cognition.

The formulation of the IWFC as a robust game in the sense described by Aghassi and Bertsimas (2006) is basically a *max–min problem*, in which each user tries to maximize its own utility while the environment and the other users are trying to minimize that user's utility (Danskin, 1967; Basar and Bernhard, 1995); worst-case interference scenarios have been studied for *digital subscriber lines* (DSL) by Brady and Cioffi (2006). Considering an ellipsoidal uncertainty set, the IWFC problem (7.13) can be reformulated as the following robust optimization problem:

$$\max_{\mathbf{p}^i} \left[ \min_{\|\Delta \mathbf{I}^i\| \leq \varepsilon} \sum_{k=1}^{n} \log_2 \left( 1 + \frac{p_k^i}{\overline{I}_k^i + \Delta I_k^i} \right) \right] \tag{7.27}$$

subject to

$$\left. \begin{array}{ll} \sum_{k=1}^{n} p_k^i \leq p_{\mathrm{max}}^i & \\ \max(p_k^i + \overline{I}_k^i + \Delta I_k^i) \leq \mathrm{CAP}_k & \forall k \notin PS \\ \|\Delta \mathbf{I}^i \leq \varepsilon\| & \\ p_k^i = 0 & \forall k \in PS \\ p_k^i \geq 0 & \end{array} \right\}. \tag{7.28}$$

A large $\varepsilon$ in objective function of (7.27) and the second constraint of (7.28) account for large perturbations. Basically, the new set of constraints in (7.28) guarantees that the permissible interference power level will not be violated for any perturbation from the uncertainty set considered.

Stochastic optimization guarantees some level of performance on average, but sometimes the desired quality of service may not be achieved, which, in effect, means a lack of reliable communication. On the other hand, robust optimization guarantees an acceptable level of performance under the worst-case conditions, but it is a conservative approach. We say so because real-life systems are not always in their worst behavior, but, from a reliability perspective, it can provide seamless communication even in the worst situations. Recognizing the dynamic nature of a cognitive radio network and the delay introduced by the feedback channel, the statistics of interference that are used by the transmitter to adjust its power may not represent the current situation of the network. In such situations, robust optimization is equipped with the means to prevent permissible interference power-level violation by taking into account the worst-case uncertainty in the presence of interference and noise. Therefore, sacrificing optimality for robustness seems to be a reasonable proposition in practice, and very much in the spirit of cognition. As a possible refinement, the adoption of a predictive model may make it possible for the user to choose the uncertainty set adaptively according to changes in environmental conditions and, therefore, may lead to a less conservative design; however, this design gain will have been achieved at the expense of increased system complexity.

It is worth noting that a cognitive radio can benefit from two predictive models. One is part of the DSM that makes predictions about primary users' behavior, which determines the available subcarriers and durations of their availability. These, in turn, provide estimates of the dimension of the optimization in (7.27) and (7.28) and of the control horizon respectively. The other predictive model is part of the TPC and makes predictions about secondary users' behavior and provides an estimate of the size of the uncertainty set. As will be explained later, these two predictive models work on different time scales.

## 7.13.1    Additional cost of robustness

In addition to conservatism, there is yet another price to be paid for achieving robustness. Although the constrained IWFC problem formulated in (7.13) and 7.14 is a convex optimization problem, appearance of the perturbation term $\Delta I$ in the denominator of the signal-to-(interference plus noise) ratio (SINR) in the objective function of the robust iterative water-filling algorithm problem (7.27) makes it a nonconvex optimization problem. A robust optimization technique is proposed by Teo (2007) for solving nonconvex and simulation-based problems. The proposed method is based on the assumption that the cost and constraints, as well as their gradient values, are available. The required values can even be provided by numerical simulation subroutines. It operates directly on the surface of the objective function and, therefore, does not assume any specific structure for the problem. In this method, the robust optimization problem is solved in two steps, which are applied repeatedly in order to achieve better robust designs.

- *Neighborhood search.* The algorithm evaluates the worst outcomes of a decision by obtaining knowledge of the cost surface around a specific design.
- *Robust local move.* The algorithm excludes neighbors with high costs and picks an updated design with lower estimated worst-case cost. Thereby, the decision is adjusted in order to counterbalance the undesirable outcomes.

The linearity of constraints of the robust optimization problem described in (7.27) and (7.28), especially the second set of constraints in (7.28) that involve the perturbation terms, improves the computational efficiency of the algorithm.

### 7.13.2   Two levels of robustification

In describing the design of a noncooperative cognitive radio network based on the material covered in Sections 7.9–7.13, we have introduced two levels of robustification into the network design:

(1) *Robustification at the user level*

   Examining (7.6), we see the inclusion of an SNR gap denoted by $\Gamma$ in this equation. As pointed out previously in Section 7.10, the gap $\Gamma$ is included to assure reliable communication under varying operating conditions all the time. To be more precise, this gap is included in the formula for calculating the data transmission rate of a secondary user so as to account for the "gap" between the performance of a practical coding-modulation scheme and the theoretical value of channel capacity.

(2) *Robustification at the network level*

   Next, examining the second constraint introduced in (7.28), we see by using a large enough $\varepsilon$ that we account for the possibility of large perturbations that may arise in the cognitive radio network due to uncertainties in the network.

In building this *two-level robustification strategy* into the network design, one at the local level and the other at the global level, we have provided the means needed to assure reliable communication across the cognitive radio network despite the possible presence of uncertainties that defy modeling. Reliability of the network, however, has been accomplished at the expense of suboptimality that manifests itself in a reduced data transmission performance of the network. Simply put, there is no free lunch.

## 7.14   Transient behavior of cognitive radio networks, and stability of equilibrium solutions

Although, over a short time scale, components of the cognitive radio network may remain essentially unchanged in complex and large-scale networks, the general behavior of the network can change drastically over time. If the SINR of a communication link drops below a specified threshold for a relatively long time, the connection between the transmitter and receiver for that link will be lost. For this reason, in addition to the equilibrium resource allocation, which was discussed in previous sections, the transient behavior of the network also deserves attention. In other words, studying the equilibrium states in a dynamic framework by methods that provide information about the disequilibrium behavior of the system is critical, which is the focus of this section.

In previous sections, the IWFC was proposed as an approach to find an equilibrium solution for the resource-allocation problem in cognitive radio networks. Following the approach of Luo and Pang (2006), the IWFC is reformulated as a *variational inequality (VI) problem*.[10] To explain this problem, let $X$ be a nonempty subset of the $n$-dimensional real space $R^n$, but let $\mathbf{F}$; denote a mapping from $R^n$ into itself, where $n$ denotes the number of subcarriers in the OFDM. According to Harker and Pang (1990), the variational inequality problem, denoted by VI($X$, $\mathbf{F}$), is to find a vector $\mathbf{x}^*$ in $X$ such that the condition

$$\mathbf{F}^{\mathrm{T}}(\mathbf{x}^*)(\mathbf{y} - \mathbf{x}^*) \geq 0 \qquad \text{for all } \mathbf{y} \in X \tag{7.29}$$

holds. Having formulated the IWFC as a VI problem, *projected dynamic systems (PDS) theory*, described by Nagurney and Zhang (1996), can be utilized to associate an ordinary differential equation (ODE) to the VI problem. A projection operator, which is discontinuous, appears on the right-hand side of the ODE to incorporate the feasibility constraints of the VI problem into the dynamics. This ODE provides a dynamic model for the competitive system whose equilibrium behavior is described by the VI. Also, the stationary (equilibrium) points of the ODE coincide with the set of solutions of the VI, which are the equilibrium points. Using the procedure just described, the equilibrium problem can be studied in the context of a dynamic framework. This *dynamic model* enables us not only to study the transient behavior of the network, but also to predict it.[11]

The stability of a cognitive radio network is an important issue and deserves special attention. It can be interpreted as the ability of the network to maintain or restore its equilibrium state against external perturbations. In other words, *network stability is linked to network sensitivity to perturbations.*

## 7.15    Case study II: robust IWFC versus classic IWFC

Simulation results are now presented to support the theoretical discussions of the previous sections. In the simulations reported by Setoodeh and Haykin (2009), the background noise levels $\sigma_k^i$, the normalized interference gains $\alpha_k^{ij}$, and the power budgets $p_{max}^i$ are chosen randomly from the intervals $\{0, 0.1/(m-1)\}$, $\{0, 1/(m-1)\}$, and $\{n/2, n\}$ respectively, with uniform distributions; $n$ is the number of subcarriers in the OFDM and $m$ is the number of users (communication links) in the cognitive radio network. The randomly chosen values for normalized interference gains $\alpha_k^{ij}$, which are less than $1/(m-1)$, guarantee that the tone matrices in the OFDM will be strictly diagonally dominant. For scenarios with time-varying delay in the control loops, the delays are chosen randomly.

In a cognitive radio network, when a spectrum hole disappears, users may have to increase their transmit powers at other spectrum holes, and this increases the interference. Also, when new users join the network, current users in the network experience more interference. Therefore, the joining of new users or the disappearance of spectrum holes makes the interference condition worse. Also, the cross-interference between users is time-varying because of the mobility of users. Results related to two typical but extreme scenarios are presented here to show the superiority of the robust IWFC described in (7.27) over the classic IWFC described in (7.13).

The first scenario, considered in the experiments, addresses a network with $m = 5$ communication links (users) and $n = 2$ available subcarriers, and all of the users simultaneously update their transmit powers using the interference measurements from the previous time step. At the fourth time step, two new users join the network, which increases the interference. The interference gains are also changed randomly at different time instants to consider mobility of the users. Figures 7.4 and 7.5 show the transmit powers of three users (users one, four, and seven) at two different subcarriers for the classic IWFC and robust IWFC respectively. At the second subcarrier, the classic IWFC is not able to reach an equilibrium. Data rates achieved by the chosen users are

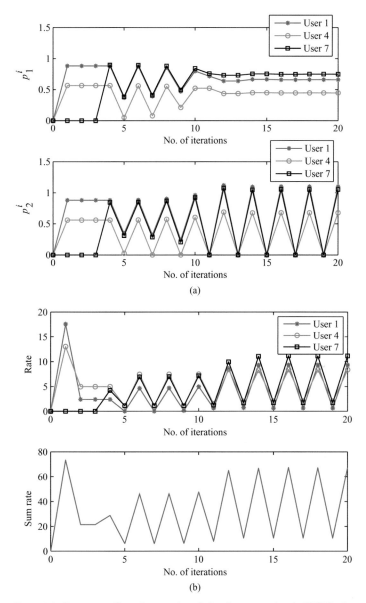

**Figure 7.4.** Resource-allocation results of simultaneous classic IWFC when two new users join a network of five users and interference gains are changed randomly to address the mobility of the users. (a) Transmit powers of three users at two subcarriers. (b) Data rates of three users and the total data rate in the network. Reproduced, with permission, from "Robust transmit power control for cognitive radio," P. Setoodeh and S. Haykin, 2009, *Proc. IEEE*, **97**, 915–939.

also shown. Also, the total data rate in the network is plotted against time, which is a measure of spectral efficiency. Moreover, although the average sum rate achieved by the classic IWFC is close to the average sum rate of the robust IWFC, it fluctuates, and in some time instants the data rate is very low, which indicates lack of spectrally efficient communication. Moreover, although the oscillation occurs mainly because of using a

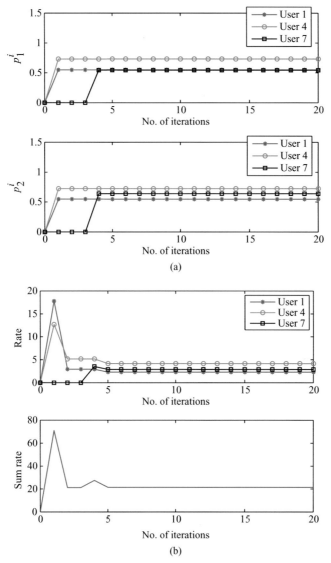

**Figure 7.5.** Resource-allocation results of simultaneous robust IWFC when two new users join a network of five users and interference gains are changed randomly to address mobility of the users. (a) Transmit powers of three users at two subcarriers. (b) Data rates of three users and the total data rate in the network. Reproduced, with permission, from "Robust transit power control for cognitive radio networks," P. Setoodeh and S. Haykin, 2009, *Proc. IEEE*, **97**, 915–939.

simultaneous update scheme, it also highlights the practical effectiveness of the robust IWFC.

In the second scenario, a network with $m = 5$ users and $n = 4$ available subcarriers is considered. Again, at the fourth time step, two new users join the network, but at the eigth time step the third subcarrier is no longer available (i.e. a spectrum hole disappears). The results are presented in Figures 7.6 and 7.7, which again show the

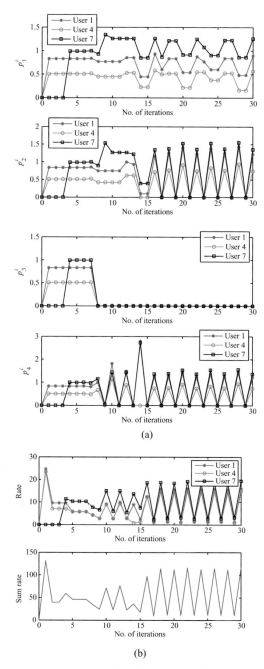

**Figure 7.6.** Resource allocation results of simultaneous classic IWFC when two new users join a network of five users, a subcarrier disappears, and interference gains are changed randomly to address mobility of the users. (a) Transmit powers of three users at four subcarriers. (b) Data rates of three users and the total data rate in the network. Figure reproduced, with permission, from the paper entitled "Robust transmit power control for cognitive radio," P. Setoodeh and S. Haykin, 2009, *Proc. IEEE*, **97**, 915–939.

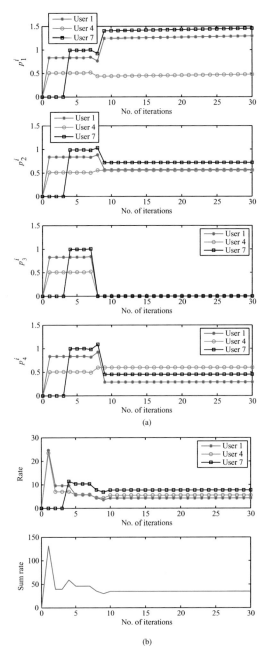

**Figure 7.7.** Resource allocation results of simultaneous robust IWFC when two new users join a network of five users, a subcarrier disappears, and interference gains are changed randomly to address the mobility of the users. (a) Transmit powers of three users at four subcarriers. (b) Data rates of three users and the total data rate in the network. Figure reproduced, with permission, from the paper entitled "Robust transmit power control for cognitive radio," P. Setoodeh and S. Haykin, 2009, *Proc. IEEE*, **97**, 915–939.

superiority of the robust IWFC. For classic IWFC, immediately after the disappearance of the third subcarrier, power in the fourth subcarrier starts to oscillate. After changing the interference gains randomly, we observe the same behavior in other subcarriers. In contrast to the robust IWFC, the classic IWFC fails again to achieve an equilibrium.

As mentioned previously, sporadic feedback introduces a time-varying delay in the transmit-power control loop, which causes different users to update their transmit powers based on outdated statistics. For instance, when the network configuration and, therefore, interference pattern changes, some users receive the related information after a delay. If the interference at a subcarrier increases and the transmitter is not informed immediately, it will not reduce its transmit power and may violate the permissible inter-ference power level for a while until it receives updated statistics of the interference in the forward channel. Similarly, this may happen to some users that update their transmit powers at lower rates compared with others. In the third scenario, a new user joins a network of three users who are competing for utilizing two subcarriers. Each user's transmitter receives statistics of the interference plus noise with a time-varying delay. Figure 7.8a shows the randomly chosen time-varying delays introduced by each user's feedback channel. The sum of transmit power and interference plus noise at the second subcarrier in the receiver of each user is plotted in Figure 7.8b and c for classic IWFC and robust IWFC respectively. Dashed lines show the limit imposed by the permissible interference power level. Although the classic IWFC is less conservative, it is not as successful as the robust iterative water-filling algorithm at preventing violations of the permissible interference power level. Similar results are obtained when users update their transmit powers with different frequencies.

Evan though the results presented in Figures 7.4–7.8 are for simple cognitive radio scenarios, they do demonstrate the practical importance of a robust iterative water-filling controller over the classic one.

## 7.16    Self-organized dynamic spectrum management

Having covered TPC, we now turn to DSM which is about distributing the available spec-trum holes among cognitive radio users and it is one of the most challenging problems in cognitive radio for several reasons (Haykin, 2005a; Khozeimeh and Haykin, 2009):

(1) DSM is equivalent to the *graph-coloring problem*,[12] which is an NP-complete problem.
(2) It is a time-varying problem.
(3) The dimensionality of the problem can assume a relatively high value, depending on the density of cognitive radio users (units).

For these practical reasons, the traditional centralized approaches that try to solve the DSM problem for the whole cognitive radio network are not practical, compelling us to find a *decentralized* approach that can achieve a satisfactory suboptimal assignment. In other words, *suboptimality of the DSM is traded off for scalability*.

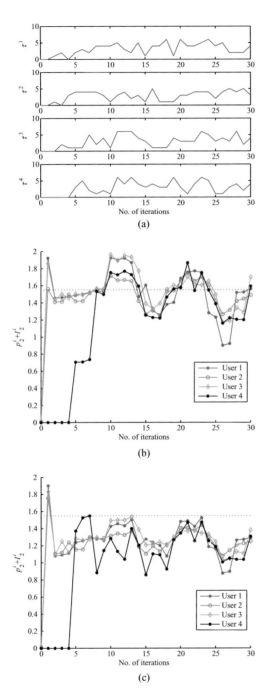

**Figure 7.8.** Resource allocation results of IWFC when interference gains change randomly with time and users use outdated information to update their transmit powers. (a) Time-varying delays introduced by each user's feedback channel. Sum of transmit power and interference plus noise for the users achieved by (b) classic IWFC and (c) robust IWFC. Dashed lines show the limit imposed by the permissible interference power level. Figure reproduced, with permission, from the paper entitled "Robust transmit power control for cognitive radio," P. Setoodeh and S. Haykin, 2009, *Proc. IEEE*, **97**, 915–939.

Inspired by the human brain, the self-organized DSM approach described herein tries to find the best channels to use for each cognitive radio unit by applying the idea of *self-organizing maps*. Self-organizing maps are a specific class of neural networks whose main goal is to adaptively transform an incoming signal-pattern of arbitrary dimension into a one- or two-dimensional discrete map in a *topologically* ordered manner. This kind of approach to self-organization relies on a form of *Hebbian learning rule* to extract patterns or features of the data and match the map to the patterns; this learning rule was discussed in Chapter 2. Recapping on the material presented therein, the Hebbian postulate of learning (Hebb, 1949) is one of the oldest learning rules. It states that changes in biological synaptic weights are proportional to the correlation between presynaptic (input) and postsynaptic (output) signals. The learning process extracts information about the radio environment and stores it in the synaptic weights of each neuron in the map. It follows, therefore, that at each step of the learning process an appropriate adjustment is applied to each synaptic weight of the neuron under consideration. The general form of weight adjustment in Hebbian learning is defined by

$$\Delta w_j(n) = F(y(n), x_j(n)), \qquad (7.30)$$

where $F(y(n), x_j(n))$ is a function of correlation between the neuronal output $y(n)$ and $j$th input signal $x_j$; the symbol $n$ used to denote discrete time in (7.30) should not be confused with the number of subcarriers in OFDM. The key point for (7.30) to work is that there must be *correlation* or *redundancy* in the input signals. Given this requirement, which is typically a practical reality, the neural network tries to find the correlation of the input and use the correlation for future use.

In self-organized DSM, each cognitive radio unit continuously monitors its own environment to extract the radio activity pattern in its surrounding neighborhood, using the Hebbian learning rule. Specifically, it tries to estimate the presence of a legacy user in each subband and when a new communication link is created, it is established on a common spectrum hole between the transmit- and receive-cognitive radio link that has the lowest probability of there being a legacy user. Simply stated:

In order to increase spectrum utilization using the self-organized DSM technique, the cognitive radio network tries to complement the legacy network's spectrum occupancy pattern by matching its own spectrum usage pattern to the pattern of channels which have had the least or no legacy activity in the recent past.

Although it is impossible to predict exactly how the spectrum holes come and go and for how long they remain available, they do *not* appear to have a completely random behavior. Rather, if we look at the radio spectrum in a specific location in a time window of interest, we do see that there is correlation between the presence of legacy users and the radio activity in their channels. For example, we rarely see activity in those channels that have no legacy users around them, whereas, on the other hand, we do see activity from time to time on those channels that do have legacy users around them. In addition to the pattern attributed to the physical presence of legacy users forming a spatial pattern, evaluation over time can give rise to patterns appearing in the spectrum holes. For example, we may see no activity in a cellphone channel during the night even though some users are present. The important point to note here is that the self-organized DSM scheme has the built-in capability to capture both patterns by employing the Hebbian learning rule, which extracts useful patterns from the incoming radio data in an *adaptive manner*.

Thus, in a cognitive radio network employing self-organized DSM, the cognitive radio units form a *common control channel* with their neighboring cognitive radio units to exchange sensing information and control data with each other; in other words, *cooperation* is established among cognitive radio users. Sharing the spectrum-sensing information, in the manner just described, significantly improves the ability of these units to discover what the legacy users are doing.

To summarize, the self-organized DSM algorithm works as follows:

- Each cognitive radio unit creates a list of synaptic weights, with each weight being associated with one of the spectrum subbands.
- The weights are continuously updated using Hebbian learning, based on spectrum-sensing information obtained by the cognitive radio unit and information received from neighboring units.
- When a new communication link needs to be established, a common spectrum hole with a high weight in both the transmitter and receiver cognitive radio units is used to form the link.
- If a legacy user is detected on a link under use, the cognitive radio units *immediately* stop using the link and try to find another common spectrum hole with high weights to use.

To conclude, the self-organized DSM algorithm is suitable for solving the dynamic spectrum management problem for three compelling reasons:

(1) The Hebbian learning process is a time-varying and highly local learning rule; this, therefore, provides a good match for DSM, which is also a time-varying and local problem in its own way.
(2) The algorithm is computationally very *simple* to implement.
(3) Most importantly, the network is decentralized and its complexity depends on the *density* of cognitive radio units, not the total number of them; therefore, it is *scalable*.

## 7.17    Cooperative cognitive radio networks

The study of a *decentralized and self-organized cognitive radio network* may also be approached using *cooperative game theory*, albeit for a different application. To be more specific, we look to a special form of cooperative game known as a *coalitional game* (Saad *et al.*, 2009). To elaborate, a *coalitional game* involves a group of *rational players* who have agreed among themselves to work together as a single entity with the objective of each player being that of strengthening its own position in the game. Thus, the set of players, denoted by $\mathcal{M} = \{1, 2, \ldots, m\}$, constitutes the first fundamental concept in a coalitional game. The second fundamental concept in a coalitional game is the *coalition value*, denoted by $v$, which quantifies the worth of the coalition carried out in the game. Thus, in mathematical terms, a coalitional game is described by the doublet $\{\mathcal{M}, v\}$.

On the basis of how the coalition value $v$ is defined, we may identify two types of coalitional game, as described here:

(1) *Coalitional games of the characteristic form, having transferable utility* (TU). The utility is said to be transferable in the sense that the coalition value depends solely on the players, regardless of how they are individually structured (von Neumann and Morgenstern, 1944). The total utility, represented by a real number, can be divided between the coalition players in a *flexible* manner. More specifically, the coalition values in TU games represent *payoffs*, which can be distributed among players in the game by using an appropriate *fairness rule*. For example, the fairness rule could be simply that of having the total utility distributed equally among all the coalition players.

(2) *Coalitional goals having nontransferable utility* (NTU). In this second type of coalitional game, there may exist *fixed* restrictions imposed on how the total utility is to be distributed among the coalition players (Autmann and Peleg, 1960). In particular, there may be a subgroup of players in the coalitional game who would prefer to operate in a noncooperative manner, yet remain members of the coalitional game. In this second type of coalitional game, the coalitional value $v$ is no longer a function of the players over the real line. Rather, we now have a set of *playoff vectors* to deal with. To elaborate, let $S$ denote the set of noncooperative players, with $S \subseteq M$. Then, each element $x_i$ of a vector $\mathbf{x} \in v(S)$ represents a payoff that player $i$ can obtain within coalition $S$. It is presumed that, for this payoff, player $i$ has selected a certain noncooperative strategy for itself while remaining a member of the coalition subset $S$.

From the classification of coalitional games into the two types just described, it is apparent that the TU type may be viewed as a special case of the NTU type. Of particular interest is the fact that these two types constitute the most important class of coalitional games encountered in practice.

## 7.17.1 Application: cooperative spectrum sensing for cognitive radio networks

Saad *et al.* (2009) proposed the idea of *cooperative spectrum sensing* (CSS) for cognitive radio networks, with the aim of providing a tradeoff between two probabilities:

- The *probability of miss*; that is, the probability of missing the detection of a white space.
- The *probability of false alarm*; that is, the probability of detecting a subband occupied by a primary user and declaring it to be a white space.

With this tradeoff in mind, the CSS problem is modeled as a *dynamic coalition-formation game* between secondary users in a cognitive radio network. Bearing in mind the following factors for a given coalition:

- imposition of an upper bound on false alarm,
- increased false alarm with increasing size of the coalition, and
- distances between users in the coalition,

Saad *et al.* (2009) show that a *grand coalition* is seldom formed. The term "grand coalition" refers to whether the cooperation of all users in the coalition can be taken for granted.

So, to get around this difficulty, Saad *et al.* go on to propose a coalition-formation algorithm that consists of three phases:

*Phase I.* In this phase, the secondary users in a coalition perform *local sensing* of the radio environment.

*Phase II.* In this second phase, the secondary users engage in adaptive coalition formation, based on a set of *merge and split rules* (Apt and Witzel, 2006). Let $\mathcal{M}$ denote a set of secondary users and $\{S_i\}_{i=1}^{l}$ denote any collection of disjoint coalitions, where $S_i \subset \mathcal{M}$ for all $i$. The $S_i$ agree to merge into a single coalition $G = \cup_{i=1}^{l} S_i$ if this new coalition is preferred by all the secondary users over the previous state of affairs. Similarly, a coalition $S \subset \mathcal{M}$ agrees to *split* into a number of smaller coalitions if the resulting set $\{S_i\}_{i=1}^{l}$ is preferred by the secondary users over the current set $S$.

*Phase III.* Once the coalitions are formed, each secondary user reports its sensing information about the radio environment to its own coalition "head" that makes the final decision on whether the subband in the radio spectrum is occupied by a primary user or not.

This three-phase coalition-formation algorithm has been tested for a toy experiment on a collaborative game involving a relatively small number of secondary users with interesting results; the interested reader is referred to Saad *et al.* (2009) for more details.

## 7.18     Emergent behavior of cognitive radio networks

The cognitive radio environment is naturally time varying. Most importantly, it exhibits a unique combination of characteristics (among others): adaptivity, awareness, cooperation, competition, and exploitation. Given these characteristics, we may wonder about the emergent behavior of a cognitive radio network in light of what we know on two relevant fields: *self-organizing systems* and *evolutionary games*.

First, we note that the emergent behavior of a cognitive radio network viewed as a game is influenced by the *degree of coupling* that may exist between the actions of different players operating in the game. The coupling may have the effect of *amplifying* local perturbations, and if they are left unchecked, the amplifications of local perturbations may ultimately lead to *instability*. From the study of self-organizing systems, we know that competition among the constituents of such a system can act as a stabilizing force (Haykin, 2009). By the same token, we expect that competition among the users of cognitive radio for limited resources (e.g. spectrum holes) may have the influence of a *stabilizer*.

For additional insight, we next look to evolutionary games. The idea of evolutionary games, developed for the study of ecological biology, was first introduced by Maynard Smith in 1974. In his landmark work, Maynard Smith (1974, 1982) wondered whether the theory of games could serve as a tool for modeling conflicts in a population of animals. In specific terms, two critical insights into the emergence of so-called *evolutionary stable strategies* were presented by Maynard Smith, as succinctly summarized by Glimcher (2003) and Schuster (2001):

- The animals' behavior is stochastic and unpredictable, when it is viewed at the microscopic level of individual acts.
- The theory of games provides a plausible basis for explaining the complex and unpredictable patterns of the animals' behavior.

Two key issues are raised here:

(1) *Complexity.*[13] The emergent behavior of an evolutionary game may be *complex*, in the sense that a change in one or more of the parameters in the underlying dynamics of the game can produce a dramatic change in behavior. Note that the dynamics must be nonlinear for complex behavior to be possible.
(2) *Unpredictability.* Game theory does not require that animals be fundamentally unpredictable. Rather, it merely requires that the individual behavior of each animal be *unpredictable with respect to its opponents* (Maynard Smith, 1982; Glimcher, 2003).

From this brief discussion on evolutionary games, we may conjecture that the emergent behavior of a cognitive radio network is explained by the possible unpredictable action of each user, as seen individually by the other secondary users.

Moreover, given the conflicting influences of cooperation, competition, and exploitation on the emergent behavior of a cognitive radio environment, we may identify two possible end results (Weisbunch, 1991):

(1) *Positive emergent behavior*, which is characterized by *order* and, therefore, a harmonious and efficient utilization of the radio spectrum by all users of the cognitive radio. (The positive emergent behavior may be likened to Maynard Smith's evolutionary stable strategy.)
(ii) *Negative emergent behavior*, which is characterized by *disorder* and, therefore, a culmination of traffic jams, chaos,[14] and unused radio spectrum.

From a practical perspective, what we first need is a reliable criterion for the early detection of negative emergent behavior (i.e. disorder) and, second, corrective measures for dealing with this undesirable behavior. With regard to the first issue, we recognize that cognition, in a sense, is an exercise in assigning probabilities to possible behavioral responses, in light of which we may say the following.

In the case of positive emergent behavior, predictions are possible with nearly complete confidence. On the other hand, in the case of negative emergent behavior, predictions are made with far less confidence. We may thus think of a *likelihood function* based on predictability as a criterion for the onset of negative emergent behavior.

In particular, we may envision a *maximum-likelihood detector*, the design of which is based on the predictability of negative emergent behavior.

Network measures should be sought that provide estimates about the global behavior of the network based on local observations made by users. This highlights the importance of analytical models that enable us to predict the future. Based on the knowledge so obtained, we can engineer the future to improve network robustness against potential disruptions.

## 7.19     Provision for the feedback channel

A discussion of cognitive radio networks would be incomplete without describing how we can provide for the *feedback channel*, which plays a critical role in the cognitive information-processing cycle of each user in the network. The question is how this provision can be satisfied in an effective and practical manner.

To be specific, the feedback channel can be established in three ways:

(1) *A dedicated universal channel for cognitive radio.* In this approach, a specific spectrum band in the radio spectrum is licensed and reserved for the feedback channel. This solution has the advantage of a simple system design and reliability. However, it is expensive (due to spectrum licensing) and also it is hard to find a worldwide common-free channel for this purpose due to different spectrum utilization policies in different countries. Furthermore, the cognitive radio can be easily interrupted by "jamming" the feedback channel. Finally, dedicating such a spectrum band to cognitive radio contradicts one of the main goals of cognitive radio, which is that of increasing the radio spectrum utilization; the dedicated spectrum band in this approach would be wasted whenever cognitive radio units are not in use.

(2) *Using available spectrum holes.* In this second approach, cognitive radio can use spectrum holes both for data transmission and as a feedback channel. Using spectrum holes is more flexible and efficient in terms of spectrum utilization than using a dedicated channel. However, the cognitive radio network cannot always be established because sometimes it is possible that there is no spectrum hole available in the radio environment to tap on. When there are no spectrum holes available, there is no feedback channel and cognitive radio users lose communication and synchronization. Thus, the moment some spectrum holes become available, they cannot immediately start data transmission and must wait until the necessary synchronization and negotiations are established. Furthermore, the radio spectrum is a highly dynamic environment and the spectrum holes may change in time. Therefore, every time the feedback channel becomes unavailable, cognitive radio users lose synchronization and need to stop data transmission until the feedback channel is established again.

(3) *Using unlicensed bands.* In this third and last approach, cognitive radio users can readily establish their own feedback channel using "unlicensed" bands. In this case, the feedback channel is always available and cognitive radio users never lose synchronization. The cognitive radio network can always be established even when there is no spectrum hole available in the environment. However, in adopting the use of unlicensed bands for the provision of feedback channel, the cognitive radio users may need to combat a high level of noise and interference due to other radios working in the unlicensed bands. In this scenario, the underlying theory and design of the cognitive radio network would have to be expanded to account for *noise in the feedback channel*. This is an issue that has not received the attention it deserves in the literature.

Using the unlicensed bands for establishing the feedback channel appears to be a good choice for the following reasons:

(1) The cognitive radio network is always established and users in the network never lose synchronization. Furthermore, even when there is no spectrum hole available in the environment, the user may employ the unlicensed bands to transmit data. Thus, users in the network can always provide communication and achieve one of their primary goals, which is providing reliable communication whenever and wherever needed (Haykin, 2005a).

(2) Even when there is no spectrum hole, cognitive radio users can always cooperate and share their radio-scene analysis information, which results in better and faster detection of spectrum holes.

(3) When a spectrum hole used by a communication link (user) becomes unavailable, the users can negotiate through the feedback channel to find another common spectrum hole and change their data channel momentarily.

There is, therefore, much to be said for adopting an unlicensed band in the radio spectrum as the medium for establishing the feedback channel for cognitive radio networks (Khozeimeh and Haykin, 2009).

## 7.20   Summary and discussion

### 7.20.1   Communication link as secondary user

In this chapter, we focused on the second application of cognition, namely cognitive radio. For a clear understanding of how cognitive radio works, Figure 7.1 pictures two cognitive radio units in communication. Each such unit is equipped with its own transceiver, made up of two parts: transmitter and receiver. Moreover, each cognitive radio unit is also equipped with its own radio scene analyzer (RSA) in the receiver part of the transceiver, and dynamic spectrum manager (DSM) and transmit-power controller (TPC) in the transmitter part. Examining the directed information flow in Figure 7.1, we clearly see that the perception–action cycle embodies the following four components:

- RSA of the cognitive radio unit on the right;
- feedback channel;
- DSM and TPC of the cognitive radio unit on the left; and
- data channel (i.e. radio environment).

In terms of terminology, the figure teaches us that, when we discuss a cognitive radio network, the term "user" refers to the "communication link" that connects the transmitter of the cognitive radio unit on the left to the receiver of the cognitive radio unit on the right.

### 7.20.2   Spectrum sensing

The next issue discussed in the chapter was spectrum sensing, which is another way of referring to radio scene analysis. Needless to say, spectrum sensing is one of the primary tasks involved in the design of cognitive radio. Among the many techniques on spectrum sensing described in the literature, we picked power spectrum estimation as the method of choice. In particular, we revisited the MTM, described in Chapter 3, as a mathematically rigorous procedure for accomplishing this task.

We also presented highlights of a contest on spectrum sensing for wireless microphones, which is more difficult than spectrum sensing of ATSC-DTV signals for cognitive radio. The contest was carried out by Qualcomm, San Diego, CA, in 2010. What is truly satisfying is the fact that all three top winners of the contest used power-spectrum estimation, in one form or another, for solving this difficult spectrum-sensing problem; the MTM was among those three winners. This contest confirms a point that was made in the influential paper by Haykin (2005a). Nonparametric power-spectrum estimation is the best procedure for spectrum sensing in cognitive radio applications.

## 7.20.3    Transmit-power control

Another important task in cognitive radio is that of transmit-power control, which is performed in the transmitter. To satisfy this requirement, we looked to ideas in game theory, information theory, optimization theory, and control theory. Under game theory, we focused on the Nash equilibrium that is a predictive concept well suited for modeling nonstationary processes, involving rational players (users). Under information theory, we focused on the iterative water-filling algorithm, the attractive features of which include: relatively fast rate of convergence, implementation in a decentralized manner, and the efficient use of OFDM. Under optimization theory, we focused on robustification of the iterative-water filling algorithm to guard against uncertainties of a physical nature. Under control theory, we focused on the variational inequality problem and projected dynamic systems to transform the IWFC into a new mathematical representation: an ODE framework for cognitive radio networks that is convenient for analysis of the network behavior. We then put all these ideas together and performed experiments to demonstrate the utility of the robust IWFC as a practical tool for solving the transmit-power control problem in a noncooperative cognitive radio network built on OFDM.

## 7.20.4    Self-organizing networks

*Self-organizing networks* are emerging as one of the most promising solutions for wireless networks in general by offering reliable service, improved energy efficiency, and ease of implementation. For their design, we may look to one of two approaches:

(1) *The human brain for inspiration*

The human brain is a highly complex dynamic system, yet it is capable of self-organization in ways that can only be termed as truly remarkable. In this context, Hebb's 1949 book entitled *The Organization of Behavior* has played a significant role in the study of *self-organizing systems* over the past six decades. This classic book introduced what is now commonly referred to as *Hebb's postulate of learning* that inspires us to this day. Most importantly, the book laid out a general framework for relating behavior to organization at the localized synaptic level through the dynamics of neural networks (Seung, 2000).

The approach taken by Khozeimeh and Haykin (2010) in the study of self-organized dynamic spectrum management for cognitive radio networks is inspired by Hebb's postulate of learning. Rephrasing this postulate in the context of wireless communications, we may make the following statement:

> The weighting assigned to a link, connecting the transmitter at one end to a receiver at the other end in a communication network, varies in accordance with the correlation between the input and output signals of the link.

For obvious reasons, some form of constraint has to be added to this postulate to assure stability of the link (Khozeimeh and Haykin, 2010).

(2) *Game theory for inspiration*

Coalitional game theory–a specialized class of cooperative game theory – provides another approach for the design of self-organized cognitive radio networks. With the secondary users of a cognitive radio network assuming the role of rational players, coalitional cognitive radio networks are categorized into two types:

- *TU coalitional networks*, in which the total utility is divided arbitrarily among the secondary users, subject to a fairness rule.
- *NTU coalitional networks*, in which arbitrary division of the total utility among the secondary users is prevented by a subset of the users who adopt a noncooperative strategy of their own.

The TU type may be viewed as a special case of the NTU type.

Saad *et al.* (2009) applied the principles of coalitional networks to spectrum sensing in a cognitive radio network. To this end, the network is modeled as a *dynamic coalition-formation game* between secondary users. A distinctive step in the algorithm so developed is an adaptive merge-and-split phase that converges to a partitioning of the secondary users with an optimal payoff allocation, whenever this partition exists.

## 7.20.5   Cognitive femtocell networks

With the ever-increasing growth of data-hungry wireless devices, exemplified by smart phones and net books, serious concern is being raised about the issues of *data overload* and *outage* in cellular networks. To tackle these practical issues, network providers are looking into the use of *femtocells* to increase network capacity and improve spectrum utilization. (Femtocells were mentioned previously in Section 7.4.) From a wireless-communication point of view, femtocells may be viewed as *low-power, indoor base stations* that are connected to cellular networks through *high-speed Internet links* typically available in most homes and small offices. Thus, following Ortiz (2008), we say:

> Femtocells are mini-cell towers located inside homes or small offices for the purpose of providing reliable wireless communication with the outside world.

Insofar as indoor users are concerned, the advantages offered by the deployment of femtocells include the following:

- improved indoor wireless reception;
- increased rates of data transmission;

- reduced power consumption and lengthened battery life brought about by the use of relatively short indoor wireless links.

By the same token, the network providers gain advantages of their own through the deployment of femtocells:

- improved spectrum utilization, thereby serving more customers without having to invest in building more towers and buying expensive radio bands;
- savings in network infrastructure realized through use of the Internet linking homes and small offices to the outside world; and
- addressing the issue of user dissatisfaction resulting from electromagnetic propagation loss, which is achieved by exploiting the fact that femtocells are able to provide indoor wireless communication at maximum speed.

However, for indoor users and network providers to gain the advantages just described, two practical issues would have to be resolved:

(1) *Interference.* With femtocells communicating among themselves as well as macrocells in the outside world, novel transmit-power control and dynamic spectrum management algorithms will have to be found to mitigate the interference problem.
(2) *Heterogeneous network support.* In the context of femtocell networks, the term "heterogeneous" applies to the varying size of femtocells from one location to another and the different capabilities of the base stations.

To tackle both of these problems, femtocell networks will have to be *self-organizing* as well as *self-optimizing*.

To become so, they have to be aware of the underlying wireless communication environment, *learn* from it, and *adapt* to it on a *continuous-time basis*, so as to provide *reliable communication whenever and wherever needed*. What we have just summarized here is nothing but the definition of cognitive radio. In other words, for femtocells to achieve their goals, they would have to be *cognitive* with one difference; whereas in traditional cognitive radio networks we speak of primary (legacy) users and secondary users, in the new generation of *cognitive femtocell networks* there are no secondary users. Rather, there is uniformity in the network, with only legacy users of varying needs.

Despite this difference, the fundamental principles of traditional cognitive radio covered in this chapter, namely spectrum sensing in the receiver, feedback linkage from the receiver to the transmitter, and transmit-power control and dynamic spectrum management in the transmitter, are all applicable to cognitive femtocell networks subject to the inherent realities of femtocells.

## Notes and practical references

1. *Impulsive noise in wireless communications*
   The most common natural source of noise encountered at the front end of communication receivers is *thermal noise*, which is justifiably modeled as *AWGN*.

By far the most important artificial source of noise in mobile communications is *man-made noise*, which is radiated by different kinds of electrical equipment across a frequency band extending from about 2 MHz to about 500 MHz (Parsons, 2000). Unlike thermal noise, man-made noise is *impulsive* in nature; hence the reference to it as *impulsive noise*. In urban areas, the impulsive noise generated by motor vehicles is a major source of interference to mobile communications.

With the statistics of impulsive noise being radically different from the Gaussian characterization of thermal noise, the modeling of noise in a white space due to the combined presence of Gaussian noise and impulsive noise in urban areas may complicate procedures for identifying spectrum holes.

2. *The FFT algorithm*

Assuming that the Slepian tapers (windows) are precomputed for a prescribed size $K$, each eigencoefficient in the MTM can be computed using the FFT algorithm. With routines such as FFTW, described by Frigo and Johnson (2005), the computation is fast to begin with and "tricks" could be used to speed up the process even further. To be specific, the FFTW is not tuned to a fixed machine; rather, it uses a *planner* that adapts its algorithms to maximize performance.

When it comes to implementation, it is ironic, but not surprising, that the FFT algorithm plays such a fundamental role not only in implementing the MTM for spectrum sensing, but also the IWFC for transmit-power control based on OFDM.

3. *Experimental results of case study I*

The experimental results on ATSC-DTV signals presented in Section 7.6 are a shortened version of material on this same topic in Haykin *et al.* (2009). For more detailed results and related discussions, the reader is referred to that paper.

4. *The 2008 FCC report*

In its 2008 report and order on TV white space (FCC, 2008), the FCC established rules to allow new wireless devices to operate in the white space (unoccupied spectrum) of the broadcast TV spectrum on a secondary basis. It is expected that this will lead to new innovative products, especially broadband applications. To avoid causing interference, the new devices will incorporate geolocation capability and the ability to access over the Internet a database of primary spectrum users, such as locations for TV stations and cable systems' headends. The database itself is insufficient to offer interference guarantees, given the ability to predict propagation characteristics accurately while efficiently using the spectrum. The portability of low-power primary users, such as wireless microphones, makes the database approach infeasible, motivating the need for sensitive and accurate spectrum-sensing technology to include these devices.

5. *The 2010 Qualcomm competition on wireless microphones*

With the need to shed more light on spectrum sensing for wireless microphones, the Qualcomm Company, San Diego, conducted an open contest in 2010 for the best technique for this challenging application of cognitive radio. The power level of wireless-microphone signals is in the range of −100 to −110 dBm, where dBm refers to power expressed in decibels with 1 mW of power as the reference. No further information about the test data was provided to the contestants. For each signal in the test data, a file was supplied listing whether a wireless microphone was present or not and at which specific frequency within the channel. The top three submissions selected by Qualcomm (in alphabetical order) were: Queen's University, Canada; University of California at Los Angeles (UCLA); and University of Illinois at Urbana-Champaign (UIUC). The overall winner was the submission from UIUC. The criteria for evaluating the submissions were twofold: results of the computations and paper organization.

The top three submissions, which are illuminating in their individual ways, and summarized below.

(1) *MTM (Burr, 2010, Queen's University, Canada)*

In the submission made by Queen's University, the MTM was used, with access to two statistical tests:

- *harmonic F-test* for identifying the significance of a potential nearby periodic signal against a background of noise; and
- *magnitude-squared coherence* (MSC) between two antenna sources.

Both tests were used in the overall MTM-based technique, in addition to the measurement of SNR of possible signals being present in the test data. Both statistical tests were discussed in Chapter 3.

Work was done only on three sets of a possible 10 sets of test data due to time limitation. Each of these sets was considered as separate antennas. The antenna sets were subdivided yet again into 50 independent sub-blocks, each of length 120 μs. On each of these sub-blocks, an MTM spectrum estimation and accompanying harmonic *F*-test were computed. A histogram-based binning scheme was used to determine the best five candidates for a wireless-microphone signal across the antenna set, and the results returned to the *base station*. For each candidate frequency, 50 total MSCs were computed between the two antennas, one for each sub-block, and the minimum of the set was taken.

This procedure was repeated for the three data sets and the results passed through a multistage logical testing procedure to determine which candidates were truly wireless-microphone signals. The testing procedure also allowed for strong and weak interference signals to be identified to aid in the possible (temporary) allocation of subbands to wireless microphones. The results were returned with indications of the minimum MSC estimate between the two antennas across all three sets for the candidate, as well as individual antenna results. Careful examination showed that certain interference signals were detected at great strength on one antenna and weakly or not at all on the other. This result confirmed the intuition that using *multiple antennas may well have provided better detection possibilities* than a single antenna, especially through use of the MSC and normal coincidence results. (This observation on the use of multiple antennas made in the submission by Burr is interesting. In Chapter 3 on spectrum sensing, expansion of the MTM for space–time processing was discussed, wherein the use of multiple antennas was identified as a method for estimating the directions of arrival (DoA) of incoming RF signals.)

(2) *Two-stage spectrum estimation procedure (Dulmange et al., UCLA)*

In the submission made by UCLA, a *two-stage spectrum estimation procedure* was used. The first stage of the procedure was coarse sensing, which divided the channel into 400 kHz subbands with 100 kHz overlap. The division of the channel into sub-bands was accomplished using a 64-point FFT, which was used to estimate the power spectrum. Coarse sensing identified candidate sub-bands that were likely to contain a wireless microphone. This was accomplished by comparing the power in a given subband with the power in the adjacent and second adjacent subbands. Subbands with sufficiently high power compared with neighboring subbands were processed further by a fine-sensing stage.

Each candidate subband selected by the coarse sensing stage was filtered and decimated by a factor of 16. The power spectrum of the decimated signal was then calculated using a 512-point FFT to provide a high-resolution spectrum estimate of the 400 kHz subband. Multiple power spectral estimates were averaged to increase the SNR. A frequency-domain correlator was used to match the spectrum from the training data. There was one frequency-domain template from the training set, where the audio signal input to the wireless microphone was music. The second frequency-domain template was from the training data, when there was no audio signal into the wireless microphone, so the audio signal was silent.

The fine-sensing stage found the largest correction to the two frequency-domain templates and then used that as its test statistic. If the test statistic was larger than a specified threshold, then the subband was declared to contain a wireless microphone; if, on the other hand, the test statistic was less than the threshold, then the subband was declared not to contain a wireless microphone.

(3) *High-resolution power-spectrum estimation using Hamming window (Sreckanth et al., 2010, UIUC)*

The UIUC contest submission began by calculating a high-resolution power spectral density using the FFT and Hamming window. Multiple power-spectral estimates were then averaged to increase the SNR. Next, it was observed that wireless microphones must be on multiples of 25 kHz as measured from the lower band edge of the TV channel. They included a frequency tolerance of ±5 kHz to allow for transmitter and sensing-receiver LO frequency errors. Since wireless microphones must be entered on a multiple of 25 kHz, that meant within a 6 MHz TV channel there are 240 small subbands centered on multiples of 25 kHz on which to search for the wireless microphones. Thus, only those subbands were considered; over each of them, the power spectral estimate over a 20 kHz subband was multiplied by a window that weighted the middle of the subband more than the edges. Next, the resulting windowed power spectral estimator over the 20 kHz subband was accumulated to represent the power in that subband. Then, again, in order to eliminate interference, the median of the power in the 240 subbands was calculated and used as an estimate of the noise floor power. Subsequently, for each subband the bandwidth was calculated as the number of power spectral estimator bins, within 6 kHz, that exceeded the noise floor; this procedure was used to eliminate narrowband interference.

Next, for each subband, a nonlinear combination of the signal power and the bandwidth was calculated. The nonlinear function combined the signal power and bandwidth into test statistics for each subband. The maximum of each of these test statistics was calculated, providing the final test statistic, which was compared with a prescribed threshold. If the test statistic exceeded the threshold, then a wireless microphone was declared to be present in the channel; on the other hand, if it did not exceed the threshold, then the channel was declared empty.

*The message to take from the Qualcomm test*

Although, indeed, the three approaches described above for tackling the spectrum-sensing problem for wireless microphones are quite different, they do, however, share a common point: all three approaches appeal to power-spectrum estimation for their implementations.

6. *Game theory*

In a historical context, the formulation of game theory may be traced back to the pioneering work of John von Neumann in the 1930s, which culminated in the publication of the co-authored book entitled *Theory of Games and Economic Behavior* (von Neumann and Morgenstern, 1944). For modern treatments of game theory, see the books by Fundenbergh and Levine (1999) and Basar and Olsder (1999).

7. *Classic papers on the Nash equilibrium*

The Nash equilibrium is named for Nobel Laureate John Nash; the Nash (1950, 1951) papers are classics.

8. *Origin of the competitive optimality criterion*

The competitive optimality criterion is discussed in Chapter 4 of Yu's doctoral dissertation (Yu, 2002). In particular, Yu developed an iterative water-filling algorithm for a suboptimal solution to the multi-user DSL environment, viewed as a noncooperative game.

9. *Shannon's information capacity formula*

Consider a band-limited, power-limited Gaussian channel. The channel output is perturbed by AWGN of zero mean and power spectral density $N_0/2$. A sample of this noise process, therefore, is Gaussian distributed with zero mean and noise variance $N_0W$, where $W$ is the channel bandwidth. According to Shannon's information theory, the information capacity of the channel is defined by

$$C = W \log_2\left(1+\frac{P}{N_0W}\right) \text{ bits per second}$$

or, equivalently,

$$C = \frac{1}{2}\log_2\left(1+\frac{P}{N_0W}\right) \text{ bits per channel use,}$$

where the logarithm is to base 2. According to this formula, it is easier to increase the information capacity of a communication channel by expanding the channel bandwidth than by increasing the average transmitted power $P$ for a prescribed noise variance.

10. *Variational inequality*

Let $\chi$ be a nonempty subset of $R^n$ and let $\mathbf{F}$ be a mapping from $R^n$ into itself. The VI problem, denoted by $VI(\chi, \mathbf{F})$ is to find a vector $\mathbf{x}^* \in \chi$ such that

$$\mathbf{F}^T(\mathbf{x}^*)(\mathbf{y}-\mathbf{x}^*) \geq 0 \qquad \text{for all } \mathbf{y} \in \chi.$$

Loosely speaking, this VI states that the vector $\mathbf{F}(\mathbf{x}^*)$ must be at an acute angle with all *feasible vectors* emanating from $\mathbf{x}^*$. Formally, $\mathbf{x}^* \in \chi$ is a solution to $VI(\chi, \mathbf{F})$ if, and only if, $\mathbf{F}(\mathbf{x}^*)$ is inward normal to $\chi$ at $\mathbf{x}^*$.

11. *Sensitivity considerations in cognitive radio networks*

For more detailed theoretical considerations of the dynamic model of a cognitive radio network, the reader is referred to Setoodeh and Haykin (2009). Therein, the network stability is linked to network sensitivity perturbations in theoretical terms.

12. *Graph-coloring problem*

Graph-coloring refers to the problem of coloring the vertices of a given graph with a minimum number of colors. To explain, consider a unidirectional graph denoted by $G = (V, E)$ when $V$ is the set of $|v| = n$ vertices and $E$ is the set of edges. A $k$-coloring of the graph $G$ is said to be a mapping $\phi: V \rightarrow \Gamma$, where $\Gamma = \{1, 2, \ldots, k\}$ is the set of $|\Gamma| = k$ integers, each of which represents a specific color. The graph-coloring is *valid* if, and only if,

$$\phi(v) \neq \phi(u) \qquad \text{for all } [u, v] \in E.$$

Otherwise, the graph-coloring is said to be invalid. In other words, for a graph-coloring to be valid, any two vertices connected by an edge have different colors.

  For further details on graph-coloring for DSM and pertinent references, the reader is referred to Khozeimeh and Haykin (2009).

13. *Complexity*

The new sciences of complexity (whose birth was assisted by the Santa Fe Institute, New Mexico) may well occupy much of the intellectual activities in the 21st century (Stein, 1989). In the context of complexity, it is perhaps less ambiguous to speak of complex behavior rather than complex systems (Nicolis and Prigogine, 1989). A nonlinear dynamic system may be complex in computational terms but be incapable of exhibiting complex behavior. By the same token, a nonlinear system can be simple in computational terms but its underlying dynamics can be rich enough to produce complex behavior.

14. *Emergent behavior characterization*

The possibility of characterizing negative emergent behavior as a chaotic phenomenon needs some explanation. Idealized chaos theory is based on the premise that dynamic noise in the state-space model (describing the phenomenon of interest) is zero (Haykin *et al.*, 2002). However, it is unlikely that this highly restrictive condition is satisfied by real-life physical phenomena. So, the proper thing to say is that it is feasible for a negative emergent behavior to be *stochastic chaotic*.

# 8  Epilogue

## 8.1  The perception–action cycle

We begin this final chapter of the book by reemphasizing the basic ideas that bind the study of cognitive dynamic systems to the human brain, where the networks dealing with perception are collocated with those dealing with action.

For a very succinct statement on the overall function of a cognitive dynamic system made up of actuator and perceptor, we simply say:

Perception for control via feedback information

Elaborating on perception performed in the perceptor and action performed in the actuator, we may go on to say:

(1) *Cognitive perception* addresses optimal estimation of the state in the perceptor by processing incoming stimuli under *indirect control* of the actuator.
(2) *Cognitive control* addresses optimal decision-making in the actuator under *feedback guidance* from the perceptor.

These two closely related functions *reinforce* each other continually from one cycle of directed information processing to the next. The net result is the *perception–action cycle*, discussed in detail in Chapter 2. This *cyclic directed information flow* is a cardinal characteristic of every cognitive dynamic system, be it of a neurobiological or artificial kind. Naturally, the exact details of the perception–action cycle will depend on the application of interest. In any event, a practical benefit of the perception–action cycle is *information gain* about how the state of the environment is inferred by the system, with the gain increasing from one cycle to the next.

When we speak of the perception–action cycle in a cognitive dynamic system, we immediately think of four other basic cognitive processes: memory, attention, intelligance, and language; all five processes, collectively, define a cognitive dynamic system.

Memory manifests itself in three parts:

(1) *Perceptual memory*, which is reciprocally coupled to the environmental scene analyzer in the perceptor. Its function is to *store* knowledge gained about the environment by having processed past data, and have that knowledge *updated* in light of information contained in new environmental data; this updating is continued from one cycle to the next. Simply stated, the perceptual memory uses the

incoming stimuli in two ways. First, past stimuli are used to learn a model of the environment. Second, the model is adaptively updated in an on-line manner by exploiting new information contained in the stimuli. With such a model of the environment available on each cycle, it is left to the environmental scene analyzer to compute an optimal estimate of the environmental state and update it from cycle to cycle.

(2) *Executive memory*, which is reciprocally coupled to the environmental scene actuator. In a manner similar to the perceptual memory, the executive memory uses *feedback information* from the perceptor in two ways. First, given past feedback information on the state of the environment sent by the perceptor, the executive memory learns to *categorize* the decision-making process performed by the environmental scene actuator. That categorization is updated from one cycle to the next by exploiting newly received feedback information. In so doing, the environmental scene actuator is enabled to continually improve its decision-making process and initiate the next cycle.

(3) *Working memory* reciprocally couples the perceptual and executive memories by mediating between them on every cycle. The net result of this mediation is twofold. First, the cognitive dynamic system, as a whole, assumes the form of a *synchronous, self-organized correlative-learning machine*. Second, the working memory provides predictive information on the consequences of model selection made in the perceptor and action taken by the actuator on or in the environment.

Turning next to *attention* as the third process essential for cognition: its function is the prioritized allocation of computational resources. In a manner similar to memory, we have *perceptive attention* in the perceptor and *executive attention* in the actuator. Both of these attentional mechanisms look after their respective resource-allocation domains. However, unlike perception and memory, the attentional mechanism does not have a physical location of its own. Rather, it builds on the perception–action cycle and memory to exercise its function by using algorithmic mechanisms.

Finally, we come to *intelligence*, which is the most complex and most powerful of all the four cognitive processes. It exercises information-processing power by building on perception, memory, and attention through the combination of local and global feedback loops distributed throughout the system. As we increase the hierarchical depth of the memory, the number of such loops is enlarged exponentially, thereby making intelligence that much more powerful in decision-making in the face of environmental uncertainties; hence the profound importance of intelligence.

Language was not discussed in the book as it is outside its scope.

## 8.2    Summarizing remarks on cognitive radar and cognitive radio

What we have just summarized on the perception–action cycle, viewed particularly in the context of the visual brain, applies appropriately to a cognitive radar whose actuator and perceptor are collocated.

However, in the case of cognitive radio, there are some practical differences that need careful attention. First, cognitive radio is intended for wireless communications, whereas cognitive radar is intended for remote-sensing applications. Moreover, unlike cognitive radar, the actuator and perceptor of cognitive radio are separately located. Thus, although the perception–action cycle applies equally well to cognitive radio, we have to use *engineering ingenuity* as to how the four cognitive processes, namely perception, memory, attention, and intelligence, are actually implemented.

To illustrate how, indeed, this implementation was done in Chapter 7 on cognitive radio, we recall the following steps:

(1) With spectrum sensing as the basic function of the perceptor, aimed at the identification of spectrum holes (i.e. underutilized subbands of the radio spectrum), the need for modeling the environment may no longer be a requirement, provided a *nonparametric* (i.e. *model-free approach*) is adopted. For example, the *MTM* is a method of choice that is used to estimate the power spectrum by directly processing the incoming RF stimuli; we may therefore eliminate the need for modeling the radio environment.

(2) Turning next to the actuator, we may look to the human brain for ideas on *self-organization* to implement the function of DSM so as to distribute the available spectrum holes among competing secondary users in a cognitive radio network. This kind of thinking leads us to build a *self-organizing map based on Hebbian learning*, whereby information on the spectrum holes is stored in the synaptic weights of the map. Provision for memory is thereby made. Moreover, the self-organizing map develops a *predictive-modeling capability* to do two things:
   • pay attention to the continually changing user-communication patterns in the network and
   • predict the duration-availability of spectrum holes.

There is one other function that needs to be performed: transmit-power control. To implement this second function, we may look to ideas in game theory, then information theory, and optimal control, as described in detail in Chapter 7. In particular, as explained by Glimcher (2003), the use of game theory does provide a basis for learning. Moreover, the control mechanism has a built-in capability for adaptation. So, the integration of DSM and transmit-power control provides the actuator of cognitive radio the desired capability for intelligent decision-making in the face of environmental uncertainties (e.g. when the spectrum holes become available and when they disappear from availability on demands made by legacy users, with both events happening in uncertain manner).

### 8.2.1    Lessons learned from cognitive radar and cognitive radio

We may now summarize the first lesson learned from the two applications, cognitive radar and cognitive radio, as follows:

• When an application of interest closely follows human cognition in conceptual terms, we may look to the perception–action cycle explained in detail in Chapter 2 for the structural design of a cognitive dynamic system to satisfy the practical requirements of that particular application.

- When, on the other hand, the application of interest lends itself to the use of cognition but does *not* closely follow human cognition in conceptual terms, we may look to the human brain for inspiration with the objective of identifying those learning tools that are relevant to that application. Moreover, we have to look to engineering ingenuity for other ideas on how to complete the structuring of a cognitive dynamic system for the prescribed application.

There is one other important lesson learned from the study of cognitive radar and cognitive radio. As different as they are, they do share three practical issues:

(1) Perception is an *ill-posed inverse problem*, in that one or more of Hadamand's three conditions for well-posedness is violated; hence the need for *regularization*.
(2) The environment is *nonstationary*; hence the need for *adaptation* to deal with statistical variations of environmental stimuli.
(3) The presence of *environmental uncertainties* is equally unavoidable; hence the need for *intelligent choices* in the decision-making mechanism.

The use of cognition provides practical solutions to all three issues.

## 8.3 Unexplored issues

In this final section of the final chapter of the book, we briefly discuss a list of topics that are important to the study of cognitive dynamic systems, but they would have taken us much too far afield or else they are in their early stages of development.

### 8.3.1 Network of cooperative cognitive radars (Haykin, 2005b)

In radar applications for remote sensing, the traditional approach has been to go for a single radar, powerful enough for the application at hand. However, there are certain remote-sensing applications where there is merit for a new approach, based on a network of low-cost radars working in a cooperative manner and thereby reinforcing each other over time. One such application is dense networks of cooperative radars for weather hazard forecasting and warning (McLaughlin *et al.*, 2009).

#### 8.3.1.1 Dense networks of cooperative radars for weather forecasting
In today's operational networks (e.g. NEXRAD in the USA) for hazard weather forecasting and warning, the emphasis has been on a small number of highly powerful coherent radars with large antennas (e.g. 9 m diameters) spaced hundreds of kilometers apart from each other. In contrast, in the dense networks of cooperative radars described by McLaughlin *et al.* (2009), the antennas are expected to be 1 m diameter in size, and the radars are spaced tens of kilometers apart. Figure 8.1 illustrates the dense-network

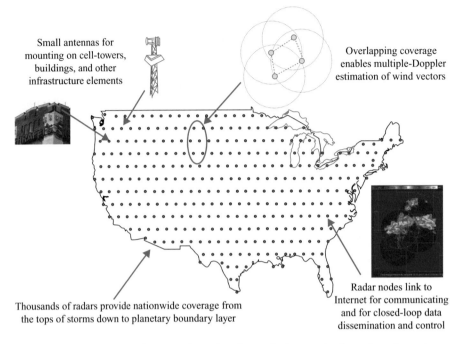

Small antennas for
mounting on cell-towers,
buildings, and other
infrastructure elements

Overlapping coverage
enables multiple-Doppler
estimation of wind vectors

Thousands of radars provide nationwide coverage from
the tops of storms down to planetary boundary layer

Radar nodes link to
Internet for communicating
and for closed-loop data
dissemination and control

**Figure 8.1.** Dense network of inexpensive radars for predicting weather hazards and warning across the entire USA. (© Copyright 2011 American Meteorological Society (AMS)).

concept, blanketing the contiguous USA. It is envisioned that the inexpensive radars used in the network would require less than 100 W of average actuator power, yet they would be capable of mapping storms with about 0.5 km spatial resolution throughout the entire atmosphere.

An important advantage of a dense network of low-cost radars over large radars in operational sensors is the following (McLaughlin *et al.*, 2009).

The short range of radars in the dense network defeats the Earth curvature-blocking problem, thereby enabling the network to comprehensively map damaging winds and heavy rainfall from tops of the storms down to the boundary layer beneath the view of today's operational radar networks.

Moreover, the short-range operation of radars in the dense network offers the potential for significant improvements in spatial resolution and forecasting-update times. These improvements, in turn, enable a better characterization of storm morphology and analysis. The net result of all these improvements is more accurate weather-hazard forecasting and warning, compared with operational networks of large radars in current use.

### 8.3.1.2    Advantage of cognitive radars over traditional ones

In Chapter 6, the practical advantages of cognitive radars over traditional radars were described in detail, supported by computer simulations. With a dense network of cooperative radars for weather forecasting as the application of interest, it is apropos that we reiterate those advantages:

(1) *Information gain* about the state of the environment, which is achieved through the use of *global feedback* from the perceptor to the actuator; by virtue of the *perception–action cycle*; this information gain is increased from one cycle to the next.

(2) *Prediction of consequence of actions taken by the radar*, which is realized by distributing *multiscale memory* across each low-cost radar system.

(3) With *attention* naturally based on perception and memory, the cognitive radar acquires the ability to *allocate computational resources* in order of practical importance.

(4) Finally, with *intelligence* based on perception, memory, and attention, the coherent radar's decision mechanism is provided with the means to select *intelligent choices* in the face of environmental uncertainties that are highly likely to arise in practice.

These important practical benefits realized through the use of cognition are compelling reasons for the adoption of cognitive coherent low-cost radars in a dense network for weather forecasting in place of their traditional counterparts. So much so, when it comes to hazard-weather forecasting and warning, the network of cognitive radars acquires the information-processing power that would enhance the likelihood of outperforming the traditional networks of large radars that much more.

## 8.3.1.3 Self-organizing networking

The current thinking behind the dense-network concept is to link radar nodes in the network to the Internet for the purposes of communication, closed-loop data dissemenation, and control. Here again, there is the potential for further improvements to deployment of the dense network by exploiting the self-organizing capability of the human brain.

The human brain is a *highly powerful, parallel distributed information-processing, correlative learning machine* (Haykin, 2009). Most importantly, the brain is capable of *self-organization* through a mechanism known as *Hebbian learning*, which is one of the oldest methods of learning (Hebb, 1949). The idea behind this learning process is based on *correlation* between the presynaptic (input) signal applied to a synaptic weight in a neural network and the corresponding postsynaptic (output) signal produced as response to the input. The synaptic weight refers to a linkage (i.e. synapse) with an adjustable weight connecting one computational unit (i.e. neuron) to an adjacent one.

In Chapter 7 we described the application of Hebbian learning to solve the DSM problem in a cognitive radio network, catering to the communication needs of a large number of users. In effect, through Hebbian learning, the network assumes the form of a *self-organizing map* with built-in memory and short-time predictive-modeling capability. Emboldened by the successful application of Hebbian learning to cognitive radio networks, it is our belief that much may also be gained by applying it to a dense network of cooperative radars.

To sum up, there is much to be gained by expanding the scope of the dense-network concept for weather forecasting through the combined use of cognition at the elemental radar level and self-organization at the network level. Putting it altogether, we envision a *self-organized network of cooperative cognitive radars* based on the use of low-cost coherent radars, each of which is capable of range and Doppler estimation. With accurate weather forecasting and warning as the objective, such a network has the potential for outperforming the traditional network of large radars with operational improvements on several fronts.

## 8.3.2     Double-layer network dynamics (Setoodeh, 2010)

There are two worlds of wireless communications: the legacy (old) wireless world and the cognitive (new) wireless world. Spectrum holes are the medium through which the two worlds interact. Releasing subbands by primary users allows the cognitive radio users to perform their normal tasks and, therefore, to survive. In other words, the old world affects the new world through appearance and disappearance of the spectrum holes and there is a *master–slave relationship* between them. Hence, the two worlds of wireless communications are going on side by side. This makes a cognitive radio network a multiple-time-scale dynamic system: a large-scale time in which the activities of primary users change and a small-scale time in which the activities of secondary users change accordingly. Such systems are called *double-layer dynamic systems*. A theoretical framework must, therefore, be developed to capture the multiple-time-scale nature of cognitive radio networks and lay the groundwork for further research. This topic is similar to an uncharted territory and has a great deal of potential, both in theoretical and practical terms.

### 8.3.2.1     Two-time-scale behavior

A cognitive radio network, which is a system of systems, is a goal-seeking system. The following classes of problems are involved in developing a cognitive radio network:

(1) Specifying the goal that the system is pursuing (i.e. efficient spectrum utilization and ubiquitous network connectivity).
(2) Discriminating between the available alternatives based on the meaning of a desirable decision.
(3) Choosing a desirable action based on a decision-making process.

By the same token, every subsystem in the network (i.e. every cognitive radio) is a goal-seeking system too.

Owing to the master–slave relationship between the legacy and the cognitive wireless worlds, the spectrum supply chain network has a hierarchical structure. Figure 8.2 depicts this hierarchical structure for $S$ secondary users and $L$ spectrum legacy owners, each of which owns spectrum subbands including $M_l$ subcarriers ($l = 1, \ldots, L$) and provides a service to a number of primary users. In an open spectrum regime, the activities of the cognitive radio users should not affect the performance of primary users. In other words, the existence of the cognitive radio users in a spectrum legacy owner's band should not be noticed by the primary users that receive service from that legacy owner. While the primary customers of the legacy owners do not need to know anything about the secondary users, secondary users should be quite cautious about the activities of the primary users.

A cognitive radio network is a dynamic system formed by a group of interacting subsystems (i.e. cognitive radios). Since the activities of the primary users determine the available subbands for secondary usage, the role of the primary users in a cognitive radio network can be interpreted as the role of a high-level-network controller that decides which resources can be used by cognitive radio users. Then, the resource-allocation algorithm used by each cognitive radio user determines the share of that user from the available resources.

The resource-allocation algorithms play the role of local controllers that control the corresponding subsystems in a decentralized manner. Regarding the master–slave relationship between the two wireless worlds, a decentralized hierarchical control structure can be considered for a cognitive radio network as depicted in Figure 8.2. Actions of the high-level controller, which are discrete events, and actions of the local controllers, which are continuous, are associated with slow and fast dynamics of the network and lead to its two-time-scale behavior.

The resource-allocation problem is solved in two stages, regarding discrete events and continuous states. Therefore, the local controllers in Figure 8.3 are two-level controllers. The corresponding two-level control scheme is shown in Figure 8.4. The supervisory-level (i.e. the higher level) controller is, in effect, an event-driven controller that deals with appearance and disappearance of spectrum holes. The radio scene analyzer will inform the supervisory-level controller if it detects a change in the status of the available spectrum holes. In that case, the supervisory-level controller calls for reconfiguration of the actuator in order to adapt the transmitting parameters to the new set of available channels. The field-level (i.e. the lower level) controller is a state-based controller that adjusts the transmit power over the set of available channels chosen by the supervisory-level controller according to the interference level in the radio environment. A cognitive radio may build an internal model for the external world. This model is used to predict the availability of certain subbands, the duration of their availability, and the approximate interference level in those subbands. This information will be critical for providing seamless communication in the dynamic wireless environment. Both the

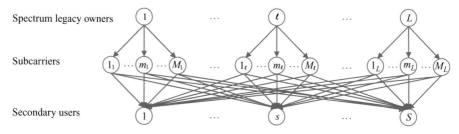

**Figure 8.2.** The spectrum supply chain network.

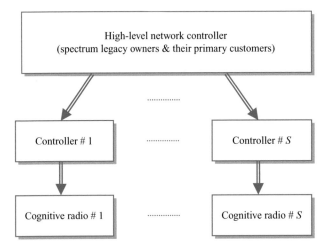

**Figure 8.3.** Decentralized hierarchical control structure in a cognitive radio network.

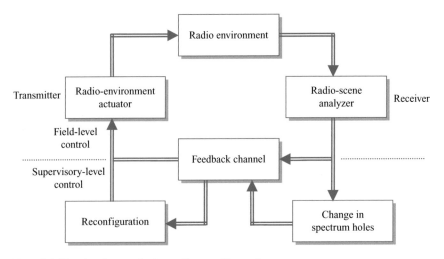

**Figure 8.4.** Two-level control scheme for cognitive radio.

supervisory-level and the field-level controllers will benefit from a predictive model, which determines the control horizon, to plan ahead.

### 8.3.3    Security in cognitive radio networks (Clancy and Goergen, 2008)

A discussion of cognitive radio networks would be incomplete without devoting this last part of the Epilogue to security in these networks.

Recognizing that, by definition, a cognitive radio has a built-in ability to adapt to the environment, it is essential that the cognitive radio be equipped to select optimal and secure means of communications. It follows, therefore, that attention ought to be focused on *attacks* that are fundamental to the cognitive radio, leaving issues such as data integrity and confidentiality to higher layer cryptographic techniques (Clancy and Goergen, 2008).

Clancy and Goergen (2008) define three classes of attacks, all of which pertain to the Physical (PHYS)-layer of the network:

(1) *Sensory manipulation attacks*

In this class of attacks, all that an attacker needs to do, for example, is to create a modulated waveform sufficiently similar to that of the primary user, thereby triggering a "false" positive in the spectrum-sensing algorithm.

(2) *Belief-manipulation attacks*

To optimize multigoal objective functions, a cognitive radio would have to be equipped with adaptive-filtering algorithms that are designed to realize such objective functions. In this second class of attacks, the attacker essentially tries to interfere with the adaptive-filtering algorithm so as to "poison" whatever beliefs that the cognitive radio has learned.

The multigoal objective function manipulation is just one example of belief manipulation attacks. In a generic sense, this second class of attacks includes any attack that is aimed at manipulating the state of an adaptive-filtering algorithm that jeopardizes the long-term behavior of a cognitive radio.

(3) *Malicious behavior attacks*

In the first two classes of attacks, the attacker's aim is to make the cognitive radio act in a suboptimal manner. In the third class of attacks, the attacker's aim is to "teach" the radio to become unknowingly malicious in a self-propagating manner. For example, suppose a primary user employs a narrowband modulation scheme and the secondary user employs OFDM equipped with a few pilot tones superimposed on a few subcarriers for the purposes of perceptor channel estimation and synchronization. In such a scenario, all that an attacker has to do is to transmit a carrier on these pilot tones, thereby blocking any useful communication among secondary users of the network.

To mitigate the effectiveness of attacks, various approaches are described by Clancy and Goergen (2008); a summary of the approaches follows:

(1) *Robust sensory input*

To reduce the gullibility of a cognitive radio to attacks, the radio would need to be equipped with the means to discriminate between ambient perceptor noise and RF interference, thereby being able to distinguish between natural and man-made RF events. This is where the MTM with the harmonic $F$-test for spectral line components can come to the rescue (Haykin *et al.*, 2009).

(2) *Mitigation in individual radios*

In this second approach, attacks against individual radios are mitigated by embedding "common sense" in the design of each radio. To illustrate what we mean by common sense, consider the following example (Newman, personal communication, 2010). Suppose that a digital radio has been trained to believe that the DTV signals of interest are 6 MHz wide. Given this knowledge, the attacker slowly causes the cognitive radio to start believing that the DTV signals are 6.5 MHz wide instead of 6 MHz wide by emulating a bunch of signals that look like DTV signals except they are 6.5 MHz wide. In reality, this kind of attack depends on the attacker having the ability to adapt to noisy conditions through the use of adaptive-filtering algorithms of its own. This example attack may be viewed as the "teaching" attack. As such, belief-manipulation attacks are the real threat to cognitive radio users, which conventional wireless systems do not have to worry about. In any event, once the cognitive radio is taught to misclassify the real DTV signals, the radio may go on to transmit in those bands and thereby jam all nearly DTV perceptors. This scenario may, in turn, teach other cognitive radios in the same network that it is apropos to transmit in the misclassified band, thereby spreading havoc across the entire cognitive radio network.

Even though the ideal cognitive radio has the notion of autonomy and is able to adapt to environmental changes, there should be some form of anchor so as to avoid the cognitive radio from being led far off the correct path by a malicious user. To this end, there could be "common sense" rules built into design of the cognitive radio, such as primary users will never have a bandwidth larger than 6.4 MHz or the ambient noise will never be larger than −50 dBm (Newman, personal communication, 2010).

(3) *Mitigation in networks*

In this third approach, it is presumed that there is control-channel connectivity between cognitive radios, which, in turn, suggests the use of *swarm intelligence*; swarm intelligence is a set of algorithms that are designed to mimic animal behaviors. One such technique of particular applicability to security is the so-called *particle swarm optimization*. To elaborate briefly, each cognitive radio in a network represents a "particle," with each particle having its own hypothesis about what the best behavior is in a particular situation, and with that particular hypothesis being a weighted average of all the hypotheses in the network.

# Glossary

## A

| ADP | approximate dynamic programming |
| AR | autoregressive |
| ARMA | autoregressive moving average |
| ATSC | Advanced Television Systems Committee |
| ATSC-DTV | Advanced Television Systems Committee–Digital Television |
| AWGN | additive white Gaussian noise |

## B

| BS | base station |

## C

| CD-CKF | continuous-discrete cubature Kalman filter |
| CD-EKF | Continuous-discrete extended Kalman filter |
| CD-UKF | continuous-discrete unscented Kalman filter |
| CFDP | cycle-frequency domain profile |
| CKF | cubature Kalman filter |
| CPE | customer premise equipment |
| CRLB | Cramér–Rao lower bound |
| CSS | cooperative spectrum sensing |

## D

| dB | decibel |
| dBi | DBm referred to isotropic antenna |
| dBm | decibel (with respect to one) milliwatt |
| DEKF | decoupled extended Kalman filter |
| DFT | discrete Fourier transform |
| DoA | direction of arrival |
| DoF | degrees of freedom |
| DP | dynamic programming |
| DSM | dynamic spectrum management |
| DTV | digital television |

## E

| | |
|---|---|
| EA-RMSE | ensemble-averaged root mean-square error |
| EKF | extended Kalman filter |
| EM | executive memory |
| EM-U1 | executive memory – unsupervised 1 |
| EM-U2 | executive memory – unsupervised 2 |
| EM-S | executive memory – supervised |

## F

| | |
|---|---|
| FAR | Fore-active radar |
| FCC | Federal Communications Commission (in the USA) |
| FFT | fast Fourier transform |
| FFTW | fastest Fourier transform in the west |

## G

| | |
|---|---|
| GHA | generalized Hebbian algorithm |
| GQ | gradient $Q$, where the $Q$ refers to $Q$-learning |

## H

| | |
|---|---|
| Hz | hertz |

## I

| | |
|---|---|
| iid | independently and identically distributed |
| IEEE | Institute of Electronics and Electrical Engineers |
| IWFC | iterative water-filling controller |

## L

| | |
|---|---|
| LFM | linear frequency modulation |
| LO | local oscillator |

## M

| | |
|---|---|
| MA | moving average |
| MAP | maximum a posteriori (probability) |
| MDP | Markov decision process |
| MFBLP | modified forward–backward linear prediction |
| ML | maximum likelihood |
| MLP | multilayer perceptron |
| MMSE | minimum mean-square error |
| MPEG | Moving Picture Expert Group |
| MPEG-2TS | Moving Picture Expert Group-2 Transport System |
| MTM-SVD | multitaper method–singular value decomposition |
| MUSIC | multiple signal classification |
| MVDR | minimum-variance distortionless response |

## N

| NEXRAD | next-generation radar |
| NLOS | near line of sight |
| NTSC | National Television System Committee |
| NTU | nontransferable utility |

## O

| ODE | ordinary differential equation |
| OFDM | orthogonal frequency-division multiplexing |

## P

| PAC | perception–action cycle |
| PAC-DP | perception–action cycle-dynamic programming |
| PM | perceptual memory |
| PM-S | perceptual memory–supervised |
| PM-U1 | perceptual memory–unsupervised 1 |
| PM-U2 | perceptual memory–unsupervised 2 |
| PSD | predictive sparse decomposition |

## R

| RF | radio frequency |
| RMLP | recursive multilayer perceptron |
| RMSBE | root–mean-square Bellman error |
| RMSE | root mean-square error |
| RMSPBE | root-mean-square projected Bellman error |
| RSA | radio scene analyzer |
| RX | receiver |
| RX-CR | receiver-cognitive radio |

## S

| SDR | software-defined radio |
| SESM | sparse encoding symmetric machine |
| SINR | signal-to-interference plus noise ratio |
| SMC | sequential Monte Carlo |
| SNR | signal-to-noise ratio |
| SVD | singular value decomposition |

## T

| TAR | traditional active radar |
| TD | temporal difference |
| TF-MSC | two-frequency magnitude-squared coherence |
| TPC | transmit-power controller |
| TU | transferrable utility |
| TV | television |

| | |
|---|---|
| TX | transmitter |
| TX-CR | transmitter-cognitive radar |

## U

| | |
|---|---|
| UCLA | University of California at Los Angeles |
| UHF | ultrahigh frequency |
| UIUC | University of Illinois at Urbana-Champaign |
| UKF | unscented Kalman filter |

## V

| | |
|---|---|
| VI | variational inequality |
| VSB | vestigial side-band |

## W

| | |
|---|---|
| WOSA | weighted overlapped segment averaging |

# References

Aghassi, M. and Bertsimas, D. (2006). Robust game theory. *Mathematical Programming, Series B*, **107**, 231–73.

Aleksander, I. and Morton, H. (1990). *An Introduction to Neural Computing*. London: Chapman and Hall.

Anderson, J. (1995). *An Introduction to Neural Networks*. Cambridge, MA: MIT Press.

Anderson, B. D. O. and Moore, J. B. (1979). *Linear Optimal Control*. Englewood Cliffs, NJ: Prentice-Hall.

Annastasio, T. J. (2003). Vestibulo-occular reflex. In M. A. Arbib, ed., *The Handbook of Brain Theory and Neural Networks*, second edition. Cambridge, MA: MIT Press, pp. 1192–96.

Apt, K. R. and Witzel, A. (2006). A generic approach to coalition formation. *International Game Theory Review*, **11**, 347–67.

Arasaratnam, I. and Haykin, S. (2009). Cubature Kalman filters. *IEEE Transactions on Automatic Control*, **54**, 1254–69.

Arasaratnam, I., Haykin, S., and Hurd, T. R. (2010). Cubature Kalman filtering for continuous-discrete systems: theory and simulations. *IEEE Transactions on Signal Processing*, **58**, 4977–4993.

Athans, M., Wishner, R. P., and Bertolini, A. (1968). Suboptimal state estimation for continuous-time nonlinear systems from discrete noise measurements. *IEEE Transactions on Automatic Control*, **AC13**, 504–14.

Autmann, R. J. and Peleg, B. (1960). Von Neumann–Morgenstern solutions to cooperative games without side payments. *Bulletin of the American Mathematical Society*, **6**, 173–9.

Baddeley, A. (2003). Working memory: looking back and looking forward. *Nature Reviews Neuroscience*, **4**, 829–39.

Baird, L. C. (1999). Reinforcement learning through gradient descent. Ph.D. thesis, Carnegie-Mellon University, May.

Barlow, H. (1961). The coding of sensory images. In W. H. Thorpe and O. L. Zangwill, eds, *Current Problems in Animal Behaviour*. Cambridge: Cambridge University Press, pp. 331–60.

Barlow, H. (2001). Redundancy reduction revisited. *Network: Computational Neural Systems*, **12**, 241–53.

Bar-Shalom, Y., Li, X., and Kirburajan, T. (2001). *Estimation with Applications to Tracking and Control*. Wiley.

Barto, A. G., Sutton, R. S., and Anderson, C. W. (1983). Neuronlike elements that can solve difficult learning control problems. *IEEE Transactions on Systems, Man, and Cybernetics*, **13**, 835–46.

Basar, T. and Bernhard, P. (eds) (1995). *H∞-Optimal Control and Related Minimax Design Problems: A Dynamic Game Approach*, second edition, Boston. MA: Birkhäuser.

Basar, T. and Olsder, G. J. (1999). *Dynamic Noncooperative Game Theory*. SIAM.

Bellman, R. E. (1957). *Dynamic Programming*. Princeton, NJ: Princeton University Press.

Bellman, R. E. (1961). *Adaptive Control Processes: A Guided Tour*. Princeton, NJ: Princeton University Press.

Bellman, R. E. and Dreyfus, S. E. (1962). *Applied Dynamic Programming*. Princeton, NJ: Princeton University Press.

Bengio, Y. and LeCun, Y. (2007). Scaling learning algorithms toward AI. In L. Bottou, O. Chapelle, D. DeCoste, and J. Weston, eds, *Large Scale Kernel Machines*. Cambridge, MA: MIT Press. pp. 321–59.

Bennett, B. Hoffman, D. Nichola, J., and Prokash, C. (1989). Structure from two orthographic views of rigid motion. *journal of the Optical Society of America*, **A.6**, 1052–69.

Bernardo, J. M. and Smith, A. F. M. (1998). *Bayesian Theory*. Wiley.

Bertsekas, D. P. (2005). *Dynamic Programming and Optimal Control*, vol. 1, third edition. Athena Scientific.

Bertsekas, D. P. (2007). *Dynamic Programming and Optimal Control*, vol. 2, third edition. Athena Scientific.

Bertsekas, D. P. and Tsitsiklis, J. N. (1996). *Neuro-Dynamic Programming*. Belmont, MA: Athena Scientific.

Bertsekas, D. P. and Tsitsiklis, J. N. (2008). *Introduction to Probability*, second edition. Belmont, MA: Athena Scientific.

Bishop, C. M. (2006). *Pattern Recognition and Machine Learning*. Springer.

Boyd, S. and Vandenberghe, L. (2004). *Convex Optimization*. Cambridge University Press.

Bradtke, S. J. and Barto, A. G. (1996). Linear least-squares algorithms for temporal difference learning. *Machine Learning*, **22**, 33–57.

Brady, M. H. and Cioffi, J. M. (2006). The worst-case interference in DSL systems employing dynamic spectrum management. *EURASIP Journal on Advanced Signal Processing*, 1–11.

Bronez, T. P. (1992). On the performance advantage of multitaper spectral analysis. *IEEE Transactions on Signal Processing*, **40**, 2941–46.

Buddhiko, M. M. (2007). Understanding dynamic spectrum access: models taxonomy, and challenges. In *Proceedings of IEEE DYSPAN*, April.

Cappé, O., Moulines, E., and Ryden, T. (2005). *Inference in Hidden Markov Models*. Springer.

Churchland, P. S. and Sejnowski, T. J. (1992). *The Computational Brain*. Cambridge, MA: MIT Press.

Clancy, T. C. and Goergen, N. (2008). Security in cognitive radio networks: threats and mitigation. In *International Conference on Cognitive Radio Oriented Wireless Networks and Communucations (Crowncom)*, May, pp. 1–8.

Cohen, L. (1995). *Time–Frequency Analysis*. Englewood Cliffs, NJ: Prentice-Hall.

Cools, R. (1997). Constructing cubature formulae: the science behind the art. *Acta Numerica*, 6, 1–54.

Cover, T. M. and Thomas, J. A (2006). *Elements of Information Theory*, second edition. New York: Wiley.

Danskin, J. (1967). *The Theory of Max–Min*. Springer-Verlag.

Dayan, P. and Abott, L. (2001). *Theoretical Neuroscience: Computational and Mathematical Modeling of Neural Systems*. Cambridge, MA: MIT Press.

Donoho, D. L. and Elad, M. (2003). Optimally sparse representation in general (nonorthogonal) dictionaries via $l^1$ minimization. *Proceedings of the National Academy of Sciences of the United States of America*, **100**(5), 2197–202.

Doya, K., Ishi, S., Pouget A., and Rao, R. (2007). *Bayesian Brain: Probabilistic Approaches to Neural Coding*. Cambridge, MA: MIT Press.

Drosopoulos, A. and Haykin, S. (1992). Adaptive radar parameter estimation with Thomson's multiple-window method. In S. Haykin and A. Steinhardt, eds, *Radar Detection and Estimation*. New York: Wiley.

Dupuy, J.-P. (2009). *On the Origins of Cognitive Science: The Mechanization of the Mind*. Cambridge, MA: MIT Press.

Elliott, R. J. and Haykin, S. (2010). A Zakai equation derivation of the extended Kalman filter, *Automatica*, **46**, 620–4.

FCC. (2002). Spectrum Policy Task Force, Report ET Docket No. 02-135, Federal Communications Commission, November.

FCC. (2008). *Second Report and Order and Memorandum Opinion and Order In the Matter of Unlicensed Operation in the TV Broadcast Bands, Additional Spectrum for Unlicensed Devices Below 900 MHz and in the 3 GHz Band*. Docket number 08-260, Federal Communication Commission, November.

FCC. (2010). *Second Memorandum Opinion and Order In the Matter of Unlicensed Operation in the TV Broadcast Bands, Additional Spectrum for Unlicensed Devices Below 900 MHz and in the 3 GHz Band*. Docket number 10-174, Federal Communication Commission, September.

Fisher, R.A. (1912). On an absolute criteria for fitting frequency curves. *Messenger of Mathematics*, **41**, 155–60.

Fisher, R. A. (1922). On the mathematical foundation of theoretical statistics. *Philosophical Transactions of the Royal Society of London, Series A*, **222**, 309–68.

Frigo, M. and Johnson, S. G. (2005). The design and implementation of FFTW3. *Proceedings of the IEEE*, **93**, 216–31.

Fukushima, M. (2007). Stochastic and robust approaches to optimization problems under uncertainty. *Proceedings of International Conference on Informatics Research for Development of Knowledge Society Infrastructure (ICKS)*, pp. 87–94.

Fundenberg, D. and Levine, D. K. (1999). *The Theory of Learning in Games*. Cambridge, MA: MIT Press.

Fuster, J. M. (2003). *Cortex and Mind: Unifying Cognition*. Oxford, UK: Oxford University Press.

Gardner, W. A. (1988). Signal interception: a unifying theoretical framework for feature detection. *IEEE Transactions on Communications*, **36**, 897–906.

Gardner, W. A. ed. (1994). *Cyclostationarity in Communications and Signal Processing*. New York: IEEE Press.

Geman, S. and Geman, D. (1984). Stochastic relaxation, Gibbs distributions, and the Bayesian restoration of images. *IEEE Transactions on Pattern Analysis and Machine Intelligence*, **PAMI-6**, 721–41.

Giannakis, G. B. and Serpedin, E. (1998). Blind identification of ARMA channels with periodically modulated inputs. *IEEE Transactions on Signal Processing*, **46**, 3099–104.

Gjessing, D. T. (1986). *Target Adaptive Matched Illumination Radar: Principles and Applications*. Peter Peregrinus Ltd on Behalf of the Institution of Electrical Engineers, London, UK.

Glimcher, P. W. (2003). *Decisions, Uncertainty, and the Brain: The Science of Neuroeconomics*. Cambridge, MA: MIT Press.

Golub, G. H. and Van Loan, C. F. (1996). *Matrix Computations*, third edition. Johns Hopkins University Press.

Grenander, U. (1976–1981). *Lectures in Pattern Theory I, II and III: Pattern Analysis, Pattern Synthesis and Regular Structures*. Springer-Verlag.

Gross, B. (1964). *The Managing of Organizations: The Administrative Struggle*. New York: Free Press of Glencoe.

Grossberg, S. (1988). *Neural Networks and Natural Intelligence*. Cambridge, MA: MIT Press.

Guerci, J. R. (2010). *Cognitive Radar: The Knowledge-Aided Fully Adaptive Approach*. Artech House.

Habibi, S. R. and Burton, R. (2003). The variable structure filter. *Journal of Dynamic Systems Measurement and Control*, **125**, 287–93.

Hanzo, L., Akhtman, Y., Wang, L., and Jiang, M. (2010). *MIMO-OFDM for LTE, WiFi and WiMAX: Coherent versus Non-Coherent and Cooperative Turbo-Transceivers*. Wiley–IEEE.

Harker, P. T. and Pang, J.-S. (1990). Finite-dimensional variational inequality and nonlinear complementarity problems: a survey of theory, algorithms and applications. *Mathematical Programming*, **48**, 161–220.

Hastie, T., Tishbirani, R., and Friedman, J. (2001). *The Elements of Statistical Learning: Data Mining, Inference, and Prediction*. Canada: Springer.

Haykin, S. (2000). *Communication Systems*, third edition. New York: Wiley.

Haykin, S. (2002). *Adaptive Filter Theory*, fourth edition. Prentice-Hall.

Haykin, S. (2005a). Cognitive radio: brain-empowered wireless communications. *IEEE Journal on Selected Areas in Communications*, **23**, 201–20.

Haykin, S. (2005b). Cognitive radar networks. In *1st IEEE Workshop on Computational Advances in Multi-sensor Adaptive Processing*, Jalisco State, Mexico.

Haykin, S. (2006a). Cognitive dynamic systems, point-of-view article. *Proceedings of the IEEE*, **94**, 1910–11.

Haykin, S. (2006b). Cognitive radar: a way of the future. *IEEE Signal Processing Magazine*, **23**, 30–41.

Haykin, S. ed. (2007). *Adaptive Radar Signal Processing*. Wiley.

Haykin, S. (2009). *Neural Networks and Learning Machines*. Upper Saddle River, NJ: Prentice-Hall.

Haykin, S. and Thomson, D. J. (1998). Signal detection in a nonstationary environment reformulated as an adaptive pattern classification problem. *Proceedings of the IEEE*, **86**, 2325–44.

Haykin, S., Bakker, R., and Currie, B. (2002). Uncovering nonlinear dynamics: the case study of sea clutter. *Proceedings of the IEEE*, **90**, 860–81.

Haykin, S., Thomson, D. J., and Reed, J. H. (2009). Spectrum sensing for cognitive radio. *Proceedings of the IEEE*, **97**, 849–77.

Haykin, S., Zia, A., Xue, Y., and Arasaratnam, I. (2011). Control-theoretic approach to tracking radar: first step towards cognition. *Digital Signal Processing* **21**, 576–85.

Haykin, S. and Xue, Y. (2012). Cognitive Radar. To be published.

Hebb, D. O. (1949). *The Organization of Behavior: A Neurosychological Theory*. New York: Wiley.

Hecht-Nielsen, R. (1995). Replicator neural networks for universal optimal source and coding. *Science*, **269**, 1860–63.

Ho, Y. C. and Lee, R. C. K. (1964). A Bayesian approach to problems in stochastic estimation and control. *IEEE Transactions on Automatic Control*, **AC-9**, 333–9.

Hurd, H. L. and Miamee, A. (2007). *Periodically Correlated Random Sequences*. New York: Wiley.

Julier, S. J. and Uhlmann, J. K. (2004). Unscented filtering and nonlinear estimation. *Proceedings of the IEEE*, **92**, 401–422.

Julier, S. J., Uhlmann, J. K., and Durrant-Whyte, H. F. (2000). A new method for nonlinear transformation of means and covariances in filters and estimators. *IEEE Transactions on Automatic Control*, **45**, 472–82.

Kalman, R. (1960). A new approach to linear filtering and prediction problems. *Transactions of the ASME, Journal of Basic Engineering, Series D*, **82**, 35–45.

Kavukcuoglu, K., Ranzato, M., and LeCun, Y. (2008). Fast inference in sparse coding algorithms with applications to object recognition. Technical Report, CBLL, Courant Institute, NYU, CBLL-TR-2008-12-01.

Kershaw, D. J. and Evans, R. J. (1994). Optimal waveform selection for tracking systems. *IEEE Transactions on Information Theory*, **40**, 1536–50.

Kersten, D. (1990). Statistical limits to image understanding. In C. Blakemore, ed., *Vision: Coding and Efficiency*. Cambridge, UK: Cambridge University Press.

Khalil, H. K. (2002). *Nonlinear Systems*, third edition. Upper Saddle River, NJ: Prentice-Hall.

Khozeimeh, F. and Haykin, S. (2009). Dynamic spectrum management for cognitive radio: an overview. *Wireless Communications and Mobile Computing*, **9**, 1447–59.

Khozeimeh, F. and Haykin, S. (2010). Self-organizing dynamic spectrum management for cognitive radio networks. In *The 8th Conference on Communication Networks and Services Research (CNSR-2010)*, pp. 1–8.

Knill, D. C. and Pouget, A. (2004). The Bayesian brain: the role of uncertainty in neural coding and computation for action. *Trends in Neuroscience*, **12**, 712–19.

Knill, D. C. and Richards, W. eds. (1996). *Perception as Bayesian Inference*. Cambridge, UK: Cambridge University Press.

Krishnamurthy, V. and Djonin, D. V. (2009). Optimal threshold policies for POMDPs in radar resource management. *IEEE Transactions on Signal Processing*, **57**, 3954–69.

Kumarasan, R. and Tufts, D. W. (1983). Estimating the angles of arrival of multiple plane waves. *IEEE Transactions on Aerospace and Electronic Systems*, **AES-19**, 134–9.

Li, Y. and Stüber, G. (2006). *Orthogonal Frequency Division Multiplexing for Wireless Communications*. Springer.

Lo, J. T. and Bassu, D. (2001). Adaptive vs. accommodative neural networks for adaptive system identification. In *Proceedings of the International Joint Conference on Neural Networks*, Washington, DC, July, pp.1279–1284.

Lo, J. T. and Yu, L. (1995). Adaptive neural filtering by using the innovations approach. In *Proceedings of the 1995 World Congress on Neural Networks*, vol. 2, July, pp. 29–35.

Loève, M. (1946). Fonctions aléatoires de second ordrer. *Revue Scientifique Paris*, **84**, 195–206.

Loève, M. (1963). *Probability Theory*. Van Nostrand.

Luo, Z. and Pang, J. (2006). Analysis of iterative waterfilling algorithm for multiuser power control in digital subscriber lines. *EURASIP Journal of Applied Signal Processing*, **2006**, ID 24012, 1–10.

Maei, H. R. and Sutton, R. S. (2010). GQ($\lambda$): a general gradient algorithm for temporal difference prediction learning with eligibility traces. In E. Baum, M. Hutter, and E. Kitzelmann, eds, *AGI 2010*. Atlantis Press, pp. 91–6.

Maei, H. R., Szepesva'ri, Cs., Bhatnagar, S., and Sutton, R. S. (2010). Toward off-policy learning control with function approximation. In *Proceedings of the 27th International Conference on Machine Learning*, Haifa, Israel.

Marple, L. S. L. (1987). *Digital Spectral Analysis with Applications*. Englewood Cliffs, NJ: Prentice-Hall.

Maybeck, P. S. (1982). *Stochastic Models, Estimation, and Control*, vol. 2. New York: Academic Press.

Maynard Smith, J. (1974). The theory of games and the evolution of animal conflicts. *Journal of Theoretical Biology*, **47**, 209–21.

Maynard Smith, J. (1982). *Evolution and the Theory of Games*. Cambridge, UK: Cambridge University Press.

McLaughlin, D., Payne, D., Chandrasekar, V., Philips, B., Kurose, J., Zink, M., *et al.* (2009). Short-wavelength technology and the potential for distributed networks of small radar systems. *Bulletin of the American Meteorological Society*, **90**, 1797–1817.

Mitola, J. (2000). Cognitive radio: an integrated agent architecture for software defined radio. Doctor of Technology Dissertation, Royal Institute of Technology, Sweden.

Mitola, J. and McGuire, G. Q. (1999). Cognitive radio: making software radios more personal. *IEEE Personal Communications*, **6**(4), 13–18.

Molisch, A. F. (2005). *Wireless Communications*. Chichester, UK: IEEE Press/Wiley.

Molisch, A. F., Greenstein, L. J., and Shafi, M. (2009). Propogations issues for cognitive radio, *Proceeding of the IEEE*, **97**, 787–804.

Mooers, C. N. K. (1973). A technique for the cross-spectrum analysis of pairs of complex-valued time series, with emphasis on properties of polarized components and rotational invariants. *Deep-Sea Research*, **20**, 1129–41.

Morrone, C. and Burr, D. (2009). Visual stability during saccadic eye movements. In M. S. Gazzaniga, editor-in-chief, *The Cognitive Neurosciences*, fourth edition. MIT Press, pp. 511–24.

Morse, P. M. and Feshbach, H. (1953). *Methods of Theoretical Physics, Part I*. New York: McGraw-Hill.

Nagurney, A. and Zhang, D. (1996). *Projected Dynamical Systems and Variational Inequalities with Applications*. Springer.

Narendra, K. S. and Parthasarathy, K. (1990). Identification and control of dynamical systems using neural networks. *IEEE Transactions on Neural Networks*, **1**, 4–27.

Nash, J. F. (1950). Equilibrium points in $n$-person games. *Proceedings of the National Academy of Sciences of the United States of America*, **86**, 48–9.

Nash, J. F. (1951). Non-cooperative games. *Annals of Mathematics*, **54**, 286–95.

Nicolis, G. and Prigogine, I. (1989). *Exploring Complexity: An Introduction*. W. H. Freeman.

Olshausen, B. A. (2011). Personal communication.

Olshausen, B. A. and Field, D. J. (1996). Emergence of simple-cell receptive field properties by learning a sparse code for natural images. *Nature*, **381**, 607–9.

Ortiz, S. (2008). The wireless industry begins to embrace femtocells. *Computer*, **41**, 14–17.

Parsons, J. (2000). *The Mobile Radio Propagation Channel*. New York: Wiley.

Percival, D. B. and Walden, A. T. (1993). *Spectral Analysis for Physical Applications*. Cambridge, UK: Cambridge University Press.

Picinbone, B. (1996). Second-order complex random vectors and normal distributions. *IEEE Transactions on Signal Processing*, **44**, 2637–40.

Posner, M., ed. (1989). *Foundations of Cognitive Science*. Cambridge, MA: MIT Press.

Powell, W. B. (2007). *Approximate Dynamic Programming: Solving the Curses of Dimensionality*. Hoboken, NJ: Wiley.

Press, W. and Teukolsky, S. (1990). Orthogonal polynomials and Gaussian quadrature with non-classical weighting functions. *Computers in Physics*, **4**, 423–26.

Puskorius, G. V. and Feldkamp, L. A. (2001). Parameter-based Kalman filter training: theory and implementation. In S. Haykin, ed., *Kalman Filtering and Neural Networks*. Wiley.

Pylyshyn, Z. (1984). *Computation and Cognition: Toward a Foundation for Cognitive Science*. Cambridge, MA: MIT Press.

Ranzato, M., Boureau, Y., and LeCun, Y. (2007). Sparse feature learning for deep belief networks. In *Neural Information Processing Systems* (NIPS), Vancouver, B.C., Canada.

Reber, A. (1995). *Dictionary of Psychology*. London: Penguin Books.

Ristic, B., Arulampalam, S., and Gordon, N. (2004). *Beyond the Kalman Filter: Particle Filters for Tracking Applications*. Boston, MA: Artech House.

Robbins, H. and Monro, S. (1951). A stochastic approximation method. *Annals of Mathematical Statistics*, **22**, 400–7.

Robert, C. P. (2007). *The Bayesian Choice: From Decision-Theoretic Foundations to Computational Implementations*, second edition. Springer.

Robert, C. P. and Casella, G. (2004). *Monte Carlo Statistical Methods*, second edition. Springer.

Rogers, T. and McLelland, J. L. (2004). *Semantic Cognition: A Parallel Distributed Processing Approach*. Cambridge, MA: MIT Press.

Ross, S. M. (1983). *Introduction to Stochastic Dynamic Programming*. New York: Academic Press.

Rumelhart, D. E. and McLelland, J. L. eds. (1986). *Parallel Distributed Processing: Explorations in the Microstructure of Cognition*, vol. 1. Cambridge, MA: MIT Press.

Rumelhart, D. E., Hinton, G. E., and Williams, R. J. (1986). Learning internal representations by error propagation. In D. E. Rumelhart and J. L. McLelland, eds, *Parallel Distributed Processing: Explorations in the Microstructure of Cognition*, vol. 1. Cambridge, MA: MIT Press, pp. 318–62.

Saad, W., Han, Z., Debbah, M., Horungnes, A., and Basar, T. (2009). Coalitional game theory for communication networks. *IEEE Signal Processing Magazine*, **26**, 77–97.

Sanger, T. D. (1989). Optimal unsupervised learning in a single-layer linear feedforward neural network. *Neural Networks*, **12**, 459–73.

Sarkka, S. (2007). On unscented Kalman filtering for state estimation of continuous-time nonlinear systems. *IEEE Transactions on Automatic Control*, **52**, 1631–41.

Schmidt, R. (1981). A signal subspace approach to multiple emitter of location and spectral estimation. Ph.D. dissertation, Stanford University, Stanford, CA.

Schultz, W. (1998). Predictive reward signal of dopamine neurons. *Journal of Neurophysiology*, **80**, 1–27.

Schuster, H. G. (2001). *Complex Adaptive Systems: An Introduction*. Springer-Verlag.

Sejnowski, T. J. (2010). Personal communication.

Selfridge, O. G. (1958). Pandamonium: a paradigm for learning. In *Proceedings of a Symposium held at the National Physical Laboratory*, November. London: HMSO.

Serpedin, E., Panduru, F., Sari, I., and Giannakis, G. B. (2005). Bibliography on cyclostationarity. *Signal Processing*, **85**, 2233–303.

Setoodeh, P. (2010). Dynamic models of cognitive radio networks Ph.D. thesis, McMaster University, Ontario.

Setoodeh, P. and Haykin, S. (2009). Robust transmit power control for cognitive radio. *Proceedings of the IEEE*, **97**, 915–39.

Seung, S. (2000). Half a century of Hebb. *Nature Neuroscience*, **3**, 1166–1167.

Shannon, C. E. (1948). A mathematical theory of communication. *Bell System Technical Journal*, **27**, 379–423, 623–56.

Shellhammer, S. J. (2008). Spectrum sensing in IEEE 802.22. In *Cognitive Information Processing Workshop*, Greece, June 2008.

Shellhammer, S. J. (2010). Personal communication.

Simmons, J. A., Saillant, P. A., and Dear, S. P. (1992). Through a bat's ear. *IEEE Spectrum*, **29**(3), 46–8.

Skolnik, M. I. (2008). *Radar Handbook*. McGraw-Hill.

Slepian, D. (1965). Some asymptotic expansions for prolate spheroidal wave functions. *Journal of Mathematics and Physics*, **44**, 99–140.

Slepian, D. (1978). Prolate spheroidal wave functions, Fourier analysis and uncertainty. *Bell System Technical Journal*, **57**, 1371–1430.

Stein, D. L., ed. (1989). *Lectures in the Sciences of Complexity*. Addison-Wesley.

Stevenson, C. R., Chouinand, G., Lei, Z., Hu, W., Shellhammer, S. J., and Caldwell, W. (2009). IEEE 802.22: the first cognitive radio wireless regional area network standard. *IEEE Communications Magazine*, **47**(1), 130–38.

Stroud, A. H. (1966). *Gaussian Quadrature Formulas*. Englewood Cliffs, NJ: Prentice-Hall.

Stroud, A. H. (1971). *Approximate Calculation of Multiple Integrals*. Englewood Cliffs, NJ: Prentice-Hall.

Subramanian, A. P. and Gupta, S. H. (2007). Fast spectrum allocation in coordinated spectrum access based cellular networks. In *2nd IEEE International Symposium on New Frontiers in Dynamic Spectrum Access Networks, DySPAN 2007*, Dubin, 17–20 April, pp. 320–330.

Suga, N. (1990). Cortical computational maps for auditory imaging. *Neural Networks*, **3**, 3–21.

Sutton, R. S. (1984). Temporal credit assignment in reinforcement learning. Ph.D. dissertation, University of Massachusetts, Amherst, MA.

Sutton, R. S. and Barto, A. G. (1998). *Reinforcement Learning: An Introduction*. Cambridge, MA: MIT Press.

Sutton, R. S. (1988). Learning to predict by the methods of temporal differences. *Machine Learning*, **3**, 9–44.

Teo, K. (2007). Nonconvex robust optimization. Ph.D. thesis, Massachusetts Institute of Technology, Cambridge, MA.

Thomson, D. J. (1982). Spectrum estimation and harmonic analysis. *Proceedings of the IEEE*, **70**, 1055–96.

Thomson, D. J., (2000). Multitaper analysis of nonstationary and nonlinear time series data. In W. Fitzgerald, R. Smith, A. Walden, and P. Young, eds, *Nonlinear and Nonstationary Signal Processing*, Cambridge, UK: Cambridge University Press.

Trappenberg, T. P. (2010). *Fundamentals of Computational Neuroscience*. Oxford University Press.

Utkin, V. I. (1992). *Sliding Modes in Control and Optimization*. Springer-Verlag.

Utkin, V. I., Guldner, J., and Shi, J. (2009). *Sliding Mode Control in Electro-Mechanical Systems*, second edition. CRC Press.

Van Trees, H. L. (1968). *Detection, Estimation, and Modulation Theory,* Part 1. Wiley, New York.

Van Trees, H. L. (1971). *Detection, Estimation, and Modulation Theory,* Part III. New York: Wiley.

Von Neumann, J. and Morgenstern, O. (1944). *Theory of Games and Economic Behavior*. Princeton, NJ: Princeton University Press.

Watkins, C. I. C. H. (1989). Learning from delayed rewards. Ph.D. thesis, Cambridge University, Cambridge, UK.

Watkins, C. I. C. H. and Dayan, P. (1992). Q-learning. *Machine Learning*, **8**, 279–92.

Weisbunch, G. (1991). *Complex System Dynamics*. Addison-Wesley.

Welch, P. D. (1967). The use of fast Fourier transform for the estimation of power spectra: a method based on time-averaging over short modified periodograms. *IEEE Transactions on Audio and Electroacoustics*, **AU-15**, 70–3.

Werbos, P. (2004). ADP: goals, opportunities and principles. In J. Si, A. G. Barto, W. B. Powell, and D. Wusch II, eds *Handbook of Learning and Approximate Dynamic Programming*. Wiley.

Wicks, M. C. (2010). Waveform diversity: the way forward. Keynote Lecture. In *5th International Waveform Diversity and Design Conference*, Niagara Falls, Ontario, Canada, August.

Wicks, M. C., Mokole, E., Blunt, S., Schneible, R., and Amuso, V. (2010). *Principles of Waveform Diversity and Design*. SciTech.

Widrow, B. and Stearns, S. D. (1985). *Adaptive Signal Processing*. Prentice-Hall.

Wiener, N. (1948). *Cybernetics: Or Control and Communication in the Animal and the Machine*. Cambridge: MIT Press.

Williams, J. L., Fisher III, J. W., and Wilsky, A. S. (2007). Approximate dynamic programming for communication-constrained sensor network management. *IEEE Transactions on Signal Processing*, **55**, 4300–11.

Wolfowitz, J. (1952). On the stochastic approximation method of Robbins and Monro. *Annals of Mathematical Statistics*, **22**, 457–61.

Woodward, P. (1953). *Probability and Information Theory, with Applications to Radar*, London: Pergamon Press.

Xue, Y. (2010). Cognitive radar: theory and simulations, Ph.D. thesis, McMaster University, Ontario.

Younger, S., Hockreiter, S., and Conwall, P. (2001). Meta-learning with backpropagation. In *Proceedings of the Joint International Conference on Neural Networks*, Washington, DC, pp. 2001–6.

Yu, W. (2002). Competition and cooperation in multi-user communication environments. Doctoral dissertation, Stanford University, Stanford, CA.

Yuille, A. L. and Clark, J. (1993). Bayesian models, deformable templates and competitive priors. In L. Harris and M. Jenkins, eds, *Spatial Vision in Humans and Robots*. Cambridge, UK: Cambridge University Press.

Zeng, Y., Liang, Y.-C., Hoang, A. T., and Zhang, R. (2010). A review on spectrum sensing for cognitive radio: challenges and solutions. *EURASIP Journal on Advances in Signal Processing*, **2010**, Article ID 381465.

Zhang, Q. T. (2011). Theoretical performance and thresholds of the multitaper method for spectrum sensing, *IEEE transaction on Vehicular Technology*, **60**, 2128–38.

## Further reading

Bellman, R. E. (1971). *Introduction to the Mathematical Theory of Control Processes*, vol. II. New York: Academic Press.

Debreu, G. (1952). A social equilibrium existence theorem. *Proceedings of National Academy of Sciences of the United States of America*, **38**, 886–93.

Dreyfus, S. E. and Law, A. (1977). *The Art and Theory of Dynamic Programming*. New York: Academic Press.

Facchinei, F. and Pang, J. S. (2003). *Finite-Dimensional Variational Inequalities and Complementarity Problems*. Springer.

Jazwinski, H. (2007). *Stochastic Processes and Filtering Theory*. New York: Dover Publications.

Krazios. R. S. and. Stone, L. S (2003). Pursuit eye movements. In M. A. Arbib, ed., *The Handbook of Brain Theory and Neural Networks*, second edition, MIT Press, pp. 929–34.

Langoudakis, M. and Parr, R. (2003). Least-squares policy iteration. *Journal of Machine Learning Research*, **4**, 1107–49.

Laplace, P. (1812). *Théorie Analytique de Probabilités*, Paris: Courcier.

Mumford, D. (1996). Pattern theory: a unifying perspective. In D. C. Knill and W. Richards, eds, *Perception as Bayesian Inference*. Cambridge, UK: Cambridge University Press.

Olshausen, B. A. and Field, D. J. (1997). Sparse coding with an overcomplete basis set: a strategy employed by VI? *Vision Research*, **37**, 3311–25.

Osborne, M. and Rubenstein, A.T. (1994). *A Course in Game Theory*. Cambridge, MA: MIT Press.

Pang, J. S. and Facchinei, F. (2003). *Finite-Dimensional Variational Inequalities and Complementarity Problems*. Springer-Verlag.

Putterman, M. L. (1994). *Markov Decision Processes: Discrete Stochastic Dynamic Programming*. New York: Wiley.

# Index

Printed in the United States
by Baker & Taylor Publisher Services